SECOND EDITION

Electronics and Circuit Analysis Using MATLAB®

SECOND EDITION

Electronics and Circuit Analysis Using MATLAB®

JOHN OKYERE ATTIA

CRC PRESS

Boca Raton London New York Washington, D.C.

Library of Congress Cataloging-in-Publication Data

Attia, John Okyere.
Electronics and circuit analysis using MATLAB / John O. Attia.—2nd ed.
p. cm.
Includes bibliographical references and index.
ISBN 0-8493-1892-0 (alk. paper)
1. Electronics—Data processing. 2. Electric circuit analysis—Data processing. 3. MATLAB. I. Title.

TK7835.A88 2004
621.381′0285—dc22 2004045730

Direct all inquiries to CRC Press LLC, 2000 N.W. Corporate Blvd., Boca Raton, Florida 33431.

No claim to original U.S. Government works
International Standard Book Number 0-8493-1892-0
Library of Congress Card Number 2004045730
Printed in the United States of America 1 2 3 4 5 6 7 8 9 0
Printed on acid-free paper

Dedication

Dedicated to my family members

Christine, John II, and Angela

for

their unfailing love, support, and encouragement

Preface

MATLAB® is numeric computation software for engineering and scientific calculations. MATLAB is increasingly being used by students, researchers, practicing engineers, and technicians. The causes of MATLAB popularity are legion. Among them are its iterative mode of operation, built-in functions, simple programming, rich set of graphing facilities, possibilities for writing additional functions, and extensive toolboxes.

The goals of writing this book are:

1. To provide the reader with a simple, easy, hands-on introduction to MATLAB
2. To demonstrate the use of MATLAB for solving electronics problems
3. To show the various ways MATLAB can be used to solve circuit analysis problems
4. To show the flexibility of MATLAB for solving general engineering and scientific problems

Audience

This book can be used by students, professional engineers, and technicians. The first part of the book can be used as a primer to MATLAB. It will be useful to all students and professionals who want a basic introduction to MATLAB. Parts 2 and 3 are for electrical engineering students and electrical engineers who want to use MATLAB to explore the characteristics of semiconductor devices and the application of MATLAB for analysis and design of electrical and electronic circuits and systems.

Organization

The book is divided into three parts: Introduction to MATLAB, circuit analysis applications using MATLAB, and electronics applications with MATLAB. It is recommended that the reader work through and experiment with the examples at a computer while reading Chapters 1, 2, and 3. The hands-on approach is one of the best ways of learning MATLAB.

Part II consists of Chapters 4 to 8. This part covers the applications of MATLAB in circuit analysis. The topics covered in Part II are dc analysis, transient analysis, alternating current analysis, and Fourier analysis. In addition, two-port networks are covered. I have covered the underlying theory and concepts briefly, not with the aim of writing a textbook on circuit analysis and electronics. Selected problems in circuit analysis have been solved using MATLAB.

Part III includes Chapters 9 through 13. The topics discussed in this part are diodes, semiconductor physics, operational amplifiers, transistor circuits, and electronic data analysis. Application of MATLAB for problem solving in electronics is discussed. Extensive examples showing the use of MATLAB for solving problems in electronics are presented.

Each chapter has its own bibliography and problems. Since the text contains a large number of examples that illustrate electronics and circuit analysis principles and applications with MATLAB, the m-files of the examples in the book are available at the CRC website. The reader can run the examples without having to enter the commands. The examples can also be modified to suit the needs of the reader.

Changes in the Second Edition

Chapters 1 to 3 have been rewritten to include additional MATLAB functions and to bring the materials in those chapters up to date with changes in the MATLAB software package. New topics have been added in Chapters 7, 8, and 10. A new chapter on electronic data analysis has been added. This book has 101 solved examples, 26% more than the first edition. Furthermore, there are 134 end-of-chapter problems, which represent 58% more problems than those in the previous edition of this book.

Acknowledgments

I appreciate the suggestions and comments from a number of reviewers of the first edition of this book, including Dr. Murari Kejariwal, Dr. Reginald Perry, Dr. Richard Wilkins, Dr. Warsame Ali, Anowarul Huq, and John Abbey. Their frank and positive criticisms led to considerable improvement of this work.

Special thanks go Nora Konopka, acquisition editor at CRC Press, and to Helena Redshaw, supervisor, EPD department at CRC Press, for their support in getting this book to market. A final note of gratitude goes to my wife, Christine N. Okyere, who encouraged me to finish the book in record time.

Acknowledgments

List of Examples in Text

Contents

1

MATLAB Fundamentals

MATLAB is numeric computation software for engineering and scientific calculations. The name MATLAB stands for MATRIX LABORATORY. MATLAB is primarily a tool for matrix computations, developed by John Little and Cleve Moler of MathWorks, Inc. MATLAB is a high-level language whose basic data type is a matrix that does not require dimensioning. All computations are performed in complex-valued double-precision arithmetic to guarantee high accuracy.

MATLAB has a rich set of plotting capabilities, with graphics integrated into it. Since MATLAB is also a programming environment, a user can extend the functional capabilities of MATLAB by writing new modules.

MATLAB has a large collection of toolboxes in a variety of domains. Examples include control system, signal processing, neural network, image processing, and system identification. The toolboxes consist of functions that can be used to perform computations in a specific domain.

1.1 MATLAB Basic Operations

When MATLAB is invoked, the command window will display the prompt >>. MATLAB is then ready for entering data or executing commands. To quit MATLAB, type the command

```
exit or quit
```

MATLAB has on-line help. To see the list of MATLAB's help facility, type

```
help
```

The help command followed by a function name is used to obtain information on a specific MATLAB function. For example, to obtain information on the use of fast Fourier transform function, **fft**, one can type the command

```
help fft
```

Another way to obtain help in MATLAB is the **lookfor** command. Whereas the **help** command searches for an exact function name match, the **lookfor** command searches the quick summary information of each function for a match. The lookfor command tends to be slower than the **help command**.

The basic data object in MATLAB is a rectangular numerical matrix with real or complex elements. Scalars are thought of as 1-by-1 matrices. Vectors are considered as matrices with one row or column. MATLAB has no dimension statement or type declarations. Storage of data and variables is allocated automatically once the data and variables are used.

MATLAB statements are normally of the form:

```
variable = expression
```

Expressions typed by the user are interpreted and immediately evaluated by the MATLAB system. If a MATLAB statement ends with a semicolon, MATLAB evaluates the statement but suppresses the display of the results. MATLAB is also capable of executing a number of commands that are stored in a file. This will be discussed in Section 1.6. A matrix,

$$A = \begin{bmatrix} 1 & 2 & 3 \\ 2 & 3 & 4 \\ 3 & 4 & 5 \end{bmatrix}$$

may be entered as follows:

```
A = [1 2 3; 2 3 4; 3 4 5];
```

Note that the matrix entries must be surrounded by brackets [ ], with row elements separated by blanks or by commas. The end of each row, with the exception of the last row, is indicated by a semicolon. A matrix A can also be entered across three input lines as

```
A = [ 1  2  3
        2  3  4
        3  4  5];
```

In this case, the carriage returns replace the semicolons. A row vector B with four elements,

```
B = [ 6 9 12 15 18 ]
```

can be entered in MATLAB as

```
B = [6 9 12 15 18];
```

or

```
B = [6, 9, 12, 15, 18]
```

For readability, it is better to use spaces rather than commas between the elements. The row vector B can be turned into a column vector by **transposition**, which is obtained by typing

```
C = B'
```

The above results in

```
C =
        6
        9
        12
        15
        18
```

Other ways of entering the column vector C are

```
C =  [6
      9
      12
      15
      18]
```

or

```
C  = [6; 9; 12; 15; 18]
```

MATLAB is case sensitive in naming variables, commands, and functions. Thus b and B are not the same variable. If you do not want MATLAB to be case sensitive, you can use the command

```
casesen off
```

To obtain the size of a specific variable, type **size ( )**. For example, to find the size of matrix A, you can execute the following command:

```
size(A)
```

The result will be a row vector with two entries. The first is the number of rows in A, the second the number of columns in A.

Table 1.1 shows additional MATLAB commands to get one started on MATLAB. Detailed descriptions and usages of the commands can be obtained from the MATLAB help facility or from MATLAB manuals.

1.2 Matrix Operations

The basic matrix operations are addition (+), subtraction (–), multiplication (*), and conjugate transpose (') of matrices. In addition to the above basic operations, MATLAB has two forms of matrix division: the left inverse operator \ or the right inverse operator /.

TABLE 1.1

Some Basic MATLAB Commands

Command	Description
%	Comments; everything appearing after the % command is not executed
demo	Access on-line demo programs
length	Length of a matrix
clear	Clears the variables or functions from workspace
clc	Clears the command window during a work session
clg	Clears graphic window
diary	Saves a session in a disk, possibly for printing at a later date

Matrices of the same dimensions may be subtracted or added. Thus if E and F are entered in MATLAB as

```
E = [7 2 3; 4 3 6; 8 1 5];

F = [1 4 2; 6 7 5; 1 9 1];
```

and

```
G = E - F

H = E + F
```

then matrices G and H will appear on the screen as

```
G =

     6    -2     1
    -2    -4     1
     7    -8     4

H =

     8     6     5
    10    10    11
     9    10     6
```

A scalar (1-by-1 matrix) may be added to or subtracted from a matrix. In this particular case, the scalar is added to or subtracted from all the elements of another matrix. For example,

```
J = H + 1
```

gives

```
J =

     9     7     6
    11    11    12
    10    11     7
```

Matrix multiplication is defined provided the inner dimensions of the two operands are the same. Thus, if X is an n-by-m matrix and Y is an i-by-j matrix, X*Y is defined provided m is equal to i. Since E and F are 3-by-3 matrices, the product

```
Q = E*F
```

results as

```
Q =
        22      69      27
        28      91      29
        19      84      26
```

Any matrix can be multiplied by a scalar. For example,

```
2*Q
```

gives

```
ans =
           44     138      54
           56     182      58
           38     168      52
```

Note that if a variable name and the "=" sign are omitted, a variable named **ans** is automatically created.

Matrix division can either be the left division operator \ or the right division operator /. The right division a/b, for instance, is algebraically equivalent to

$$\frac{a}{b}$$

while the left division a\b is algebraically equivalent to

$$\frac{b}{a}$$

If $Z * I = V$ and Z is nonsingular, the left division $Z\backslash V$ is equivalent to the MATLAB expression

$$I = inv(Z) * V$$

where **inv** is the MATLAB function for obtaining the inverse of a matrix. The right division denoted by V/Z is equivalent to the MATLAB expression

$$I = V * inv(Z)$$

TABLE 1.2

Some Common MATLAB Functions

Function	Description
abs(x)	Calculates the absolute value of x
acos(x)	Determines $\cos^{-1}$x, with the results in radians
asin(x)	Determines $\sin^{-1}$x, with the results in radians
atan(x)	Calculates $\tan^{-1}$x, with the results in radians
atan2(x)	Obtains $\tan^{-1}\left(\frac{y}{x}\right)$ over all four quadrants of the circle; the results are in radians
cos(x)	Calculates cos(x), with x in radians
exp(x)	Computes e^x
log(x)	Determines the natural logarithm $\log_e(x)$
sin(x)	Calculates sin(x), with x in radians

TABLE 1.3

Some Utility Matrices

Function	Description
diag(A)	Produces a vector consisting of the diagonal of a square matrix A
eye(n)	Generates an n-by-n identity matrix
eye(n,m)	Generates an n-by-m identity matrix
ones(n)	Produces an n-by-m matrix with all the elements being unity
ones(n,m)	Produces an n-by-m matrix with all the elements being unity
zeros(n)	Generates an n-by-n matrix of zeros
zeros(n,m)	Generates an n-by-m matrix of zeros

In addition to the function **inv**, there are some other common MATLAB functions worth noting. These functions can be found in Table 1.2.

There are MATLAB functions that can be used to produce special matrices and also to initialize variables. Examples are given in Table 1.3.

Example 1.1 Voltage vs. Current Relation of Network

The voltage vs. current relationship of a network is given as

$$\begin{bmatrix} 10 \\ 15 \\ 12 \end{bmatrix} = \begin{bmatrix} 3 & 6 & 9 \\ 6 & 15 & 12 \\ 9 & 12 & 20 \end{bmatrix} \begin{bmatrix} I_1 \\ I_2 \\ I_3 \end{bmatrix}$$

Solve for the current I_1, I_2, and I_3.

Solution

We use the **inv** command to solve for the current.

MATLAB script

```
% Example 1.1
% this program determines the current
% matrix Z and voltage vector V
% Z is the impedance matrix
% V is the voltage matrix
% initialize the matrix Z and vector V
Z = [3     6     9;
     6     15    12;
     9     12    20];
V = [10   15   12]';
% solve for the loop currents
I = inv(Z)*V;
% Current is printed
I
```

MATLAB produces the following result:

```
I =

     -3.6491
      1.2807
      1.4737
```

Thus

I_1 = -3.6491 I_2 = 1.2807 I_3 = 1.4737

1.3 Array Operations

Array operations refer to element-by-element arithmetic operations. Preceding the linear algebraic matrix operations, * / \ ', by a period (.) indicates an array or element-by-element operation. Thus, the operators .* , .\ , ./, .^ , represent element-by-element multiplication, left division, right division, and raising to the power, respectively. For addition and subtraction, the array and matrix operations are the same. Thus, + and .+ can be regarded as an array or matrix addition.

If A1 and B1 are matrices of the same dimensions, then A1.*B1 denotes an array whose elements are products of the corresponding elements of A1 and B1. Thus, if

```
A1 = [2 7 6
      8 9 10];

B1 = [6 4 3
      2 3 4];
```

then

```
C1 = A1.*B1
```

results in

```
C1 =
        12     28     18
        16     27     40
```

An array operation for left and right division also involves element-by-element operation. The expressions A1./B1 and A1.\B1 give the quotient of element-by-element division of matrices A1 and B1. The statement

```
D1 = A1./B1
```

gives the result

```
D1 =
        0.3333     1.7500     2.0000
        4.0000     3.0000     2.5000
```

and the statement

```
E1 = A1.\B1
```

gives

```
E1 =
       3.0000     0.5714     0.5000
       0.2500     0.3333     0.4000
```

The array operation of raising to the power is denoted by .^. The general statement will be of the form:

q = r1.^s1

If r1 and s1 are matrices of the same dimensions, then the result q is also a matrix of the same dimensions. For example, if

```
r1 = [7 3 5];

s1 = [2 4 3];
```

then

```
q1 = r1.^s1
```

gives the result

```
q1 =
       49     81    125
```

One of the operands can be scalar. For example,

```
q2 = r1.^2
q3 = (2).^s1
```

will give

```
q2 =
        49      9     25
```

and

```
q3 =
         4     16      8
```

Note that when one of the operands is scalar, the resulting matrix will have the same dimensions as the matrix operand.

1.4 Complex Numbers

MATLAB allows operations involving complex numbers. Complex numbers are entered using function i or j. For example, a number $z = 2 + j2$ may be entered in MATLAB as

```
z = 2+2*i
```

or

```
z = 2+2*j
```

Also, a complex number za,

$$za = 2\sqrt{2}\exp[(\pi/4)j]$$

can be entered in MATLAB as

```
za = 2*sqrt(2)*exp((pi/4)*j)
```

It should be noted that when complex numbers are entered as matrix elements within brackets, one should avoid any blank spaces. For example, $y = 3 + j4$ is represented in MATLAB as

```
y = 3+4*j
```

If spaces exist around the + sign, such as

u = 3 + 4*j

MATLAB considers it as two separate numbers, and y will not be equal to u.

If w is a complex matrix given as

$$w = \begin{bmatrix} 1+j1 & 2-j2 \\ 3+j2 & 4+j3 \end{bmatrix}$$

then we can represent it in MATLAB as

```
w = [1+j  2-2*j;  3+2*j  4+3*j]
```

which will produce the result

```
w =
      1.0000 + 1.0000i   2.0000 - 2.0000i
      3.0000 + 2.0000i   4.0000 + 3.0000i
```

If the entries in a matrix are complex, then the "prime" (') operator produces the conjugate transpose. Thus,

```
wp = w'
```

will produce

```
wp =
      1.0000 - 1.0000i   3.0000 - 2.0000i
      2.0000 + 2.0000i   4.0000 - 3.0000i
```

For the unconjugate transpose of a complex matrix, we can use the point transpose (.') command. For example,

```
wt = w.'
```

will yield

```
wt =
       1.0000 + 1.0000i   3.0000 + 2.0000i
       2.0000 - 2.0000i   4.0000 + 3.0000i
```

There are several functions for manipulating complex numbers. Some of the functions are shown in Table 1.4.

TABLE 1.4

Some MATLAB Functions for Manipulating Complex Numbers

Function	Description
conj(Z)	Obtains the complex conjugate of a number Z. If Z = x + iy, then conj(Z) = x – iy
real(Z)	Returns the real part of the complex number Z
imag(Z)	Returns the imaginary part of the complex number Z
abs(Z)	Computes the magnitude of the complex number Z
angle(Z)	Calculates the angle of the complex number Z, determined from the expression `atan2(imag(Z), real(Z))`

Example 1.2 Magnitude of Input Impedance

The input impedance of a circuit is given as

$$Z = \frac{(5+j6)(4-j8)}{9-j2} + 4\angle 30^\circ \text{ Ohms}$$

Find the magnitude of the impedance Z.

Solution

MATLAB script

```
% Example 1.2
% Evaluation of Z
% the complex numbers are entered
Z1 = 5+6*j;
Z2 = 4-8*j;
Z3 = 9-2*j;
theta = (30/180)*pi;    % angle in radians
Z4 = 4*exp(j*theta);
Z_imp = (Z1*Z2/Z3)+Z4;
Z_mag = abs (Z_imp);    % magnitude of Z
Z_angle = angle(Z_imp)*(180/pi);  % Angle in degrees
disp('complex number Z in polar form, mag, phase');%
displays text inside brackets
Z_polar = [Z_mag, Z_angle]
```

MATLAB produces the following result: complex number Z in polar form, mag, and phase

```
Z_polar = 11.2039    9.7942
```

Thus the magnitude of Z is 11.2039

1.5 The Colon Symbol (:)

The colon symbol (:) is one of the most important operators in MATLAB. It can be used (1) to create vectors and matrices, (2) to specify submatrices and vectors, and (3) to perform iterations. The statement

```
t1 = 1:6
```

will generate a row vector containing the numbers from 1 to 6 with unit increment. MATLAB produces the result

```
t1 =
          1    2    3    4    5    6
```

Nonunity positive or negative increments may be specified. For example, the statement

```
t2 = 3:-0.5:1
```

will result in

```
t2 =
          3.0000    2.5000    2.0000    1.5000    1.0000
```

The statement

```
t3 = [(0:2:10);(5:-0.2:4)]
```

will result in a 2-by-6 matrix

```
t3 =
      0         2.0000  4.0000  6.0000  8.0000  10.0000
      5.0000    4.8000  4.6000  4.4000  4.2000   4.0000
```

Other MATLAB functions for generating vectors are linspace and logspace. **Linspace** generates linearly evenly spaced vectors, while **logspace** generates logarithmically evenly spaced vectors. The usage of these functions is of the form:

linspace(i_value, f_value, np)

logspace(i_value, f_value, np)

where i_value is the initial value, f_value is the final value, and np is the total number of elements in the vector.

For example,

t4 = linspace(2, 6, 8)

will generate the vector

```
t4 =
     Columns 1 through 7
     2.0000 2.5714 3.1429 3.7143 4.2857 4.8571 5.4286
     Column 8
     6.0000
```

Individual elements in a matrix can be referenced with subscripts inside parentheses. For example, t2(4) is the fourth element of vector t2. Also, for matrix t3, t3(2,3) denotes the entry in the second row and third column.

Using the colon as one of the subscripts denotes all of the corresponding row or column. For example, t3(:,4) is the fourth column of matrix t3. Thus, the statement

```
t5 = t3(:,4)
```

will give

```
t5 =

        6.0000
        4.4000
```

Also, the statement t3(2,:) is the second row of matrix t3. That is, the statement

```
t6 = t3(2,:)
```

will result in

```
t6 =

     5.0000   4.8000   4.6000   4.4000   4.2000   4.0000
```

If the colon exists as the only subscript, such as t3(:), the latter denotes the elements of matrix t3 strung out in a long column vector. Thus, the statement

```
t7 = t3(:)
```

will result in

```
t7 =

         0
    5.0000
    2.0000
    4.8000
    4.0000
    4.6000
    6.0000
    4.4000
    8.0000
    4.2000
   10.0000
    4.0000
```

Example 1.3 Power Dissipation in a Resistor

The voltage, v, across a resistance is given as (Ohm's Law) $v = Ri$, where i is the current and R the resistance. The power dissipated in resistor R is given by the expression

$$P = Ri^2$$

If $R = 10$ and the current is increased from 0 to 10 A with increments of 2 A, write a MATLAB program to generate a table of current, voltage, and power dissipation.

Solution

MATLAB script

```
% Example 1.3
% diary causes output to be written into file ex1_1.dat
% Voltage and power calculation
R=10;          % Resistance value
i=(0:2:10); % Generate current values
v=i.*R;         % array multiplication to obtain voltage
p=(i.^2)*R;   % power calculation
sol=[i v p]   % current, voltage and power values are
printed
```

MATLAB produces the following result:

```
sol =

      Columns 1 through 6
      0        2        4        6        8        10
      Columns 7 through 12
      0       20       40       60       80       100
      Columns 13 through 18
      0       40      160      360      640      1000
```

Columns 1 through 6 constitute the current values, columns 7 through 12 are the voltages, and columns 13 through 18 are the power dissipation values.

1.6 M-Files

Normally, when single line commands are entered, MATLAB processes the commands immediately and displays the results. MATLAB is also capable of processing a sequence of commands that are stored in files with extension m. MATLAB files with extension m are called m-files. The latter are ASCII text files, and they are created with a text editor or word processor. To list m-files in the current directory on your disk, you can use the MATLAB command **what**. The MATLAB command **type** can be used to show the contents of a specified file. M-files can either be script files or function files.

Both script and function files contain sequences of commands. However, function files take arguments and return values.

1.6.1 Script Files

Script files are especially useful for analysis and design problems that require long sequences of MATLAB commands. With a script file written using a text editor or word processor, the file can be invoked by entering the name of the m-file, without the extension. Statements in a script file operate globally on the workspace data. Normally, when m-files are executing, the commands are not displayed on screen. The MATLAB echo command can be used to view m-files while they are executing. To illustrate the use of a script file, a script file will be written to simplify the following complex valued expression z.

Example 1.4 Complex Number Representation

Simplify the complex number z and express it both in rectangular and polar form.

$$z = \frac{(3 + j4)(5 + j2)(2\angle 60^0)}{(3 + j6)(1 + j2)}$$

Solution

The following program shows the script file that was used to evaluate the complex number, z, and express the result in polar notation and rectangular form.

MATLAB script

```
% Example 1.4
% Evaluation of Z
% the complex numbers are entered
Z1 = 3+4*j;
Z2 = 5+2*j;
theta = (60/180)*pi;    % angle in radians
Z3 = 2*exp(j*theta);
Z4 = 3+6*j;
Z5 = 1+2*j;
% Z_rect is complex number Z in rectangular form
disp('Z in rectangular form is');% displays text inside
brackets
Z_rect = Z1*Z2*Z3/(Z4*Z5);
Z_rect
Z_mag = abs (Z_rect);    % magnitude of Z
```

```
Z_angle = angle(Z_rect)*(180/pi);   % Angle in degrees
disp('complex number Z in polar form, mag, phase');%
displays text inside brackets
Z_polar = [Z_mag, Z_angle]
```

The program is named ex1_4se.m. You can execute it by typing ex1_4se in the MATLAB command window. Observe the result, which should be

```
Z in rectangular form is

Z_rect =

   3.5546 + 0.5035i

complex number Z in polar form, mag, phase

Z_polar =

     3.5901    8.0616
```

1.6.2 Function Files

Function files are m-files that are used to create new MATLAB functions. Variables defined and manipulated inside a function file are local to the function, and they do not operate globally on the workspace. However, arguments may be passed into and out of a function file.

The general form of a function file is:

function variable(s) = function_name (arguments)
% help text in the usage of the function
%

.

.

To illustrate the usage of function files and rules for writing m-file functions, let us study the following two examples.

Example 1.5 Equivalent Resistance

Write a function file to solve the equivalent resistance of series-connected resistors, R1, R2, R3, …, Rn.

Solution

MATLAB script

```
function req = equiv_sr(r)
% equiv_sr is a function program for obtaining
%             the equivalent resistance of series
%             connected resistors
```

```
% usage:  req = equiv_sr(r)
%               r is an input vector of length n
%               req is an output, the equivalent
resistance(scalar)
%
%
n = length(r);      % number of resistors
req = sum (r);      % sum up all resistors
```

The above MATLAB script can be found in the function file equiv_sr.m.

Suppose we want to find the equivalent resistance of the series-connected resistors 10, 20, 15, 16, and 5 ohms. The following statements can be typed in the MATLAB command window to reference the function equiv_sr:

```
a = [10 20 15 16 5];
Rseries = equiv_sr(a)
```

The result obtained from MATLAB is

```
Rseries =
          66
```

Example 1.6 Quadratic Equation

Write a MATLAB function to obtain the roots of the quadratic equation

$$ax^2 + bx + c = 0$$

Solution

MATLAB script

```
function rt = rt_quad(coef)
%
% rt_quad  is a function for obtaining the roots of
%             of a quadratic equation
%
% usage: rt = rt_quad(coef)
%             coef is the coefficients a,b,c of the
quadratic
%                  equation ax*x + bx + c =0
%             rt are the roots, vector of length 2
%
% coefficient a, b, c are obtained from vector coef
a = coef(1);
b = coef(2);
c = coef(3);
int = b^2 - 4*a*c;
```

```
if int > 0
   srint = sqrt(int);
   x1= (-b + srint)/(2*a);
   x2= (-b - srint)/(2*a);
elseif int == 0
   x1= -b/(2*a);
   x2= x1;
elseif int < 0
   srint = sqrt(-int);
   p1 = -b/(2*a);
   p2 = srint/(2*a);
   x1 = p1+p2*j;
   x2 = p1-p2*j;
end
rt =[x1;
     x2];
```

The above MATLAB script can be found in the function file rt_quad.m. We can use m-file function, rt_quad, to find the roots of the following quadratic equations:

(a) $x^2 + 3x + 2 = 0$
(b) $x^2 + 2x + 1 = 0$
(c) $x^2 - 2x + 3 = 0$

The following statements, which can be found in the m-file ex1_4.m, can be used to obtain the roots:

```
% Example 1.6
ca = [1 3 2];
ra = rt_quad(ca)
cb = [1 2 1];
rb = rt_quad(cb)
cc = [1 -2 3];
rc = rt_quad(cc)

diary ex1_6.dat
ca = [1 3 2];
ra = rt_quad(ca)
cb = [1 2 1];
rb = rt_quad(cb)
cc = [1 -2 3];
rc = rt_quad(cc)
diary
```

Type in the MATLAB command window the statement ex1_6 and observe the results. The following results will be obtained:

```
ra =
        -1
        -2

rb =
        -1
        -1

rc=
        1.0000 + 1.4142i
        1.0000  - 1.4142i
```

The following is a summary of the rules for writing MATLAB m-file functions:

1. The word "function" appears as the first word in a function file. This is followed by an output argument, an equal sign, and the function name. The arguments to the function follow the function name and are enclosed within parentheses.
2. The information that follows the function, beginning with the % sign, shows how the function is used and what arguments are passed. This information is displayed if help is requested for the function name.
3. MATLAB can accept multiple input arguments and multiple output arguments can be returned.
4. If a function is going to return more than one value, all the values should be returned as a vector in the function statement. For example,

   ```
   function [mean, variance] = data_in(x)
   ```

   will return the mean and variance of a vector x. The mean and variance are computed with the function.
5. If a function has multiple input arguments, the function statement must list the input arguments. For example,

   ```
   function [mean, variance] = data(x,n)
   ```

   will return the mean and variance of a vector x of length n.

1.7 Mathematical Functions

A partial list of mathematical functions that are available in MATLAB is shown in Table 1.5. A brief description of the various functions is also given below.

TABLE 1.5

Common Mathematical Functions

Function Name	Explanation of Function
abs(x)	Absolute value or magnitude of complex number; calculates |x|
acos(x)	Inverse cosine; $\cos^{-1}(x)$, the results are in radians
angle(x)	Four-quadrant angle of a complex number; phase angle of complex number x in radians
asin(x)	Inverse sine, calculates $\sin^{-1}(x)$ with results in radians
atan(x)	Calculates $\tan^{-1}(x)$, with the results in radians
atan2(x,y)	Four-quadrant inverse; this function calculates $\tan^{-1}(y/x)$ over all four quadrants of the circle; the result is in radians in the range $-\pi$ to $+\pi$
ceil(x)	Round x to the nearest integer towards positive infinity, thus ceil(4.2) = 4; ceil(–3.3) = –3
conj(x)	Complex conjugate, i.e., x = 3 + j7; conj(x) = 3 – j7
cos(x)	Calculates cosine of x, with x in radians
exp(x)	Exponential, i.e., it calculates e^X
fix(x)	Round x to the nearest integer towards zero; fix(4.2) = 4, fix(3.3) = 3
floor(x)	Round x to the nearest integer towards minus infinity; floor(4.2) = 4 and floor(3.3) = 3
imag(x)	Complex imaginary part of x
log(x)	Natural logarithm: $\log_e(x)$
log10(x)	Common logarithm: $\log_{10}(x)$
real(x)	Real part of complex number x
rem(x,y)	Remainder after division of (x/y)
round(x)	Round towards nearest integer
sin(x)	Sine of x, with x in radians
sqrt(x)	Square root of x
tan(x)	Tangent of x

TABLE 1.6

MATLAB Predefined Values

Function	Explanation of Function
pi	Represents π
i, j	Represents the value of i ($\sqrt{-1}$)
Inf	Represents machine infinity; it is normally generated as a result of a division by zero
NaN	Stands for Not-a-Number; it is the result of an undefined mathematical operation, such as division of zero by zero
clock	Represents the current date and time in a six-element row vector containing the year, month, day, hour, minute, and second
date	Represents the current date in a character string format, such as 15-Dec-2003
eps	The short name for epsilon; it represents the floating-point precision for the computer being used; it is the smallest difference between two numbers that can be represented on the computer
ans	A special variable used to store the result of an expression if that result is not explicitly assigned to some other variable

MATLAB has a number of predefined special values. The predefined values are stored as ordinary variables, so they can be modified or overwritten by a user. A list of the most common predefined values is given in Table 1.6.

Bibliography

1. Biran, A. and Breiner, M., *MATLAB for Engineers*, Addison-Wesley, Reading, MA, 1995.
2. Chapman, S.J., *MATLAB Programming for Engineers*, Brook, Cole Thompson Learning, Pacific Grove, CA, 2000.
3. Etter, D.M., *Engineering Problem Solving with MATLAB*, 2nd ed., Prentice Hall, Upper Saddle River, NJ, 1997.
4. Etter, D.M., Kuncicky, D.C., and Hull, D., *Introduction to MATLAB 6*, Prentice Hall, Upper Saddle River, NJ, 2002.
5. Gottling, J.G., *Matrix Analysis of Circuits Using MATLAB*, Prentice Hall, Englewood Cliffs, NJ, 1995.
6. Sigmor, K., *MATLAB Primer*, 4th ed., CRC Press, Boca Raton, FL, 1998.
7. *Using MATLAB, The Language of Technical Computing, Computation, Visualization, Programming*, Version 6, MathWorks, Inc., Natick, MA, 2000.

Problems

Problem 1.1

The voltage across a discharging capacitor is

$$v(t) = 10(1 - e^{-0.2t})$$

Generate a table of voltage, $v(t)$, vs. time, t, for $t = 0$ to 50 sec with an increment of 5 sec.

Problem 1.2

Use MATLAB to evaluate the complex number

$$Z_1 = \frac{(3 + j6)(6 + j4)}{(2 + j1)j2} + 7 + j10$$

Problem 1.3

Use MATLAB to find the magnitude and phase of the complex number

$$Z_2 = \frac{(124\angle 60^0 + 100\angle 30^0)}{(60 + j80)} + 75e^{j45^0} + 25 + j36$$

Problem 1.4

Use MATLAB to simplify the expression

$$y = 0.5 + j6 + 3.5e^{j0.6} + (3 + j6)e^{j0.3\pi}$$

Problem 1.5

The voltage V is given as $V = RI$, where R and I are resistance matrix and current vector, respectively. Evaluate V given that

$$R = \begin{bmatrix} 1 & 2 & 4 \\ 2 & 3 & 6 \\ 3 & 6 & 7 \end{bmatrix}$$

and

$$I = \begin{bmatrix} 1 \\ 2 \\ 6 \end{bmatrix}$$

Problem 1.6

Use MATLAB to solve for the currents I_1, I_2, and I_3.

$$\begin{bmatrix} 20 & -10 & -15 \\ -10 & 30 & -8 \\ -15 & -8 & 65 \end{bmatrix} \begin{bmatrix} I_1 \\ I_2 \\ I_3 \end{bmatrix} = \begin{bmatrix} 10 \\ 0 \\ 0 \end{bmatrix}$$

Problem 1.7

Use MATLAB to solve for the voltages V_1, V_2, V_3, V_4, and V_5.

$$\begin{bmatrix} 4.4 & 0.5 & -0.5 & 4 & 0 \\ -0.1 & -0.2 & 0 & 0.6 & -0.2 \\ 0 & -0.5 & 0.4 & -0.2 & 0 \\ 1 & 0 & -0.8 & -0.6 & 0 \\ 0 & 0 & 0 & 0 & 1 \end{bmatrix} \begin{bmatrix} V_1 \\ V_2 \\ V_3 \\ V_4 \\ V_5 \end{bmatrix} = \begin{bmatrix} 0 \\ 0 \\ 6 \\ -12 \\ 28 \end{bmatrix}$$

Problem 1.8

Use MATLAB to find V_1, V_2, and V_3.

$$\begin{bmatrix} 0.07-j0.03 & j0.0375 & -j0.005 \\ j0.0375 & 0.02-j0.0475 & j0.0875 \\ -j0.005 & j0.0875 & 0.05-j0.04 \end{bmatrix} \begin{bmatrix} V_1 \\ V_2 \\ V_3 \end{bmatrix} = \begin{bmatrix} 0.25\angle 45^0 \\ 0 \\ 0 \end{bmatrix}$$

Problem 1.9

If $A=(x_1\ x_2\ x_3 \dots x_n)$ and $B=(y_1\ y_2\ y_3 \dots y_n)$, the dot product of A and B is given as

$$W = A.B = x_1y_1 + x_2y_2 + x_3y_3 + \dots + x_ny_n$$

(a) Write a function file to obtain the dot product of two vectors A and B.

(b) Use the function to evaluate the dot product of vectors C and D, where C = (1 5 6 8 25) and D = (2 3 8 5 7).

Problem 1.10

The equivalent resistance of resistors $R_1, R_2, R_3, \dots, R_n$ is given by

$$\frac{1}{R_{eq}} = \frac{1}{R_1} + \frac{1}{R_2} + \frac{1}{R_3} + \dots + \frac{1}{R_n}$$

(a) Write a function file that can be used to calculate the equivalent resistance of n parallel connected resistors.

(b) Find R_{eq} if $R_1 = 5$, $R_2 = 15$, $R_3 = 20$ and $R_4 = 25$. Resistances are in ohms.

Problem 1.11

Write a function file to evaluate n factorial (i.e., n!), where

$$n! = n(n-1)(n-2)\dots(2)(1)$$

Use the function to compute $x = \dfrac{7!}{3!4!}$

Problem 1.12

For a triangle with sides of length a, b, and c, the area A is given as

$$A = \sqrt{s(s-a)(s-b)(s-c)}$$

where

$$s = (a+b+c)/2$$

Write a function to compute the area given the sides of a triangle. Use the function to compute the areas of triangles with the lengths: (a) 56, 27, and 43; (b) 5, 12, and 13.

2

Plotting Functions

2.1 Graph Functions

MATLAB has built-in functions that allow one to generate x-y, polar, contour, and three-dimensional plots and bar charts. MATLAB also allows one to give titles to graphs, label the x- and y-axes, and add a grid to graphs. In addition, there are commands for controlling the screen and scaling. Table 2.1 shows a list of MATLAB's built-in graph functions. One can use MATLAB's help facility to get more information on the graph functions.

2.2 X-Y Plots and Annotations

The plot command generates a linear x-y plot. This command has three variations.

```
plot(x)
plot(x, y)
plot(x1, y1, x2, y2, x3, y3, ..., xn, yn)
```

If x is a vector, the command

```
plot(x)
```

will produce a linear plot of the elements in the vector x as a function of the index of the elements in x. MATLAB will connect the points by straight lines. If x is a matrix, each column will be plotted as a separate curve on the same graph. For example, if

```
x  = [0   3.7   6.1   6.4   5.8   3.9];
```

then **plot(x)** results in the graph shown in Figure 2.1.

If x and y are vectors of the same length, then the command

```
plot(x, y)
```

TABLE 2.1
Plotting Functions

Function	Description
axis	Freezes the axis limits
bar	Plots bar chart
contour	Performs contour plots
ginput	Puts cross-hair input from mouse
grid	Adds grid to a plot
gtext	Does mouse-positioned text
hist	Gives histogram bar graph
hold	Holds plot (for overlaying other plots)
loglog	Does log vs. log plot
mesh	Performs three-dimensional mesh plot
meshdom	Domain for three-dimensional mesh plot
pause	Wait between plots
plot	Performs linear x-y plot
polar	Performs polar plot
semilogx	Does semilog x-y plot (x-axis logarithmic)
semilogy	Does semilog x-y plot (y-axis logarithmic)
shg	Shows graph screen
stairs	Performs stair-step graph
text	Positions text at a specified location on graph
title	Used to put title on graph
xlabel	Labels x-axis
ylabel	Labels y-axis

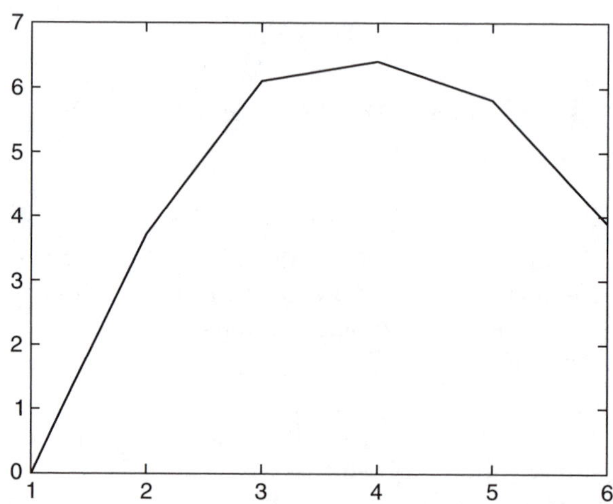

FIGURE 2.1
Graph of a row vector x.

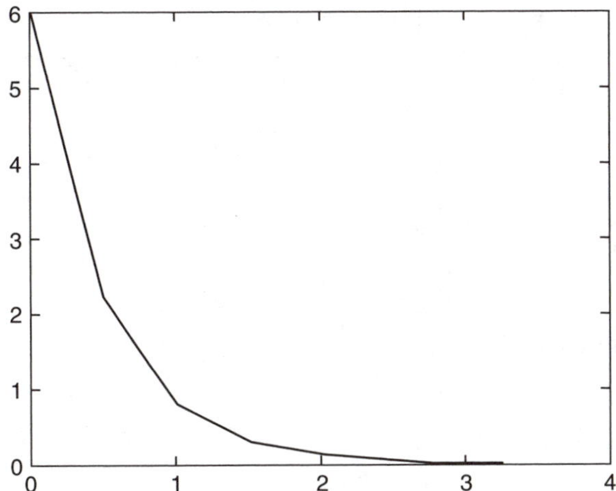

FIGURE 2.2
Graph of two vectors t and y.

plots the elements of x (x-axis) vs. the elements of y (y-axis). For example, the MATLAB commands

```
t = 0:0.5:4;
y = 6*exp(-2*t);
plot(t,y)
```

will plot the function $y(t) = 6e^{-2t}$ at the following times: 0, 0.5, 1.0, …, 4 . The plot is shown in Figure 2.2.

To plot multiple curves on a single graph, one can use the plot command with multiple arguments, such as

```
plot(x1, y1, x2, y2, x3, y3, ..., xn, yn)
```

The variables x1, y1, x2, y2, etc., are pairs of vectors. Each x-y pair is graphed, generating multiple lines on the plot. The above plot command allows vectors of different lengths to be displayed on the same graph. MATLAB automatically scales the plots. Also, the plot remains as the current plot until another plot is generated; in which case, the old plot is erased. The **hold** command holds the current plot on the screen, and inhibits erasure and rescaling. Subsequent plot commands will overplot on the original curves. The **hold** command remains in effect until the command is issued again.

When a graph is drawn, one can add a grid, a title, a label, and x- and y-axes to the graph. The commands for grid, title, x-axis label, and y-axis label are **grid** (grid lines), **title** (graph title), **xlabel** (x-axis label), and **ylabel** (y-axis label), respectively. For example, Figure 2.2 can be titled and its axes labeled with the following commands:

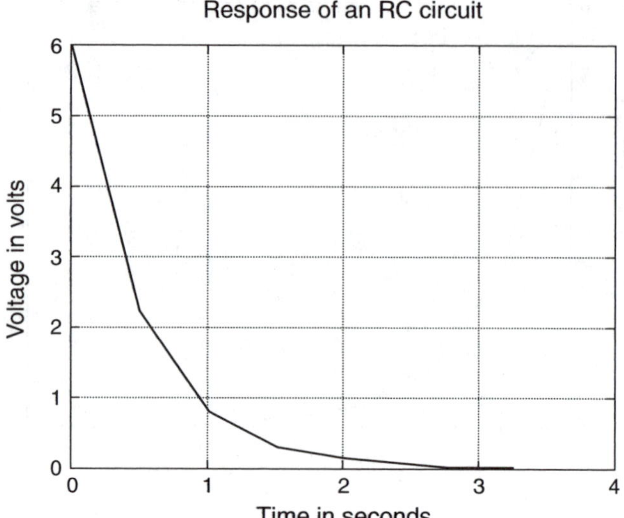

FIGURE 2.3
Graph of voltage vs. time of a response of an RC circuit.

```
t = 0:0.5:4;
y = 6*exp(-2*t);
plot(t, y)
title('Response of an RC circuit')
xlabel('time in seconds')
ylabel('voltage in volts')
grid
```

Figure 2.3 shows the graph of Figure 2.2 with title, x-axis, y-axis, and grid added.

To write text on a graphic screen beginning at a point (x, y) on the graphic screen, one can use the command

```
text(x, y, 'text')
```

For example, the statement

```
text(2.0, 1.5, 'transient analysis')
```

will write the text "transient analysis" beginning at point (2.0,1.5). Multiple text commands can be used. For example, the statements

```
plot(a1,b1,a2,b2)
text(x1,y1,'voltage')
text(x2,y2,'power')
```

will provide texts for two curves: a1 vs. b1 and a2 vs. b2. The text will be at different locations on the screen provided $x1 \neq x2$ or $y1 \neq y2$.

If the default line-types used for graphing are not satisfactory, various symbols may be selected. For example:

```
plot(a1, b1, '*')
```

draws a curve, a1 vs. b1, using star (*) symbols, while

```
plot(a1, b1, '*', a2, b2, '+')
```

uses a star (*) for the first curve and the plus (+) symbol for the second curve. Other print types are shown in Table 2.2.

For systems that support color, the color of the graph may be specified using the statement

```
plot(x, y, 'g')
```

implying "plot x vs. y using green color." Line and mark style may be added to color type using the command

```
plot(x, y, '+w')
```

The above statement implies "plot x vs. y using white + marks." Other colors that can be used are shown in Table 2.3.

The argument of the plot command can be complex. If z is a complex vector, then **plot(z)** is equivalent to **plot(real(z), imag(z)).** The following example shows the use of the plot, title, xlabel, ylabel, and text functions.

TABLE 2.2

Print Types

Line-Types	Indicators	Point Types	Indicators
Solid	-	Point	.
Dash	–	Plus	+
Dotted	:	Star	*
Dashdot	-.	Circle	o
		x-mark	x

TABLE 2.3

Symbols for Color Used in Plotting

Color	Symbol
Red	r
Green	g
Blue	b
White	w
Invisible	i

Example 2.1 Voltage and Current of an RL Circuit

For an RL circuit, the voltage $v(t)$ and current $i(t)$ are given as:

$$v(t) = 10\cos(377t)$$

$$i(t) = 5\cos(377t + 60^0)$$

Sketch $v(t)$ and $i(t)$ for t = 0 to 20 msec.

Solution

MATLAB script

```
% Example 2.1
% RL circuit
% current i(t) and  voltage v(t) are generated
%  t is time
t = 0:1E-3:20E-3;
v  = 10*cos(377*t);
a_rad = (60*pi/180);  % angle in radians
i = 5*cos(377*t + a_rad);
plot(t, v, t, v,'*',t, i, t, i,'o')
title('voltage and current of an RL circuit')
xlabel('sec')
ylabel('voltage(V) and current(mA)')
text(0.003, 1.5, 'v(t)');
text(0.009,2, 'i(t)')
```

Figure 2.4 shows the resulting graph.

2.3 Logarithmic and Polar Plots

Logarithmic and semilogarithmic plots can be generated using the commands **loglog, semilogx,** and **semilogy**. The use of these plot commands is similar to that of the plot command discussed in the previous section. The descriptions of these commands are as follows:

loglog(x, y) — generates a plot of $\log_{10}$(x) vs. $\log_{10}$(y)
semilogx(x, y) — generates a plot of $\log_{10}$(x) vs. linear axis of y
semilogy(x, y) — generates a plot of linear axis of x vs. $\log_{10}$(y)

It should be noted that since the logarithm of negative numbers and zero does not exist, the data to be plotted on the semilog axes or log-log axes should not include zero or negative values.

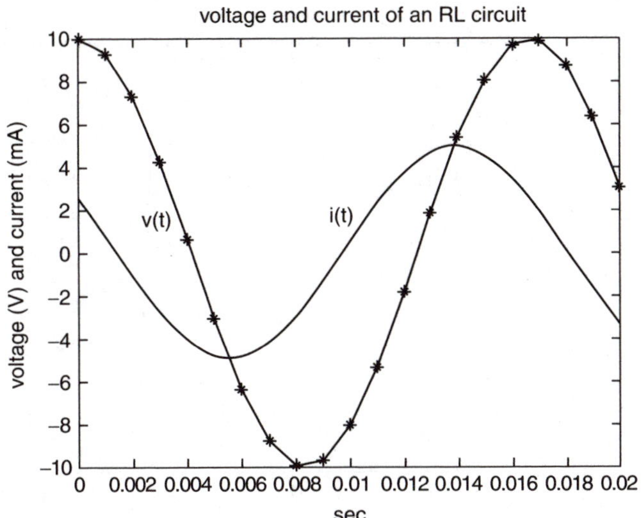

FIGURE 2.4
Plot of voltage and current of an RL circuit under sinusoidal steady-state conditions.

TABLE 2.4
Gain vs. Frequency Data for an Amplifier

Frequency (Hz)	Gain (dB)	Frequency (Hz)	Gain (dB)
20	5	2,000	34
40	10	5,000	34
80	30	8,000	34
100	32	10,000	32
120	34	12,000	30

Example 2.2 Gain vs. Frequency of an Amplifier

The gain vs. frequency of a capacitively coupled amplifier is shown Table 2.4. Draw a graph of gain vs. frequency using a logarithmic scale for the frequency and a linear scale for the gain.

Solution

MATLAB script

```
% Bode plot for capacitively coupled amplifier
f = [20 40 80 100 120 2000 5000 8000 10000 ...
     12000 15000 20000];
g = [ 5 10 30 32 34 34 34 34 32 30 10 5];
semilogx(f,g)
title('Bode plot of an amplifier')
xlabel('Frequency in Hz')
ylabel('Gain in dB')
```

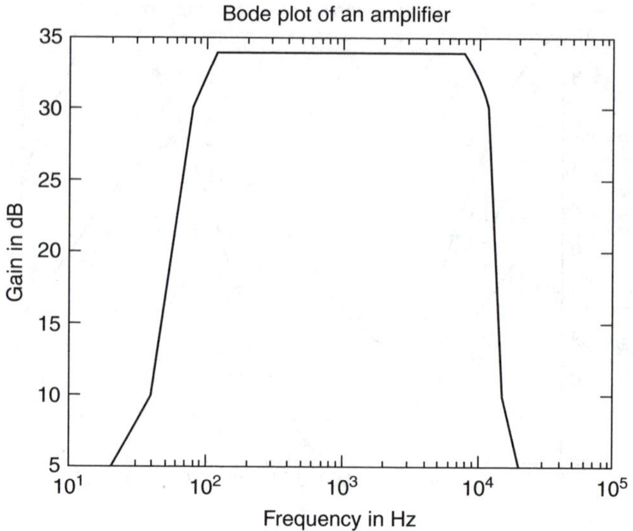

FIGURE 2.5
Plot of gain vs. frequency of an amplifier.

The plot is shown in Figure 2.5.

A **polar** plot of an angle vs. magnitude may be generated using the command

```
polar(theta, rho)
```

where theta and rho are vectors, with theta being an angle in radians and rho being the magnitude.

When the grid command is issued after the polar plot command, polar grid lines will be drawn. The polar plot command is used in the following example.

Example 2.3 Polar Plot of a Complex Number

A complex number z can be represented as $z = re^{j\theta}$. The nth power of the complex number is given as $z^n = r^n e^{jn\theta}$. If $r = 1.2$ and $\theta = 10°$, use the polar plot to plot $|z^n|$ vs. $n\theta$ for $n = 1$ to $n = 36$.

Solution

MATLAB script

```
% polar plot of z
r = 1.2;   theta = 10*pi/180;
angle = 0:theta:36*theta;   mag = r.^(angle/theta);
polar(angle,mag)
grid
title('Polar Plot')
```

The polar plot is shown in Figure 2.6.

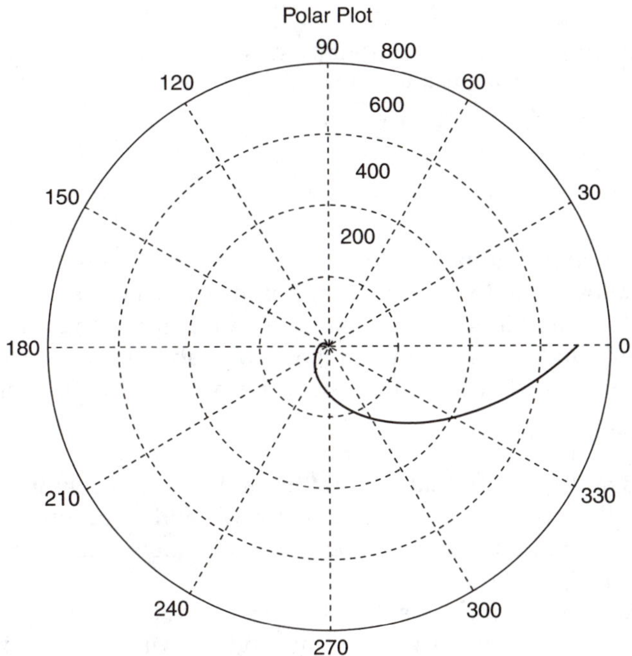

FIGURE 2.6
Polar plot of $z = 1.2^n e^{j10n}$.

The **plot3** function can be used to do three-dimensional line plots. The function is similar to the two-dimensional **plot** function. The plot3 function supports the same line size, line style, and color options that are supported by the plot function. The simplest form of the plot3 function is

plot(x, y, z)

where **x**, **y**, and **z** are equal-sized arrays containing the locations of the data points to be plotted.

2.4 Subplots and Screen Control

MATLAB has basically two display windows: a command window and a graph window. The following commands can be used to select and clear the windows:

shg shows graph window
clc clears command window
clg clears graph window
home home command cursor

TABLE 2.5

Numbering of Subwindows for Subplot Command "subplot(324)"

1	2
3	4 *(current figure)*
5	6

The graph window can be partitioned into multiple windows. The **subplot** command allows one to split the graph into two subdivisions or four subdivisions. Two subwindows can be arranged either top to bottom or left or right. A four-window partition will have two subwindows on top and two subwindows on the bottom. The general form of the subplot command is

```
subplot(i j k)
```

The digits **i** and **j** specify that the graph window is to be split into an i-by-j grid of smaller windows, arranged in *i* rows and *j* columns. The digit *k* specifies the kth window for the current plot. The subwindows are numbered from *left to right, top to bottom.*

For example, the command subplot(324) creates six subplots in the current figure and makes subplot 4 the current plotting window. This is shown in Table 2.5.

The following example illustrates the use of the subplot command.

Example 2.4 Subplots of Functions

Use MATLAB to plot (a) $y = x^2$ and (b) $z = x^3$. Plot $y = x^2$ in the top half of the graph screen and $z = x^3$ in the bottom half of the graph screen.

Solution

MATLAB script

```
%
x = -4:0.5:4;
y = x.^2; % square of x
z = x.^3; % cube of x
subplot(211), plot(x, y), title('square of x')
subplot(212), plot(x, z), title('cube of x')
```

The plots are shown in Figure 2.7.

The coordinates of points on the graph window can be obtained using the **ginput** command. There are two forms of the command:

```
[x y] = ginput

[x y] = ginput(n)
```

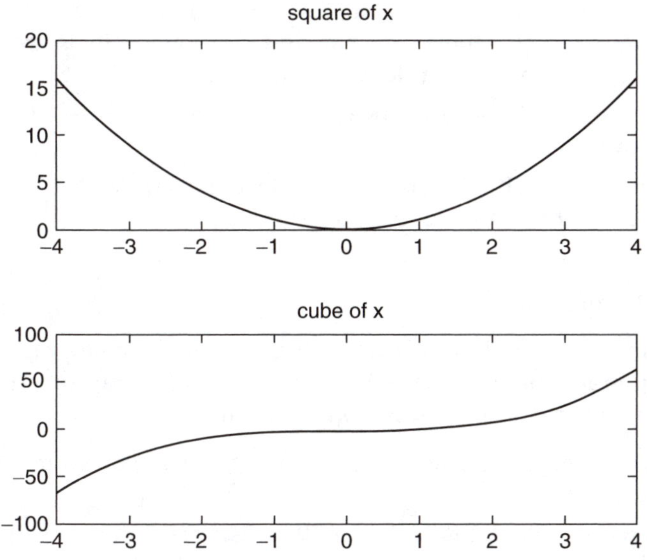

FIGURE 2.7
Plots of x^2 and x^3 using subplot commands.

- The [x y] = ginput command allows one to select an unlimited number of points from the graph window using a mouse or arrow keys. Pressing the return key terminates the input.
- The [x y] = ginput(n) command allows the selection of n points from the graph window using a mouse or arrow keys. The points are stored in vectors x and y. Data points are entered by pressing a mouse button or any key on the keyboard (except the return key). Pressing the return key terminates the input.

2.5 Other Plotting Functions

2.5.1 Bar Plots

The bar function is used to plot bar plots. The general form of the command is

```
bar(x,y)
```

This function creates a vertical bar plot, where the values in x are used to label each bar and the values in y are used to determine the height of the bar. There are other variations of the bar function, such as:

barh(x, y) — This function creates a horizontal bar plot. The values in x are used to label each bar and the values in y are used to determine the horizontal length of the bar.

bar3(x, y) — This function gives bar charts a three-dimensional appearance.

bar3h(x, y) — This function is similar to barh(x, y), but it gives the bar chart a three-dimensional appearance.

2.5.2 Hist Function

The hist function can be used calculate and plot the histogram of a set of data. A histogram shows the distribution of a set of values in a data set. The general format for using the function is as follows:

hist(x) — calculates and plots the histogram of values in a data set x by using 10 bins.

hist(x,n) — calculates and plots the histogram of values in a data set x by using n equally spaced bins.

hist(x,y) — calculates and plots the histogram of values of x using bins with centers specified by the values of the vector y.

Example 2.5 Plot of Gaussian Random Data

Random data with 5000 points, Gaussian distributed, with a mean value of zero and standard deviation of two can be generated with the equation

$$R\_data = 2*randn(500, 1) + 0.0$$

Plot the histogram of the random Gaussian data using 20 evenly spaced bins.

Solution

MATLAB script

```
% Generate the random Gaussian data
r_data = 2*randn(5000,1)+ 0.0;
hist(r_data, 20)
title('Histogram of Gaussian Data')
```

The histogram of the Gaussian random data is shown in Figure 2.8.

2.5.3 Stem Plots

The **stem** function generates a point plot with lines or stems connecting the point to the x-axis. It is normally used to plot discrete sequence data. Its usage is

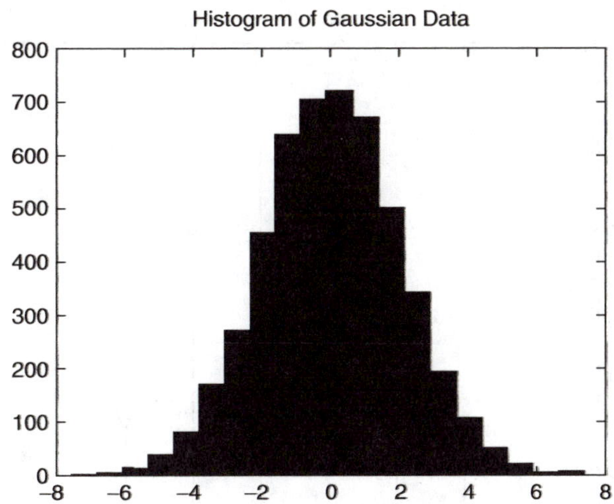

FIGURE 2.8
Histogram of Gaussian data.

`Stem(z)` — creates a plot of data points in vector z connected to the horizontal axis. An optional character string can be used to specify line style.

`Stem(x, z)` — plots the data points in z at values specified in x.

Example 2.6 Convolution between Two Discrete Data

The convolution between two discrete data X and Y is given as Z = conv(X, Y). If X = [0 1 2 3 2 1 0] and Y = [0 1 1 0], find Z. Plot X, Y, and Z.

Solution

MATLAB script

```
% The convolution function Z = conv(X, Y) is used
X = [0 1 2 3 2 1 0];
Y = [0 1 1 0];
Z = conv(X, Y);
stem(Z), title('Convolution Between X and Y')
```

The plots are shown in Figure 2.9.

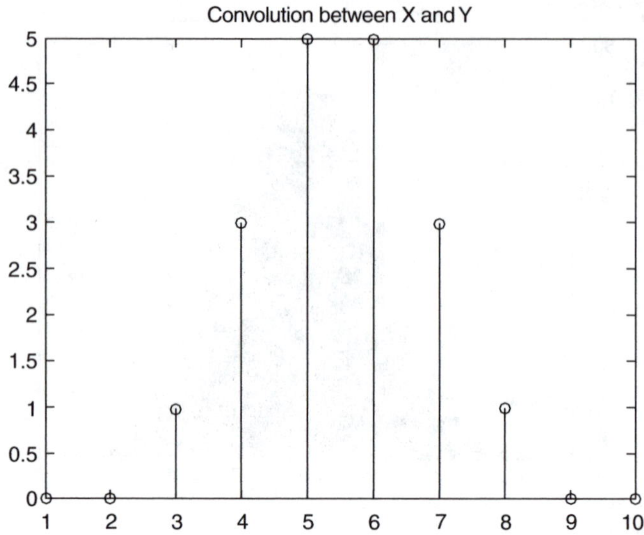

FIGURE 2.9
Convolution between X and Y.

Bibliography

1. Biran, A. and Breiner, M. *MATLAB for Engineers*, Addison-Wesley, Reading, MA, 1995.
2. Chapman, S.J., *MATLAB Programming for Engineers*, Brook, Cole Thompson Learning, Pacific Grove, CA, 2000.
3. Etter, D.M., *Engineering Problem Solving with MATLAB*, 2nd ed., Prentice Hall, Upper Saddle River, NJ, 1997.
4. Etter, D.M., Kuncicky, D.C., and Hull, D., *Introduction to MATLAB 6*, Prentice Hall, Upper Saddle River, NJ, 2002.
5. Gottling, J.G., *Matrix Analysis of Circuits Using MATLAB*, Prentice Hall, Upper Saddle River, NJ, 1995.
6. Sigmor, K., *MATLAB Primer*, 4th ed., CRC Press, Boca Raton, FL, 1998.
7. *Using MATLAB, The Language of Technical Computing, Computation, Visualization, Programming*, Version 6, MathWorks, Inc., Natick, MA, 2000.

Problems

Problem 2.1

The repulsive coulomb force that exists between two protons in the nucleus of a conductor is given as

$$F = \frac{q_1 q_2}{4\pi\varepsilon_0 r^2}$$

If $q_1 = q_2 = 1.6 \times 10^{-19}$ C, and $1/4\pi\varepsilon_0 = 8.99 \times 10^9$ Nm^2/C^2, sketch a graph of force vs. radius r. Assume a radius from 1.0×10^{-15} to 1.0×10^{-14} m with increments of 2.0×10^{-15} m.

Problem 2.2

The current flowing through a drain of a field effect transistor during saturation is given as

$$i_{DS} = k(V_{GS} - V_t)^2$$

If $V_t = 1.0$ volt and $k = 2.5$ mA/V^2, plot the current i_{DS} for the following values of V_{GS}: 1.5, 2.0, 2.5, ..., 5 V.

Problem 2.3

The voltage v and current i of a certain diode are related by the expression

$$i = I_S \exp\left[v/(nV_T)\right]$$

If $I_S = 1.0 \times 10^{-14}$ A, $n = 2.0$, and $V_T = 26$ mV, plot the current vs. voltage curve of the diode for diode voltage between 0 and 0.6 volts.

Problem 2.4

The voltage across a parallel RLC circuit is given as

$$v(t) = 5e^{-2t}\sin(1000\pi t)$$

Plot the voltage as a function of time.

Problem 2.5

Obtain the polar plot of z vs. $n\theta$, where $z = r^{-n} e^{jn\theta}$ for $\theta = 15°$ and $n = 1$ to 20. Assume that $r = 1.5$.

Problem 2.6

A function $f(x)$ is given as

$$f(x) = x^4 + 3x^3 + 4x^2 + 2x + 6$$

(a) Plot $f(x)$.
(b) Find the roots of $f(x)$.

Problem 2.7

The table below shows the grades on two examinations of ten students in a class.

Student	Exam #1	Exam #2
1	81	78
2	75	77
3	95	90
4	65	69
5	72	73
6	79	84
7	93	97
8	69	72
9	83	80
10	87	81

(a) Plot the results of each examination.
(b) Use MATLAB to calculate the mean and standard deviation of the grades on each examination.

Problem 2.8

A message signal $m(t)$ and the carrier signal $c(t)$ of a communication system are, respectively:

$$m(t) = 4\cos(120\pi t) + 2\cos(240\pi t)$$

$$c(t) = 10\cos(10{,}000\pi t)$$

A double-sideband suppressed carrier $s(t)$ is given as

$$s(t) = m(t)c(t)$$

Plot $m(t)$, $c(t)$, and $s(t)$ using the subplot command.

Problem 2.9

The closed loop gain, G, of an operational amplifier with a finite open loop gain of A is given as

$$G = \frac{-\dfrac{R_2}{R_1}}{1 + \left(\dfrac{1 + \dfrac{R_2}{R_2}}{A} \right)}$$

If $R_2 = 60$ kΩ and $R_1 = 5$ kΩ, find the closed loop gain for the following values of the open loop gain: 10^2, 10^3, 10^4, 10^5, 10^6, and 10^7. Plot the closed loop gain G vs. the open loop gain A.

Problem 2.10

The equivalent impedance of a circuit is given as

$$z(j\omega) = R + \frac{jwL}{1 - w^2 LC}$$

If L = 4 mH, C = 25 μF, and R = 120 Ω, plot the magnitude of the input impedance for w = 10, 100, 1000, 1.0e04, and 1.0e05 rad/sec.

Problem 2.11

Fifty thousand random numbers, uniformly distributed between 0 and 1, can be represented as random = rand(50000,1). Plot the histogram of the random numbers. Use 25 equally spaced bins.

Problem 2.12

Use the stem function to plot the convolution between the two discrete data x and y, if

(a) X = [0 1 1 1 0] and Y = [0 1 1 1 0]
(b) X = [0 1 2 3 4 3 2 1 0] and Y = [0 1 2 3 4 3 2 1 0]
(c) X = [0 1 1 0 –1 –1 0] and Y = [0 1 1 0]

3

Control Statements

3.1 "For" Loops

"**For**" loops allow a statement or group of statements to be repeated a fixed number of times. The general form of a for loop is

```
for index = expression
  statement group X
end
```

The expression is a matrix, and the statement group X is repeated as many times as the number of elements in the columns of the expression matrix. The index takes on the elemental values in the matrix expression. Usually, the expression is something like

```
m:n  or m:i:n
```

where m is the beginning value, n the ending value, and i is the increment.

Suppose we would like to find the squares of all the integers from 1 to 100. We could use the following statements to solve the problem:

```
sum = 0;
for i = 1:100
    sum = sum + i^2;
end
sum
```

For loops can be nested, and it is recommended that the loop be indented for readability. Suppose we want to fill a 10-by-20 matrix, b, with an element value equal to unity. The following statements can be used to perform the operation.

```
%
n = 10;% number of rows
m = 20;% number of columns
for i = 1:n
```

```
        for j = 1:m
            b(i,j) = 1; % semicolon suppresses printing in
            the loop
        end
end
b   % display the result
%
```

It is important to note that each for statement group must end with the word **end**. The following program illustrates the use of a for loop.

Example 3.1 Horizontal and Vertical Displacements

The horizontal displacement $x(t)$ and vertical displacement $y(t)$ are given with respect to time, t, as

$$x(t) = 2t$$

$$y(t) = \sin(2\pi \frac{1}{T} t)$$

where $T = 5$ msec. For $t = 0$ to 10 msec, determine the values of $x(t)$ and $y(t)$. Use the values to plot $x(t)$ vs. $y(t)$.

Solution

MATLAB script

```
%
for i= 1:100
    t = 1.0e-04;
    T = 5.0e-03;
   x(i) = 2*T*i;
   y(i) = sin(2*pi*t*i/T);
end
plot(x,y)
```

Figure 3.1 shows the plot of $x(t)$ and $y(t)$.

3.2 "If" Statements

"If" statements use relational or logical operations to determine what steps to perform in the solution of a problem. The relational operators in MATLAB for comparing two matrices of equal size are shown in Table 3.1.

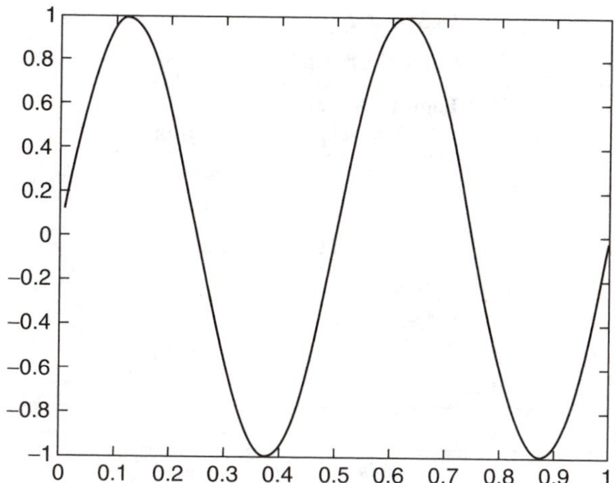

FIGURE 3.1
Plot of x vs. y.

TABLE 3.1

Relational Operators

Relational Operator	Meaning
<	Less than
<=	Less than or equal
>	Greater than
>=	Greater than or equal
==	Equal
~=	Not equal

When any of the above relational operators is used, a comparison is performed between the pairs of corresponding elements. The result is a matrix of ones and zeros, with one representing TRUE and zero FALSE. For example, if

```
a = [1 2 3 3 3 6];
b = [1 2 3 4 5 6];
a == b
```

The answer obtained is

```
ans =
         1     1     1     0     0     1
```

The ones indicate the elements in vectors *a* and *b* that are the same, and zeros represent those that are different.

There are three logical operators in MATLAB. These are shown in Table 3.2.

TABLE 3.2

Logical Operators

Logical Operator Symbol	Meaning
&	And
!	Or
~	Not

TABLE 3.3

MATLAB Logical Functions

Function Name	Description
ischar(X)	Returns a 1 if X is a character string and a 0 otherwise
isfinite (X)	Returns a 1 if the elements of X are finite and a 0 otherwise
isinf(X)	Returns 1 where the elements of X are infinite and a 0 otherwise
ismember(X, Y)	Returns 1 when the elements of X are also in Y and a 0 otherwise
isreal(X)	Returns 1 whenever the elements of X have no imaginary part and 0 otherwise

Logical operators work elementwise and are usually used on 0-1 matrices, such as those generated by relational operators. The & and ! operators compare two matrices of equal dimensions. If A and B are 0-1 matrices, then A&B is another 0-1 matrix with ones representing TRUE and zeros FALSE. The NOT(~) operator is a unary operator. The expression ~C returns 1 where C is zero and 0 when C is nonzero.

MATLAB has a number of logical functions that return a 1 whenever the test condition is true and a 0 whenever the test condition is false. The logical functions can be used to control the operations of branches and loops of a program. A list of the logical functions is shown in Table 3.3

There are several variations of the if statement:

- Simple if statement
- Nested if statement
- If-else statement
- If-elseif statement
- If-elseif-else statement

- The general form of the **simple if statement** is

```
if logical expression 1
    statement group 1
end
```

In the case of a simple if statement, if logical expression 1 is true, statement group 1 is executed. However, if the logical expression is false, statement group

1 is bypassed and the program control jumps to the statement that follows the end statement.

- The general form of a **nested if statement** is

```
if logical expression 1
        statement group 1
    if logical expression 2
        statement group 2
    end
        statement group 3
end
statement group 4
```

The program control is such that if expression 1 is true, then statement groups 1 and 3 are executed. If logical expression 2 is also true, statement groups 1 and 2 will be executed before executing statement group 3. If logical expression 1 is false, we jump to statement group 4 without executing statement groups 1, 2, and 3.

- The **if-else statement** allows one to execute one set of statements if a logical expression is true and a different set of statements if the logical statement is false. The general form of the if-else statement is

```
if logical expression 1
        statement group 1
    else
        statement group 2
end
```

In the above program segment, statement group 1 is executed if logical expression 1 is true. However, if logical expression 1 is false, statement group 2 is executed.

- The **if-elseif statement** may be used to test various conditions before executing a set of statements. The general form of the if-elseif statement is

```
if logical expression 1
        statement group 1
    elseif logical expression 2
        statement group 2
    elseif logical expression 3
        statement group 3
    elseif logical expression 4
        statement group 4
end
```

A statement group is executed provided the logical expression above it is true. For example, if logical expression 1 is true, then statement group 1 is executed. If logical expression 1 is false and logical expression 2 is true, then statement group 2 will be executed. If logical expressions 1, 2, and 3 are false and logical expression 4 is true, then statement group 4 will be executed. If none of the logical expressions is true, then statement groups 1, 2, 3, and 4 will not be executed. Only three elseif statements are used in the above example. More elseif statements may be used if the application requires them.

- The **if-elseif-else statement** provides a group of statements to be executed if other logical expressions are false. The general form of the if-elseif-else statement is

```
if logical expression 1
        statement group 1
    elseif logical expression 2
        statement group 2
    elseif logical expression 3
        statement group 3
    elseif logical expression 4
        statement group 4
    else
        statement group 5
end
```

The various logical expressions are tested. The one that is satisfied is executed. If logical expressions 1, 2, 3, and 4 are false, then statement group 5 is executed. Example 3.2 shows the use of the if-elseif-else statement.

Example 3.2 Analog-to-Digital (A/D) Converter

A three-bit A/D converter, with an analog input x and digital output y, is represented by the equation:

$$
\begin{aligned}
y &= 0 && x < -2.5 \\
&= 1 && -2.5 \le x < -1.5 \\
&= 2 && -1.5 \le x < -0.5 \\
&= 3 && -0.5 \le x < 0.5 \\
&= 4 && 0.5 \le x < 1.5 \\
&= 5 && 1.5 \le x < 2.5 \\
&= 6 && 2.5 \le x < 3.5 \\
&= 7 && x \ge 3.5
\end{aligned}
$$

Write a MATLAB program to convert analog signal x to digital signal y. Test the program by using an analog signal with the following amplitudes: –1.25, 2.57, and 6.0.

Solution

MATLAB script

```
%
y1 = bitatd_3(-1.25)
y2 = bitatd_3(2.57)
y3 = bitatd_3(6.0)

function Y_dig = bitatd_3(X_analog)
%
% bitatd_3 is a function program for obtaining
%         the digital value given an input analog
%         signal
%
% usage:  Y_dig = bitatd_3(X_analog)
%         Y_dig is the digital number
%         X_analog is the analog input
%
if X_analog < -2.5
   Y_dig = 0;
elseif X_analog >= -2.5 & X_analog < -1.5
   Y_dig = 1;
elseif X_analog >= -1.5 & X_analog < -0.5
   Y_dig = 2;
elseif X_analog >= -0.5 & X_analog < 0.5
   Y_dig = 3;
elseif X_analog >= 0.5 & X_analog < 1.5
   Y_dig = 4;
elseif X_analog >= 1.5 & X_analog < 2.5
   Y_dig = 5;
elseif X_analog >= 2.5 & X_analog < 3.5
   Y_dig = 6;
else
   Y_dig = 7;
end
Y_dig;
```

The function file, bitatd_3.m, is an m-file. In addition, the script file, ex3_2se.m, can be used to perform this example. The results obtained, when the latter program is executed, are

```
y1 =
          2
```

```
y2 =
        6

y3 =
        7
```

Example 3.3 A System with Hysteresis

The output voltage, $v_0(t)$, of a system and the input voltage, $v_S(t)$, of a system are related by the expressions

$$\begin{aligned}
v_0(t) &= -4\ V && \text{if} \quad v_S(t) \geq 0.3\ V \\
&= -4\ V && \text{if} \quad -0.3\ V < v_S(t) < 0.3\ V \quad \text{and} \quad v_0(t-1) = -4\ V \\
&= +4\ V && \text{if} \quad -0.3\ V < v_S(t) < 0.3\ V \quad \text{and} \quad v_0(t-1) = +4\ V \\
&= +4\ V && \text{if} \quad v_S(t) \leq -0.3\ V
\end{aligned}$$

If the input voltage $v_s(t)$ is a noisy signal given as

$$v_S(t) = 0.8\sin(2\pi f_0 t) + 0.4n(t)$$

where f_0 = 200 Hz and $n(t)$ is a normally distributed white noise, write a MATLAB program to find the output voltage. Plot both the input and output waveforms of the system with hysteresis.

Solution

MATLAB script

```
% vo is the output voltage
% vs is the input voltage
% Generate the sine voltage
t = 0.0:0.1e-4:5e-3;
fo=200;  % frequency of sine wave
len = length(t);
for i =1:len
   s(i) = 0.8*sin(2*pi*fo*t(i));
   % Generate a normally distributed white noise
   n(i) = 0.4*randn(1);
   % generate the noisy signal
   vs(i) = s(i) + n(i);
end
% calculation of output voltage
%
```

```
len1 = len -1;
for i=1:len1
    if  vs(i + 1) >= 0.3;
        vo(i+1) = -4;
    elseif  vs(i+1) > -0.3 & vs(i+1)< 0.3 & vo(i) == -4
            vo(i+1) = -4;
    elseif  vs(i+1) > -0.3 & vs(i+1 )< 0.3 & vo(i) == +4
            vo(i+1) = 4;
    else
            vo(i+1) = +4;
    end
end
%
% Use subplots to plot vs and vo
    subplot (211), plot (t(1:40), vs(1:40))
    title ('Noisy time domain signal')
    subplot (212), plot (t(1:40), vo(1:40))
    title ('Output Voltage')
    xlabel ('Time in sec')
```

The input and output voltages are shown in Figure 3.2.

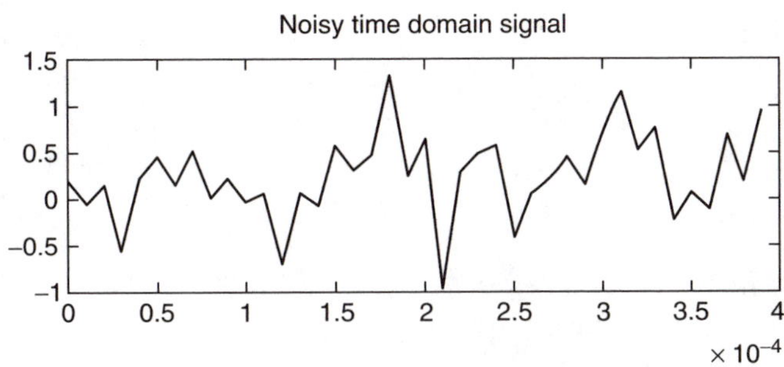

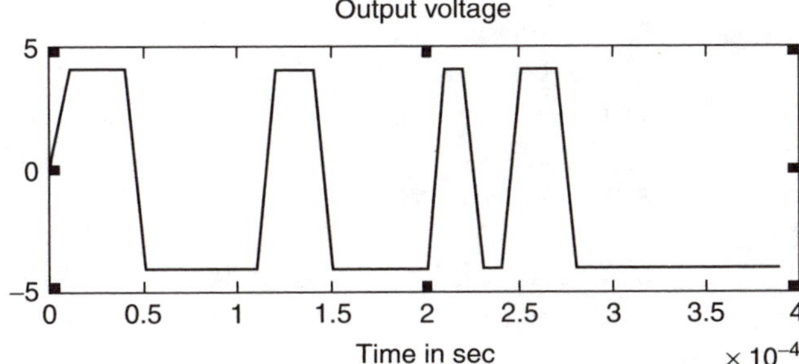

FIGURE 3.2
Input and output voltages.

3.3 "While" Loops

A "**while**" loop allows one to repeat a group of statements as long as a specified condition is satisfied. The general form of the while loop is

```
while expression 1
      statement group 1
end
statement group 2
```

When expression 1 is true, statement group 1 is executed. At the end of execution of statement group 1, expression 1 is retested. If expression 1 is still true, statement group 1 is again executed. However, if expression 1 is false, the program exits the while loop and executes statement group 2. The following example illustrates the use of the while loop.

Example 3.4 Summation of Consecutive Integers

Determine the number of consecutive integers that, when added together, will give a value equal to or just less than 210.

Solution

MATLAB script

```
% integer summation
%
int = 1;
int_sum = 0;
max_val = 210;
while int_sum < max_val
     int_sum = int_sum + int;
     int = int + 1;
end
last_int = int
if int_sum == max_val
   num_int = int - 1
   tt_int_ct =  int_sum
  elseif int_sum  > max_val
     num_int = int - 1
     tt_int_ct = int_sum - last_int
  end
end
```

The solution obtained will be

```
last_int =
                21

num_int =
                 20

tt_int_ct =
                 210
```

Thus, the number of integers starting from 1 that would add up to 210 is 20. That is,

$$1+2+3+4+...+20=210$$

3.4 Input/Output Commands

MATLAB has commands for inputting information in the command window and outputting data. Examples of input/output commands are echo, input, pause, keyboard, break, error, display, format, and fprintf. Brief descriptions of these commands are presented in Table 3.4.

Break

The **break** command may be used to terminate the execution of **for** and **while** loops. If the break command exits in an innermost part of a nested loop, the break command will exit from that loop only. The break command is useful for exiting a loop when an error condition is detected.

TABLE 3.4

Some Input/Output Commands

Command	Description
break	Exits while or for loops
disp	Displays text or matrix
echo	Displays m-files during execution
error	Displays error messages
format	Switches output display to a particular format
fprintf	Displays text and matrices and specifies format for printing values
input	Allows user input
keyboard	Invokes the keyboard as an m-file
pause	Causes an m-file to stop executing; pressing any key causes resumption of program execution

Disp

The **disp** command displays a matrix without printing its name. It can also be used to display a text string. The general form of the disp command is

```
disp(x)
disp('text string')
```

disp(x) will display the matrix x. Another way of displaying matrix x is to type its name. This is not always desirable since the display will start with a leading "x = ". **Disp ('text string')** will display the text string in quotes. For example, the MATLAB statement

```
disp('3-by-3  identity matrix')
```

will result in

```
3-by-3 identity matrix
```

and

```
disp(eye(3,3))
```

will result in

```
1     0     0
     0     1     0
     0     0     1
```

Echo

The **echo** command can be used for debugging purposes. It allows commands to be viewed as they execute. This command can be enabled or disabled.

echo on — enables the echoing of commands

echo off — disables the echoing of commands

echo — when used by itself, toggles the echo state

Error

The **error** command causes an error return from the m-files to the keyboard and displays a user written message. The general form of the command is

```
error('message for display')
```

Consider the following MATLAB statements:

```
x = input('Enter age of student');
if  x < 0
```

```
        error('wrong age was entered, try again')
    end
    x = input('Enter age of student')
```

For the above MATLAB statements, if the age is less than zero, the error message "wrong age was entered, try again" will be displayed and the user will again be prompted for the correct age.

Format

The **format** command controls the format of an output. Table 3.5 shows some formats available in MATLAB.

By default, MATLAB displays numbers in "short" format (five significant digits). The **format compact** command suppresses line feeds that appear between matrix displays, thus allowing more lines of information to be seen on the screen; **format loose** reverts to the less compact display. The format compact and format loose commands do not affect the numeric format.

Fprintf

The **fprintf** command can be used to print both text and matrix values. The format for printing the matrix can be specified, and line feed can also be specified. The general form of this command is

```
fprintf('text with format specification', matrices)
```

For example, the statements

```
cap = 1.0e-06;
fprintf('The value of capacitance is %7.3e Farads\n',
cap)
```

when executed will yield the output

```
The value of capacitance is 1.000e-006 Farads
```

TABLE 3.5

Format Displays

Command	Meaning
format short	Five significant decimal digits
format long	Fifteen significant digits
format short e	Scientific notation with five significant digits
format long e	Scientific notation with 15 significant digits
format hex	Hexadecimal
format +	+ printed if value is positive, – if negative; space skipped if value is zero

The format specifier **%7.3e** is used to show where the matrix value should be printed in the text. The 7.3e indicates that the resistance value should be printed with an exponential notation of seven digits, three of which should be decimal digits. Other format specifiers are

%c — single character
%d — decimal notation (signed)
%e — exponential notation (using a lowercase e as in 2.051e+01
%f — fixed-point notation
%g — signed decimal number in either %e or %f format, whatever is shorter

The text with format specification should end with **\n** to indicate the end of a line. However, we can also use \n to get line feeds as represented by the following example:

```
r1 = 1500;
fprintf('resistance is \n%f Ohms \n', r1)
```

The output is

```
resistance is
1500.000000 Ohms
```

Input

The **input** command displays a user-written text string on the screen, waits for an input from the keyboard, and assigns the number entered on the keyboard as the value of a variable. For example, if one types the command

```
r = input('Please enter the four resistor values');
```

when the above command is executed, the text string "Please enter the four resistor values" will be displayed on the terminal screen. The user can then type an expression such as

```
[10 15 30 25];
```

The variable r will be assigned a vector [10 15 30 25]. If the user strikes the return key, without entering an input, an empty matrix will be assigned to r.

To return a string typed by a user as a text variable, the input command may take the form

```
x = input('Enter string for prompt', 's')
```

For example, the command

```
x = input('What is the title of your graph', 's')
```

when executed, will echo on the screen, "What is the title of your graph". The user can enter a string such as "Voltage (mV) vs. Current (mA)".

Keyboard

The **keyboard** command invokes the keyboard as an m-file. When the word **keyboard** is placed in an m-file, execution of the m-file stops when the word **keyboard** is encountered. MATLAB commands can then be entered. The keyboard mode is terminated by typing the word **return** and pressing the return key. The keyboard command may be used to examine or change a variable or may be used as a tool for debugging m-files.

Pause

The **pause** command stops the execution of m-files. The execution of the m-file resumes upon pressing any key. The general forms of the pause command are

```
pause
pause(n)
```

The **pause** command stops the execution of m-files until a key is pressed. The **pause(n)** command stops the execution of m-files for n seconds before continuing. The pause command can be used to stop m-files temporarily when plotting commands are encountered during program execution. If pause is not used, the graphics are momentarily visible.

The following example uses the MATLAB **input, fprint,** and **disp** commands.

Example 3.5 Equivalent Inductance of Series-Connected Inductors

Write a MATLAB program that will accept values of inductors connected in series and find the equivalent inductance. The values of the inductors will be entered from the keyboard.

Solution

We shall use the MATLAB **input** command to accept the input of the elements, the **fprintf** command to output the result; and the **disp** command to display the text string.

MATLAB script

```
% input values of the inductors in input order
%
%
disp('Enter Inductor values with spaces between them
and  enclosed in brackets')
ind = input('Enter inductor values')
num = length(ind);     % number of elements in array ind
lequiv = 0;
```

```
 for i = 1:num
    lequiv= lequiv + ind(i);
end
%
fprintf('The Equivalent Inductance is %8.3e Henries',
lequiv)
```

If you enter the values [2 3 7 9], you get the result

```
ind =
      2 3 7 9
The Equivalent Inductance is 2.100e+001 Henries
```

Bibliography

1. Attia, J.O., *PSPICE and MATLAB for Electronics: An Integrated Approach,* CRC Press, Boca Raton, FL, 2002.
2. Biran, A. and Breiner, M., *MATLAB for Engineers,* Addison-Wesley, Reading, MA, 1995.
3. Chapman, S.J., *MATLAB Programming for Engineers,* Brook, Cole Thompson Learning, Pacific Grove, CA, 2000.
4. Etter, D.M., *Engineering Problem Solving with MATLAB,* 2nd ed., Prentice Hall, Upper Saddle River, NJ, 1997.
5. Etter, D.M., Kuncicky, D.C., and Hull, D., *Introduction to MATLAB 6,* Prentice Hall, Upper Saddle River, NJ, 2002.
6. Gottling, J.G., *Matrix Analysis of Circuits Using MATLAB,* Prentice Hall, Englewood Cliffs, NJ, 1995.
7. Sigmor, K., *MATLAB Primer,* 4th ed., CRC Press, Boca Raton, FL, 1998.
8. *Using MATLAB, The Language of Technical Computing, Computation, Visualization, Programming,* Version 6, MathWorks, Inc., Natick, MA, 2000

Problems

Problem 3.1

Write a MATLAB program to add all the even numbers from 0 to 100.

Problem 3.2

Add all the terms in the series

$$1+\frac{1}{2}+\frac{1}{4}+\frac{1}{8}+\dots$$

until the sum exceeds 1.995. Print out the sum and the number of terms needed to just exceed the sum of 1.995.

Problem 3.3

The Fibonacci sequence is given as

$$1 \quad 1 \quad 2 \quad 3 \quad 5 \quad 8 \quad 13 \quad 21 \quad 34 \quad \ldots$$

Write a MATLAB program to generate the Fibonacci sequence up to the twentieth term. Print out the results.

Problem 3.4

In the Fibonacci sequence, defined in Problem 3.3, find the sum of the first thirty terms.

Problem 3.5

A function S is given as

$$S = 1 + x + \frac{x^2}{2!} + \frac{x^3}{3!} + \quad \ldots$$

If $x = 2$, find S when 20 terms are used to evaluate the function.

Problem 3.6

The table below shows the final course grade and its corresponding relevant letter grade.

Letter Grade	Final Course Grade
A	$90 < \text{grade} \le 100$
B	$80 < \text{grade} \le 90$
C	$70 < \text{grade} \le 80$
D	$60 < \text{grade} \le 70$
F	$\text{Grade} \le 60$

For the course grades: 70, 85, 90, 97, 50, 60, 71, 83, 91, 86, 77, 45, 67, 88, 64, 79, 75, 92, and 69:

(a) Determine the number of students who attained the grades of A and F.

(b) What are the mean grade and the standard deviation?

Problem 3.7

The input $v_s(t)$ and output $v_0(t)$ of a limiter are described by the expressions

$$\begin{aligned} v_0(t) &= 0.7\ V && \text{if } v_s(t) \ge 0.7\ V \\ &= v_s(t) && \text{if } -0.7\ V < v_s(t) < 0.7\ V \\ &= -0.7\ V && \text{if } v_s(t) \le -0.7\ V \end{aligned}$$

If $v_s(t) = \cos(120\pi t)$ V, plot $v_0(t)$.

Problem 3.8

The input $v_s(t)$ and output $v_0(t)$ of a circuit are described by the expressions

$$\begin{aligned} v_0(t) &= +5\ V && \text{if } v_s(t) \ge 1\ V \\ &= +5\ V && \text{if } -0.5\ V < v_s(t) < 0.5\ V \text{ and } v_0(t-1) = +5\ V \\ &= -5 V && \text{if } -0.5\ V < v_s(t) < 0.5\ V \text{ and } v(_0 t-1) = -5\ V \\ &= -5 V && \text{if } v_s(t) \le -0.5\ V \end{aligned}$$

If $v_s(t) = 1.2 \sin(2\pi f_0 t) + 0.8n(t)$, $f_0 = 100$ Hz, and $n(t)$ is a normally distributed white noise, write a MATLAB program to find the output voltage. Plot $v_s(t)$ and output $v_0(t)$.

Problem 3.9

Write a script file to evaluate $y[1]$, $y[2]$, $y[3]$, and $y[4]$ for the difference equation given as:

$$y[n] = 2y[n-1] - y[n-2] + x[n]$$

for $n \ge 0$. Assume that $x[n] = 1$ for $n \ge 0$, $y[-2] = 2$, and $y[-1] = 1$.

Problem 3.10

The input $x[n]$ and the output $y[n]$ of a system are related by the expression

$$y[n] = x[n] + 4x[n-1] + 3x[n-2]$$

Assuming that $x[n] = 2$ for $n \ge 0$ and $x[n] = 0$ for $n < 0$, determine $y[n]$ for $n = 1$ to 10.

Problem 3.11

The equivalent impedance of a circuit is given as

$$Z_{eq}(jw) = 100 + jwL + \frac{1}{jwC}$$

If $L = 4$ H and $C = 1$ μF,

(a) Plot $|Z_{eq}(jw)|$ vs. w.
(b) What is the minimum impedance?
(c) With what frequency does the minimum impedance occur?

Problem 3.12

Write a MATLAB program that will accept values of resistors connected in parallel and find the equivalent resistance. The values of the resistors will be entered from the keyboard.

4

DC Analysis

4.1 Nodal Analysis

Kirchhoff's current law (KCL) states that for any electrical circuit, the algebraic sum of all the currents at any node in the circuit equals zero. In nodal analysis, if there are n nodes in a circuit and we select a reference node, the other nodes can be numbered from V_1 through V_{n-1}. With one node selected as the reference node, there will be n – 1 independent equations. If we assume that the admittance between nodes i and j is given as Y_{ij}, we can write the nodal equations:

$$\begin{aligned} Y_{11}V_1 + Y_{12}V_2 + \ldots + Y_{1m}V_m &= \sum I_1 \\ Y_{21}V_1 + Y_{22}V_2 + \ldots + Y_{2m}V_m &= \sum I_2 \\ Y_{m1}V_1 + Y_{m2}V_2 + \ldots + Y_{mm}V_m &= \sum I_m \end{aligned} \tag{4.1}$$

where $m = n - 1$, V_1, V_2, and V_m are voltages from nodes 1, 2, …, n with respect to the reference node, and $\sum I_x$ is the algebraic sum of current sources at node x.

Equation (4.1) can be expressed in matrix form as

$$[Y][V] = [I] \tag{4.2}$$

The solution of the above equation is

$$[V] = [Y]^{-1}[I] \tag{4.3}$$

where $[Y]^{-1}$ is an inverse of $[Y]$.

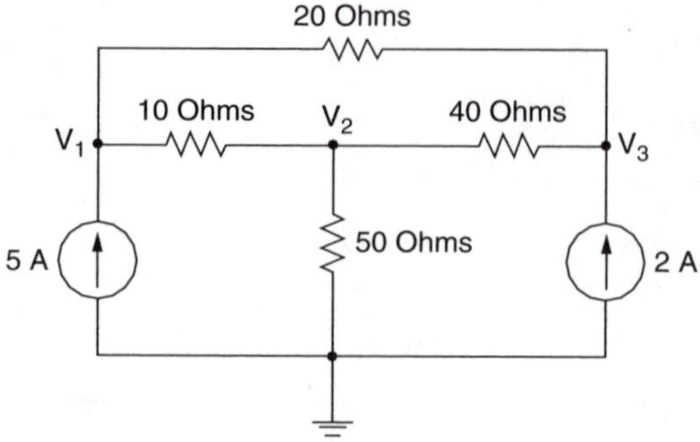

FIGURE 4.1
Circuit with nodal voltages.

In MATLAB, we can compute [V] by using the command

$$V = inv(Y) * I \tag{4.4}$$

where

inv(*Y*) is the inverse of matrix *Y*

The matrix left and right divisions can also be used to obtain the nodal voltages. The following MATLAB commands can be used to find the matrix [V]:

$$V = {}^{I}\!/_{Y} \tag{4.5}$$

or

$$V = Y \backslash I \tag{4.6}$$

The solutions obtained from Equation (4.4) to Equation (4.6) will be the same, provided the system is not ill-conditioned. The following two examples illustrate the use of MATLAB for solving nodal voltages of electrical circuits.

Example 4.1 Nodal Voltages of a Simple Circuit

For the circuit shown in Figure 4.1, find the nodal voltages V_1, V_2, and V_3.

Solution

Using KCL and assuming that the currents leaving a node are positive, we have:

For node 1,

$$\frac{V_1 - V_2}{10} + \frac{V_1 - V_3}{20} - 5 = 0$$

i.e.,

$$0.15V_1 - 0.1V_2 - 0.05V_3 = 5 \tag{4.7}$$

At node 2,

$$\frac{V_2 - V_1}{10} + \frac{V_2}{50} + \frac{V_2 - V_3}{40} = 0$$

i.e.,

$$-0.1V_1 + 0.145V_2 - 0.025V_3 = 0 \tag{4.8}$$

At node 3,

$$\frac{V_3 - V_1}{20} + \frac{V_3 - V_2}{40} - 2 = 0$$

i.e.,

$$-0.05V_1 - 0.025V_2 + 0.075V_3 = 2 \tag{4.9}$$

In matrix form, we have

$$\begin{bmatrix} 0.15 & -0.1 & -0.05 \\ -0.1 & 0.145 & -0.025 \\ -0.05 & -0.025 & 0.075 \end{bmatrix} \begin{bmatrix} V_1 \\ V_2 \\ V_3 \end{bmatrix} = \begin{bmatrix} 5 \\ 0 \\ 2 \end{bmatrix} \tag{4.10}$$

The MATLAB program for solving the nodal voltages is:

MATLAB script

```
% program computes the nodal voltages
% given the admittance matrix Y and current vector I
% Y is the admittance matrix and I is the current vector
% initialize matrix y and vector I using YV=I form
Y = [ 0.15    -0.1     -0.05;
```

```
     -0.1        0.145    -0.025;
     -0.05      -0.025     0.075];
I = [5;
     0;
     2];
% solve for the voltage
fprintf('Nodal voltages V1, V2 and V3 are \n')
v = inv(Y)*I
```

The results obtained from MATLAB are

```
Nodal voltages V1, V2 and V3
v =
      404.2857
      350.0000
      412.8571
```

Example 4.2 Circuit with Dependent and Independent Sources

Find the nodal voltages of the circuit shown in Figure 4.2.

Solution

Using KCL and the convention that current leaving a node is positive, we have:

At node 1

$$\frac{V_1}{20} + \frac{V_1 - V_2}{5} + \frac{V_1 - V_4}{2} - 5 = 0$$

Simplifying, we get

$$0.75V_1 - 0.2V_2 - 0.5V_4 = 5 \tag{4.11}$$

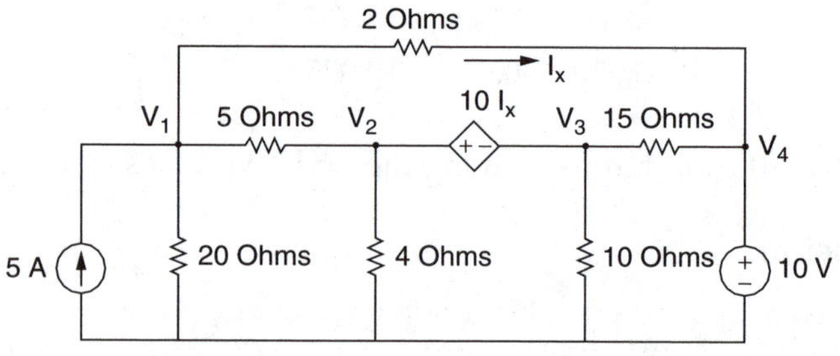

FIGURE 4.2
Circuit with dependent and independent sources.

At node 2,

$$V_2 - V_3 = 10 I_X$$

But

$$I_X = \frac{(V_1 - V_4)}{2}$$

Thus

$$V_2 - V_3 = \frac{10(V_1 - V_4)}{2}$$

Simplifying, we get

$$-5V_1 + V_2 - V_3 + 5V_4 = 0 \tag{4.12}$$

From supernodes 2 and 3, we have

$$\frac{V_3}{10} + \frac{V_2 - V_1}{5} + \frac{V_2}{4} + \frac{V_3 - V_4}{15} = 0$$

Simplifying, we get

$$-0.2V_1 + 0.45V_2 + 0.1667V_3 - 0.06667V_4 = 0 \tag{4.13}$$

At node 4, we have

$$V_4 = 10 \tag{4.14}$$

In matrix form, Equation (4.11) to Equation (4.14) become

$$\begin{bmatrix} 0.75 & -0.2 & 0 & -0.5 \\ -5 & 1 & -1 & 5 \\ -0.2 & 0.45 & 0.1667 & -0.06667 \\ 0 & 0 & 0 & 1 \end{bmatrix} \begin{bmatrix} V_1 \\ V_2 \\ V_3 \\ V_4 \end{bmatrix} = \begin{bmatrix} 5 \\ 0 \\ 0 \\ 10 \end{bmatrix} \tag{4.15}$$

The MATLAB program for solving the nodal voltages is:

MATLAB script

```
% this program computes the nodal voltages
% given the admittance matrix Y and current vector I
% Y is the admittance matrix
% I is the current vector
% initialize the matrix y and vector I using YV=I

Y = [0.75    -0.2    0    -0.5;
        -5        1   -1       5;
       -0.2    0.45   0.166666667    -0.0666666667;
        0         0    0        1];

% current vector is entered as a transpose of row vector
I = [5    0    0    10]';

% solve for nodal voltage
fprintf('Nodal voltages V1,V2,V3,V4 are \n')
V = inv(Y)*I
```

We obtain the following results.

```
Nodal voltages V1,V2,V3,V4 are

V =
         18.1107
         17.9153
        -22.6384
         10.0000
```

4.2 Loop Analysis

Loop analysis is a method for obtaining loop currents. The technique uses Kirchhoff voltage law (KVL) to write a set of independent simultaneous equations. The Kirchhoff voltage law states that the algebraic sum of all the voltages around any closed path in a circuit equals zero.

In loop analysis, we want to obtain current from a set of simultaneous equations. The latter equations are easily set up if the circuit can be drawn in planar fashion. This implies that a set of simultaneous equations can be obtained if the circuit can be redrawn without crossovers.

For a planar circuit with n-meshes, the KVL can be used to write equations for each mesh that does not contain a dependent or independent current source. Using KVL and writing equations for each mesh, the resulting equations will have the general form:

$$Z_{11}I_1 + Z_{12}I_2 + Z_{13}I_3 + \ldots Z_{1n}I_n = \sum V_1$$
$$Z_{21}I_1 + Z_{22}I_2 + Z_{23}I_3 + \ldots Z_{2n}I_n = \sum V_2 \tag{4.16}$$
$$Z_{n1}I_1 + Z_{n2}I_2 + Z_{n3}I_3 + \ldots Z_{nn}I_n = \sum V_n$$

where

$I_1, I_2, \ldots, I_n$ are the unknown currents for meshes 1 through n

$Z_{11}, Z_{22}, \ldots, Z_{nn}$ are the impedance for each mesh through which individual current flows

Z_{ij}, j # i denote mutual impedance

$\sum V_x$ is the algebraic sum of the voltage sources in mesh x

Equation (4.16) can be expressed in matrix form as

$$[Z][I] = [V] \tag{4.17}$$

where

$$Z = \begin{bmatrix} Z_{11} & Z_{12} & Z_{13} & \ldots & Z_{1n} \\ Z_{21} & Z_{22} & Z_{23} & \ldots & Z_{2n} \\ Z_{31} & Z_{32} & Z_{33} & \ldots & Z_{3n} \\ \ldots & \ldots & \cdot & \ldots & \ldots \\ Z_{n1} & Z_{n2} & Z_{n3} & \ldots & Z_{nn} \end{bmatrix}$$

$$I = \begin{bmatrix} I_1 \\ I_2 \\ I_3 \\ \cdot \\ I_n \end{bmatrix}$$

and

$$V = \begin{bmatrix} \sum V_1 \\ \sum V_2 \\ \sum V_3 \\ \ldots \\ \sum V_n \end{bmatrix}$$

The solution to Equation (4.17) is

$$[I] = [Z]^{-1}[V] \tag{4.18}$$

In MATLAB, we can compute [I] by using the command

$$I = inv(Z) * V \tag{4.19}$$

where *inv*(Z) is the inverse of the matrix Z.

The matrix left and right divisions can also be used to obtain the loop currents. Thus, the current I can be obtained by the MATLAB commands

$$I = V/Z \tag{4.20}$$

or

$$I = Z \backslash V \tag{4.21}$$

As mentioned earlier, Equation (4.19) to Equation (4.21) will give the same results, provided the circuit is not ill-conditioned. The following examples illustrate the use of MATLAB for loop analysis.

Example 4.3 Loop Analysis of a Bridge Circuit

Use the mesh analysis to find the current flowing through the resistor R_B. In addition, find the power supplied by the 10-volt voltage source. (See Figure 4.3a.)

Solution

Using loop analysis and designating the loop currents as I_1, I_2, and I_3, we obtain Figure 4.3b. Note that $I = I_3 - I_2$ and power supplied by the source is $P = 10I_1$.

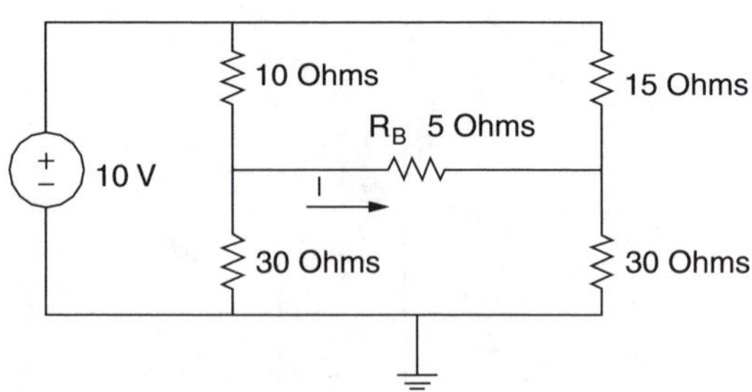

FIGURE 4.3a
Bridge circuit.

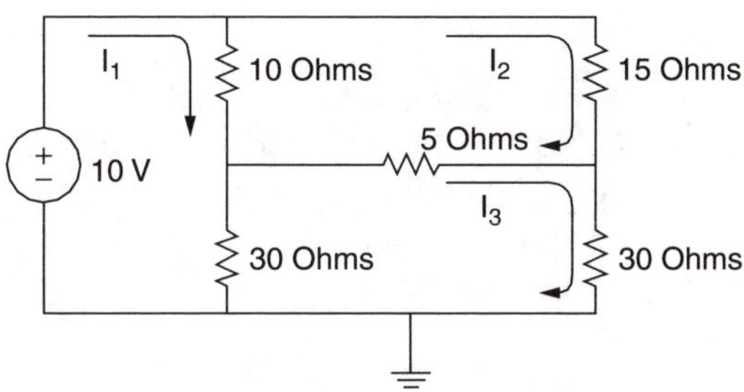

FIGURE 4.3b
Bridge circuit with loop currents.

The loop equations are:

Loop 1,

$$10(I_1 - I_2) + 30(I_1 - I_3) - 10 = 0$$

$$40I_1 - 10I_2 - 30I_3 = 10 \tag{4.22}$$

Loop 2,

$$10(I_2 - I_1) + 15I_2 + 5(I_2 - I_3) = 0$$

$$-10I_1 + 30I_2 - 5I_3 = 0 \tag{4.23}$$

Loop 3,

$$30(I_3 - I_1) + 5(I_3 - I_2) + 30I_3 = 0$$

$$-30I_1 - 5I_2 + 65I_3 = 0 \tag{4.24}$$

In matrix form, Equation (4.22) and Equation (4.23) become

$$\begin{bmatrix} 40 & -10 & -30 \\ -10 & 30 & -5 \\ -30 & -5 & 65 \end{bmatrix} \begin{bmatrix} I_1 \\ I_2 \\ I_3 \end{bmatrix} = \begin{bmatrix} 10 \\ 0 \\ 0 \end{bmatrix} \tag{4.25}$$

The MATLAB program for solving the loop currents I_1, I_2, and I_3, the current I, and the power supplied by the 10-volt source is:

MATLAB script

```
% This program determines the current
% flowing in a resistor RB and power supplied  by source
% it computes the loop currents given the impedance
% matrix Z and voltage vector V
% Z is the impedance matrix
% V is the voltage matrix
% initialize the matrix Z and vector V

Z = [40    -10    -30;
    -10     30     -5;
    -30     -5     65];

V = [10  0  0]';

% solve for the loop currents
I = inv(Z)*V;
% current through RB is calculated
IRB = I(3) - I(2);
fprintf('The current through R is %8.3f Amps \n',IRB)
% the power supplied by source is calculated
PS = I(1)*10;
fprintf('The power supplied by 10V source is %8.4f watts
\n',PS)
```

MATLAB's answers are

```
The current through R is 0.037 Amps
The power supplied by 10V source is 4.7531 watts
```

Circuits with dependent voltage sources can be analyzed in a manner similar to that of Example 4.3. Example 4.4 illustrates the use of KVL and MATLAB to solve loop currents.

Example 4.4 Power Dissipation and Source Current Determinations

Find the power dissipated by the 8-Ω resistor and the current supplied by the 10-volt source in the circuit shown in Figure 4.4a.

Solution

Using loop analysis and denoting the loop currents as I_1, I_2, and I_3, the circuit can be redrawn as shown in Figure 4.4b.

By inspection,

$$I_S = I_1 \tag{4.26}$$

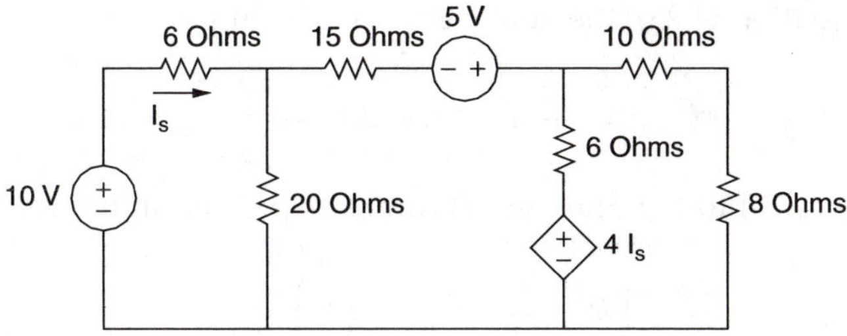

FIGURE 4.4a
Circuit for Example 4.4.

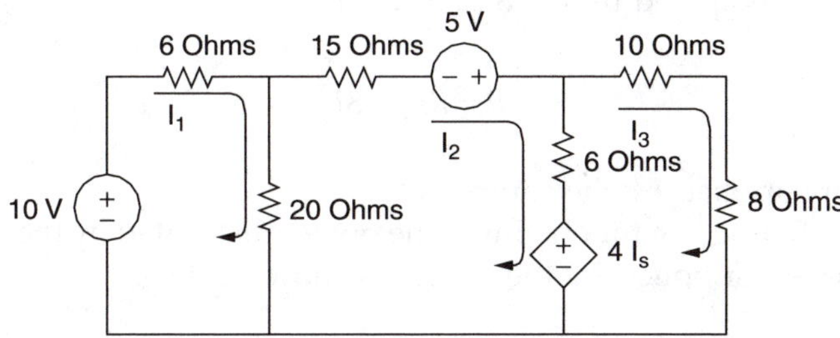

FIGURE 4.4b
Figure 4.4 with loop currents.

For loop 1,

$$-10 + 6I_1 + 20(I_1 - I_2) = 0$$

$$26I_1 - 20I_2 = 10 \tag{4.27}$$

For loop 2,

$$15I_2 - 5 + 6(I_2 - I_3) + 4I_S + 20(I_2 - I_1) = 0$$

Using Equation (4.26), the above expression simplifies to

$$-16I_1 + 41I_2 - 6I_3 = 5 \tag{4.28}$$

For loop 3,

$$10I_3 + 8I_3 - 4I_S + 6(I_3 - I_2) = 0$$

Using Equation (4.26), the above expression simplifies to

$$-4I_1 - 6I_2 + 24I_3 = 0 \tag{4.29}$$

Equation (4.25) to Equation (4.27) can be expressed in matrix form as

$$\begin{bmatrix} 26 & -20 & 0 \\ -16 & 41 & -6 \\ -4 & -6 & 24 \end{bmatrix} \begin{bmatrix} I_1 \\ I_2 \\ I_3 \end{bmatrix} = \begin{bmatrix} 10 \\ 5 \\ 0 \end{bmatrix} \tag{4.30}$$

The power dissipated by the 8-Ω resistor is

$$P = RI_3^2 = 8I_3^2$$

The current supplied by the source is $I_S = I_1$.

A MATLAB program for obtaining the power dissipated by the 8-Ω resistor and the current supplied by the source is shown below.

MATLAB script

```
% This program determines the power dissipated by
% 8 ohm resistor and current supplied by the
% 10V source
%
% the program computes the loop currents, given
% the impedance matrix Z and voltage vector V
%
% Z is the impedance matrix
% V is the voltage vector
% initialize the matrix Z and vector V of equation
% ZI=V

Z = [26    -20     0;
     -16    40    -6;
      -4    -6    24];
V = [10   5    0]';

% solve for loop currents
I = inv(Z)*V;
% the power dissipation in 8 ohm resistor is P
P = 8*I(3)^2;
% print out the results
fprintf('Power dissipated in 8 ohm resistor is %8.2f
Watts\n',P)
fprintf('Current in 10V source is %8.2f Amps\n',I(1))
```

MATLAB results are

```
Power dissipated in 8 ohm resistor is 0.42 Watts
Current in 10V source is 0.72 Amps
```

For circuits that contain both current and voltage sources, irrespective of whether they are dependent sources, both KVL and KCL can be used to obtain equations that can be solved using MATLAB. Example 4.5 illustrates one such circuit.

Example 4.5 Calculation of Nodal Voltages of a Circuit with Dependent Sources

Find the nodal voltages in the circuit shown in Figure 4.5, i.e., $V_1, V_2, \ldots, V_5$.

Solution

By inspection,

$$V_b = V_1 - V_4 \tag{4.31}$$

Using Ohm's law

$$I_a = \frac{V_4 - V_3}{5} \tag{4.32}$$

Using KCL at node 1, and supernode 1–2, we get

$$\frac{V_1}{2} + \frac{V_1 - V_4}{10} - 5V_b + \frac{V_2 - V_3}{8} = 0 \tag{4.33}$$

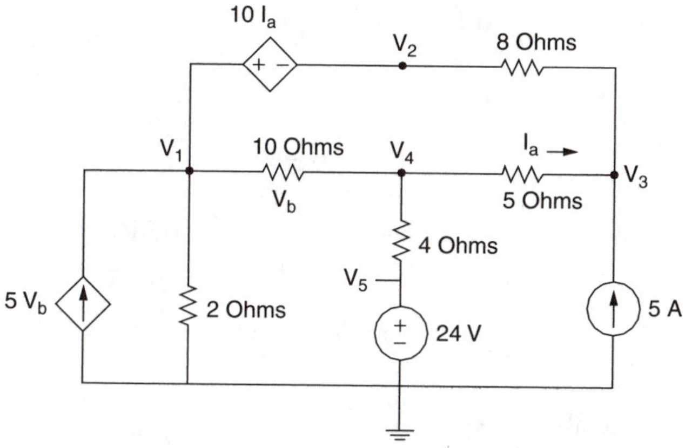

FIGURE 4.5
Circuit for Example 4.5.

Using Equation (4.31), Equation (4.33) simplifies to

$$-4.4V_1 + 0.125V_2 - 0.125V_3 + 4.9V_4 = 0 \tag{4.34}$$

Using KCL at node 4, we have

$$\frac{V_4 - V_5}{4} + \frac{V_4 - V_3}{5} + \frac{V_4 - V_1}{10} = 10$$

This simplifies to

$$-0.1V_1 - 0.2V_3 + 0.55V_4 - 0.25V_5 = 0 \tag{4.35}$$

Using KCL at node 3, we get

$$\frac{V_3 - V_4}{5} + \frac{V_3 - V_2}{8} - 5 = 0$$

which simplifies to

$$-0.125V_2 + 0.325V_3 - 0.2V_4 = 5 \tag{4.36}$$

Using KVL for loop 1, we have

$$-10I_a + V_b + 5I_a + 8(I_a + 5) = 0 \tag{4.37}$$

Using Equation (4.31) and Equation (4.32), Equation (4.37) becomes

$$-10I_a + V_b + 5I_a + 8I_a + 40 = 0$$

i.e.,

$$3I_a + V_b = -40$$

Using Equation (4.32), the above expression simplifies to

$$3\frac{V_4 - V_3}{5} + V_1 - V_4 = -40$$

Simplifying the above expression, we get

$$V_1 - 0.6V_3 - 0.4V_4 = -40 \tag{4.38}$$

By inspection,

$$V_S = 24 \tag{4.39}$$

Using Equation (4.34), Equation (4.35), Equation (4.36), Equation (4.38), and Equation (4.39), we get the matrix equation

$$\begin{bmatrix} -4.4 & 0.125 & -0.125 & 4.9 & 0 \\ -0.1 & -0.2 & 0 & 0.55 & -0.25 \\ 0 & -0.125 & 0.325 & -0.2 & 0 \\ 1 & 0 & -0.6 & -0.4 & 0 \\ 0 & 0 & 0 & 0 & 1 \end{bmatrix} \begin{bmatrix} V_1 \\ V_2 \\ V_3 \\ V_4 \\ V_5 \end{bmatrix} = \begin{bmatrix} 0 \\ 0 \\ 5 \\ -40 \\ 24 \end{bmatrix} \tag{4.40}$$

The MATLAB program for obtaining the nodal voltages is shown below.

MATLAB script

```
% Program determines the nodal voltages
% given an admittance matrix Y and  current vector I
% Initialize matrix Y and the current vector I of
% matrix equation Y V = I
Y = [-4.4  0.125  -0.125  4.9   0;
     -0.1  0       -0.2    0.55  -0.25;
      0    -0.125  0.325  -0.2   0;
      1    0       -0.6    -0.4  0;
      0    0       0       0     1];
I = [0  0  5  -40  24]';
%  Solve for the nodal voltages
fprintf('Nodal voltages V(1), V(2), ... V(5) are \n')
V = inv(Y)*I
```

The results obtained from MATLAB are

```
Nodal voltages V(1), V(2), ... V(5) are

V =

      117.4792
      299.7708
      193.9375
      102.7917
      24.0000
```

4.3 Maximum Power Transfer

Assume that we have a voltage source V_S with resistance R_S connected to a load R_L. The circuit is shown in Figure 4.6.

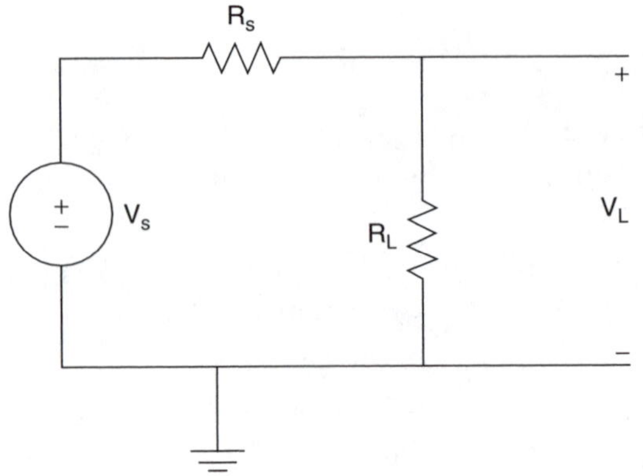

FIGURE 4.6
Circuit for obtaining maximum power dissipation.

The voltage across the load R_L is given as

$$V_L = \frac{V_s R_L}{R_s + R_L}$$

The power dissipated by the load R_L is given as

$$P_L = \frac{V_L^2}{R_L} = \frac{V_s^2 R_L}{(R_s + R_L)^2} \tag{4.41}$$

The value of R_L that dissipates the maximum power is obtained by differentiating P_L with respect to R_L, and equating the derivative to zero. That is,

$$\begin{aligned} \frac{dP_L}{dR_L} &= \frac{(R_s + R_L)^2 V_S - V_s^2 R_L (2)(R_s + R_L)}{(R_s + R_L)^4} \\ \frac{dP_L}{dR_L} &= 0 \end{aligned} \tag{4.42}$$

Simplifying the above, we get

$$(R_s + R_L) - 2R_L = 0$$

i.e.,

$$R_L = R_S \tag{4.43}$$

Thus, for a resistive network, the maximum power is supplied to a load provided the load resistance is equal to the source resistance. When $R_L = 0$, the voltage across and power dissipated by R_L are zero. On the other hand, when R_L approaches infinity, the voltage across the load is maximum, but the power dissipation is zero. MATLAB can be used to observe the voltage across and power dissipation of the load as functions of load resistance value. Example 4.6 shows the use of MATLAB to plot the voltage and display the power dissipation of a resistive circuit.

Before presenting an example of the maximum power transfer theorem, let us discuss the MATLAB functions **diff** and **find**.

4.3.1 MATLAB Diff and Find Functions

Numerical differentiation can be obtained using the backward difference expression

$$f'(x_n) = \frac{f(x_n) - f(x_{n-1})}{x_n - x_{n-1}} \tag{4.44}$$

or by the forward difference expression

$$f'(x_n) = \frac{f(x_{n+1}) - f(x_n)}{x_{n+1} - x_n} \tag{4.45}$$

The derivative of $f(x)$ can be obtained by using the MATLAB **diff** function as

$$f'(x) \cong diff(f)./\,diff(x) \tag{4.46}$$

If f is a row or column vector

$$f = [f(1) \quad f(2) \quad \ldots \quad f(n)]$$

then **diff(f)** returns a vector of difference between adjacent elements

$$diff(f) = [f(2) - f(1) \quad f(3) - f(2) \quad \ldots \quad f(n) - f(n-1)] \tag{4.47}$$

The output vector $diff(f)$ will be one element less than the input vector f.

The **find** function determines the indices of the nonzero elements of a vector or matrix. The statement

$$\text{B} = \text{find}(f) \tag{4.48}$$

will return the indices of the vector f that are nonzero. For example, to obtain the points where a change in sign occurs, the statement

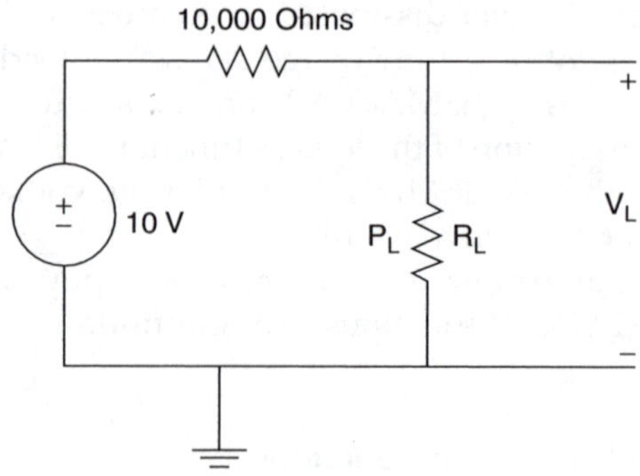

FIGURE 4.7
Resistive circuit for Example 4.6.

$$\text{Pt\_change} = \text{find(product} < 0) \tag{4.49}$$

will show the indices of the locations in *product* that are negative.

The **diff** and **find** are used in the following example to find the value of resistance at which the maximum power transfer occurs.

Example 4.6 Maximum Power Dissipation

In Figure 4.7, as R_L varies from 0 to 50 kΩ, plot the power dissipated by the load. Verify that the maximum power dissipation by the load occurs when R_L is 10 kΩ.

Solution

MATLAB script

```
% maximum power transfer
% vs is the supply voltage
% rs is the supply resistance
% rl is the load resistance
% vl is the voltage across the load
% pl is the power dissipated by the load
vs = 10;    rs = 10e3;
rl = 0:1e3:50e3;
k = length(rl); % components in vector rl
% Power dissipation calculation
for i=1:k
  pl(i) = ((vs/(rs+rl(i)))^2)*rl(i);
end
% Derivative of power is calculated using backward
difference
dp = diff(pl)./diff(rl);
rld = rl(2:length(rl)); % length of rld is 1 less than
that of rl
```

```
% Determination of critical points of derivative of
power
prod = dp(1:length(dp) - 1).*dp(2:length(dp));
crit_pt = rld(find(prod < 0));
max_power = max(pl); % maximum power is calculated
% print out results
fprintf('Maximum power occurs at %8.2f Ohms\n',crit_pt)
fprintf('Maximum power dissipation is %8.4f Watts\n',
max_power)
% Plot power versus load
plot(rl, pl, rl,pl,'o')
title('Power delivered to Load')
xlabel('Load Resistance in Ohms')
ylabel('Power in watts')
```

The results obtained from MATLAB are

```
Maximum power occurs at 10000.00 Ohms
Maximum power dissipation is 0.0025 Watts
```

The plot of the power dissipation obtained from MATLAB is shown in Figure 4.8.

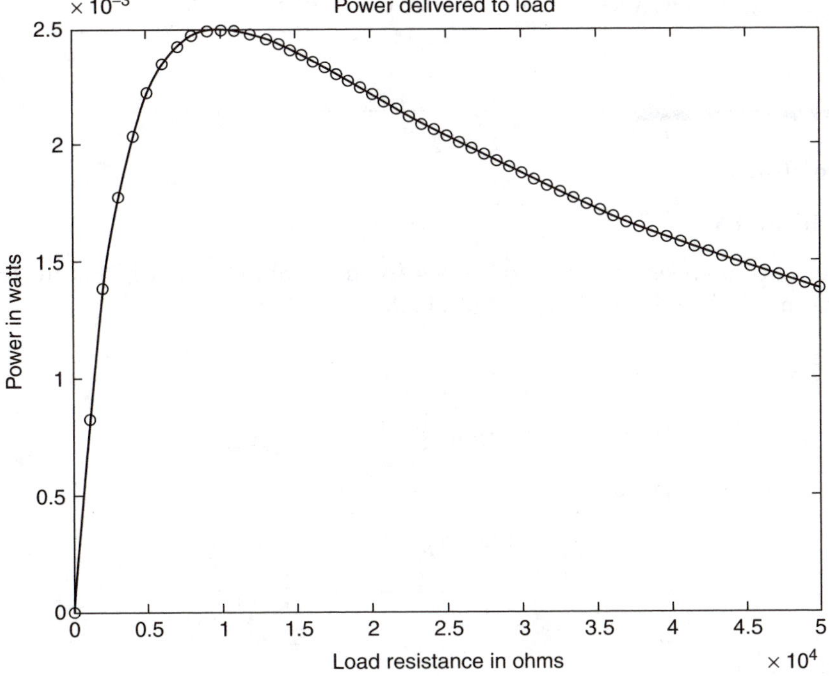

FIGURE 4.8
Power delivered to load.

Bibliography

1. Alexander, C.K. and Sadiku, M.N.O., *Fundamentals of Electric Circuits*, 2nd ed., McGraw-Hill, New York, 2004.
2. Attia, J.O., *PSPICE and MATLAB for Electronics: An Integrated Approach*, CRC Press, Boca Raton, FL, 2002.
3. Biran, A. and Breiner, M., *MATLAB for Engineers*, Addison-Wesley, Reading, MA, 1995.
4. Chapman, S.J., *MATLAB Programming for Engineers*, Brook, Cole Thompson Learning, Pacific Grove, CA, 2000.
5. Dorf, R.C. and Svoboda, J.A., *Introduction to Electric Circuits*, 3rd ed., John Wiley & Sons, New York, 1996.
6. Etter, D.M., *Engineering Problem Solving with MATLAB*, 2nd ed., Prentice Hall, Upper Saddle River, NJ, 1997.
7. Etter, D.M., Kuncicky, D.C., and Hull, D., *Introduction to MATLAB 6*, Prentice Hall, Upper Saddle River, NJ, 2002.
8. Gottling, J.G., *Matrix Analysis of Circuits Using MATLAB*, Prentice Hall, Englewood Cliffs, NJ, 1995.
9. Johnson, D.E., Johnson, J.R., and Hilburn, J.L., *Electric Circuit Analysis*, 3rd ed., Prentice Hall, Englewood Cliffs, NJ, 1997.
10. Sigmor, K., *MATLAB Primer*, 4th ed., CRC Press, Boca Raton, FL, 1998.
11. *Using MATLAB, The Language of Technical Computing, Computation, Visualization, Programming*, Version 6, MathWorks, Inc., Natick, MA, 2000.

Problems

Problem 4.1

Use loop analysis to write equations for the circuit shown in Figure P4.1. Determine the current *I* using MATLAB.

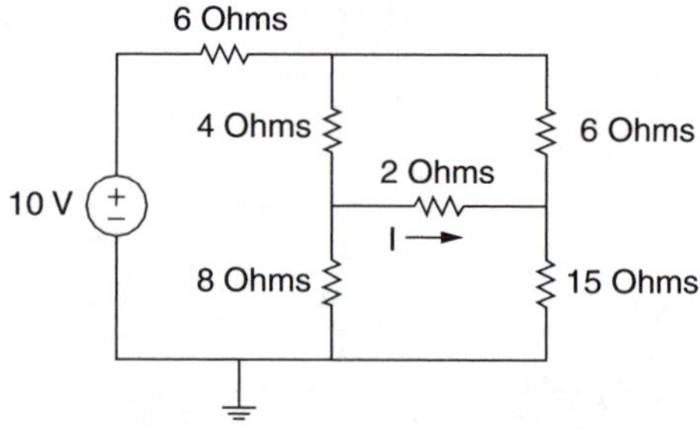

FIGURE P4.1
Circuit for Problem 4.1.

Problem 4.2

Use nodal analysis to solve for the nodal voltages for the circuit shown in Figure P4.2. Solve the equations using MATLAB.

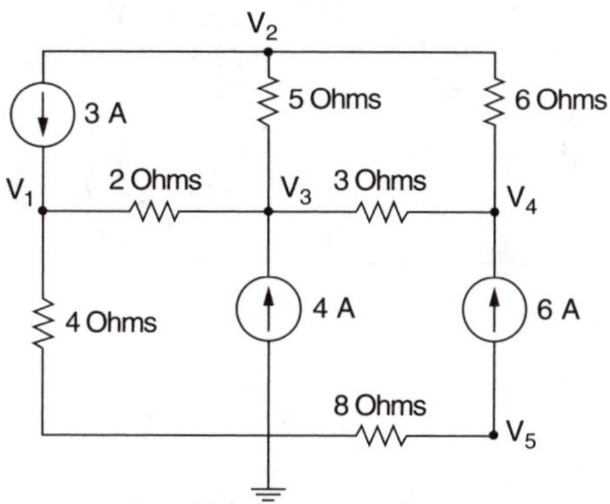

FIGURE P4.2
Circuit for Problem 4.2.

Problem 4.3

Find the power dissipated by the 4-Ω resistor shown in Figure P4.3 and the voltage V_o.

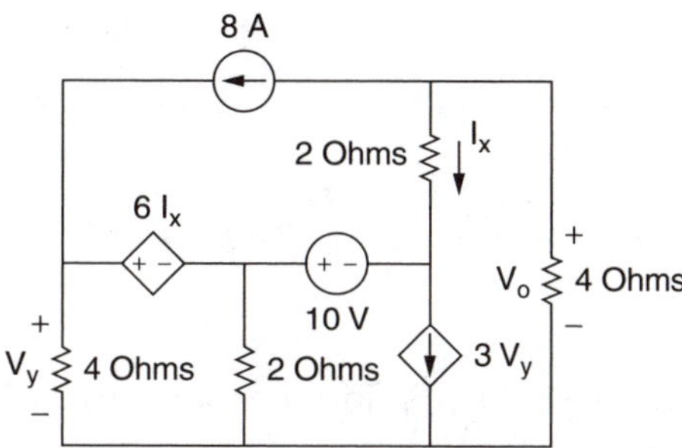

FIGURE P4.3
Circuit for Problem 4.3.

Problem 4.4

Using both loop and nodal analysis, find the power delivered by the 15-V source. Refer to Figure P4.4.

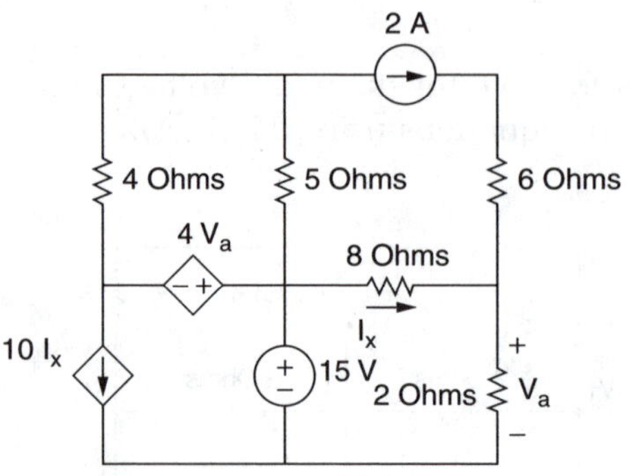

FIGURE P4.4
Circuit for Problem 4.4.

Problem 4.5

As R_L varies from 0- to 12-in increments of 2 Ω, calculate the power dissipated by R_L. Plot the power dissipation with respect to the variation in R_L. What is the maximum power dissipated by R_L? What is the value of R_L needed for maximum power dissipation? Refer to Figure P4.5.

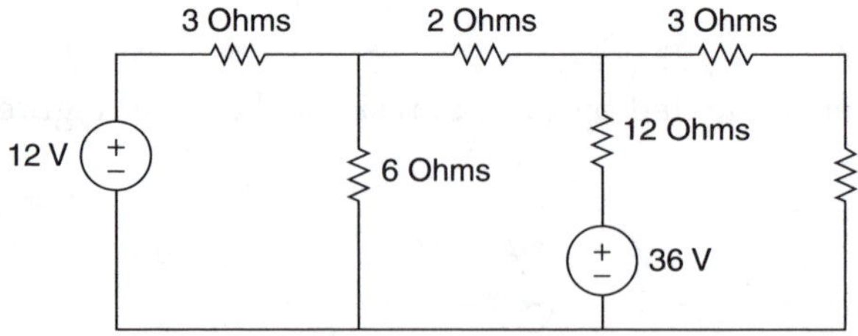

FIGURE P4.5
Circuit for Problem 4.5.

Problem 4.6

Using loop analysis and MATLAB, find the loop currents. What is the power supplied by the source? Refer to Figure P4.6.

Problem 4.7

In Figure 4.3a, find the power supplied to the 15-Ohms resistor.

Problem 4.8

In Figure P4.8, VS = 8 V, R1 = 1 kΩ, R2 = 1 kΩ, R3 = 2 kΩ, R4 = 1 kΩ, R5 = 5 kΩ, R6 = 2 kΩ, and I1 = 2 mA. Find the nodal voltages.

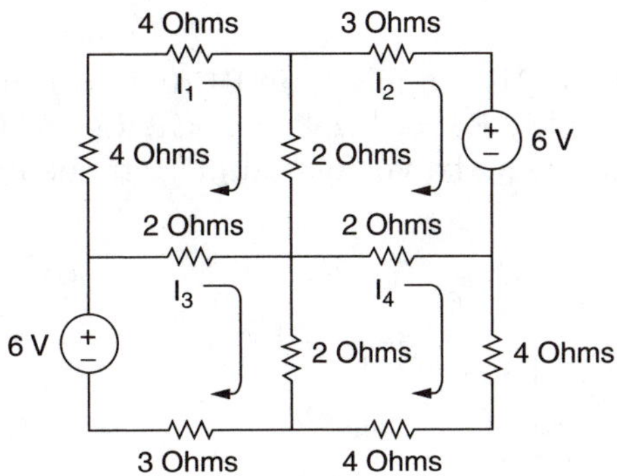

FIGURE P4.6
Circuit for Problem 4.6.

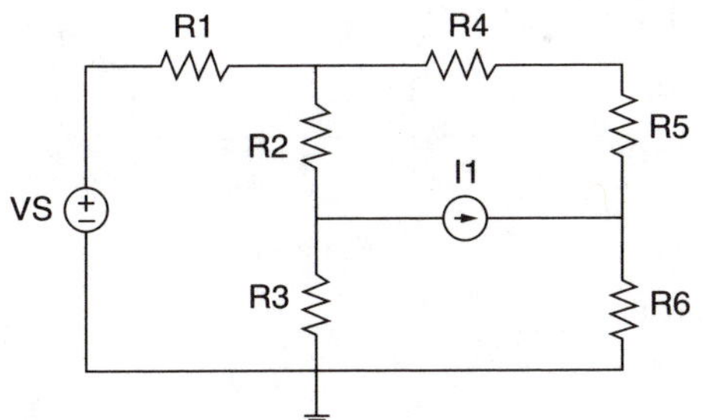

FIGURE P4.8
Resistive circuit with voltage and current sources.

Problem 4.9

In the network shown in Figure P4.9, R1 = 10 kΩ, R2 = 20 kΩ, R3 = 10 kΩ, R4 = 40 kΩ, R5 = 20 kΩ, and VS = 5 V. (a) Find the nodal voltages. (b) What is the power dissipated by R5?

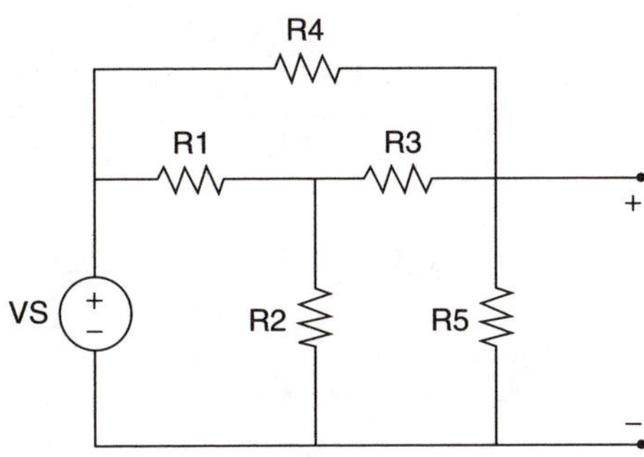

FIGURE P4.9
Bridge-T network.

Problem 4.10

For the resistive circuit shown in Figure P4.10, I1 = 1 mA, V1 = 5 V, V2 = 2 V, R1 = 1 kΩ, R2 = 2 kΩ, R3 = 2 kΩ, R4 = 5 kΩ, R5 = 5 kΩ, R6 = 10 kΩ, and R7 = 10 kΩ. (a) Find the nodal voltages. (b) Determine the voltage between nodes A and B.

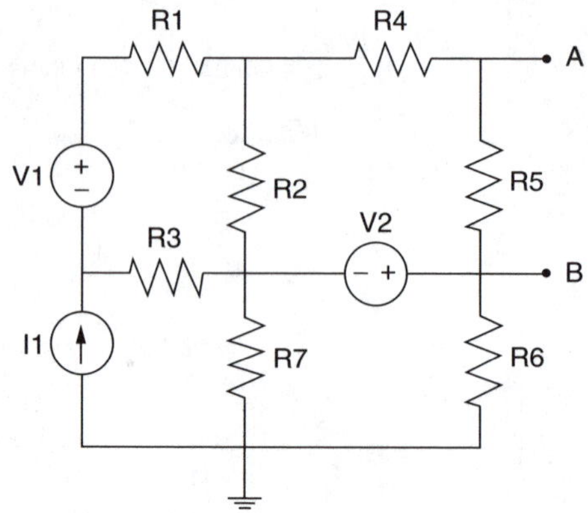

FIGURE P4.10
Multi-source resistive circuit.

5

Transient Analysis

5.1 RC Network

Considering the RC network shown in Figure 5.1, we can use Kirchhoff's current law (KCL) to write Equation (5.1).

$$C\frac{dv_o(t)}{dt}+\frac{v_o(t)}{R}=0 \tag{5.1}$$

i.e.,

$$\frac{dv_o(t)}{dt}+\frac{v_o(t)}{CR}=0$$

If V_m is the initial voltage across the capacitor, then the solution to Equation (5.1) is

$$v_0(t)=V_m e^{-\left(\frac{t}{CR}\right)} \tag{5.2}$$

where CR is the time constant.

Equation (5.2) represents the voltage across a discharging capacitor. To obtain the voltage across a charging capacitor, let us consider Figure 5.2.

Using KCL, we get

$$C\frac{dv_o(t)}{dt}+\frac{v_o(t)-V_s}{R}=0 \tag{5.3}$$

If the capacitor is initially uncharged, that is, $v_0(t) = 0$ at $t = 0$, the solution to Equation (5.3) is given as

$$v_0(t)=V_S\left(1-e^{-\left(\frac{t}{CR}\right)}\right) \tag{5.4}$$

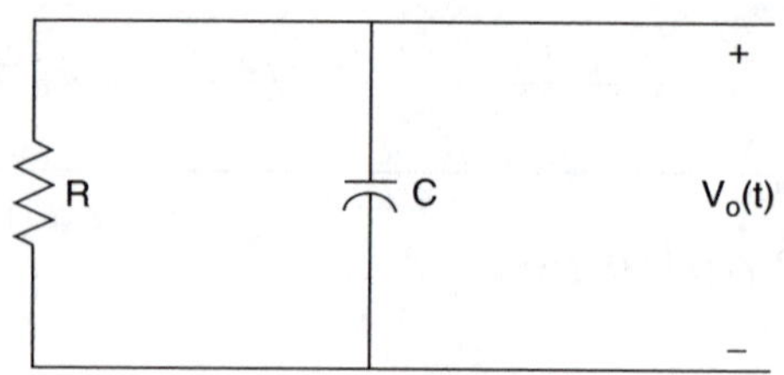

FIGURE 5.1
Source-free RC network.

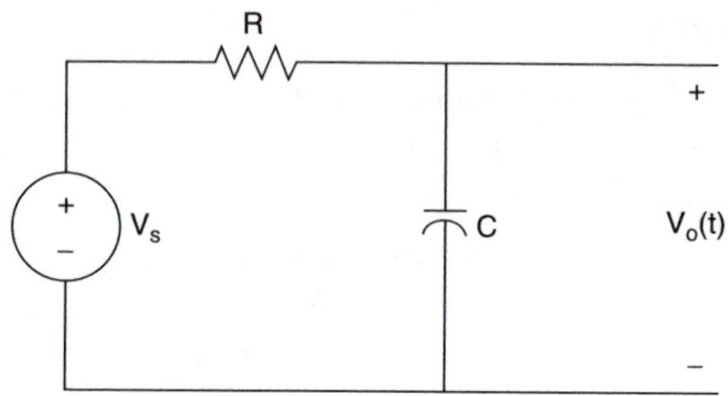

FIGURE 5.2
Charging of a capacitor.

Example 5.1 and Example 5.2 illustrate the use of MATLAB for solving problems related to RC network.

Example 5.1 RC Circuits and Time Constants

Assume that for Figure 5.2, $C = 10$ μF. Use MATLAB to plot the voltage across the capacitor if R is equal to (a) 1.0 kΩ, (b) 10 kΩ, and (c) 0.1 kΩ.

Solution

MATLAB script

```
% Charging of an RC circuit
%
c = 10e-6;
r1 = 1e3;
tau1 = c*r1;
t = 0:0.002:0.05;
v1 = 10*(1-exp(-t/tau1));
r2 = 10e3;
tau2 = c*r2;
v2 = 10*(1-exp(-t/tau2));
```

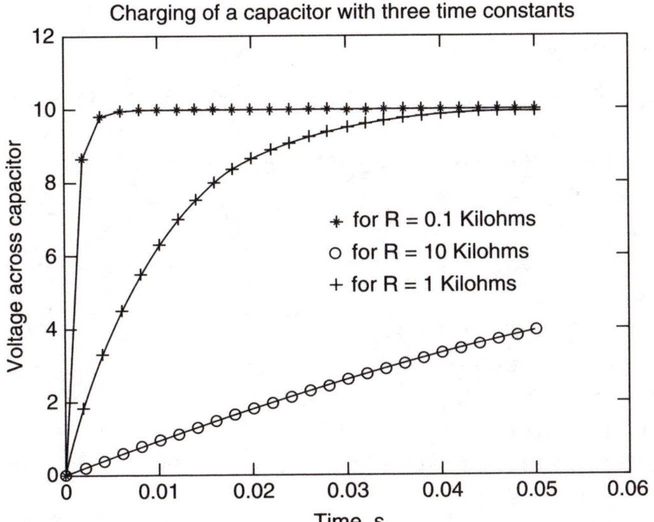

FIGURE 5.3
Charging of a capacitor with three time constants.

```
r3 = .1e3;
tau3 = c*r3;
v3 = 10*(1-exp(-t/tau3));
plot(t, v1, t, v1, '+b', t, v2, t, v2, 'ob', t, v3,
t,v3,'*b')
axis([0 0.06 0 12])
title('Charging of a capacitor with three time
constants')
xlabel('Time, s')
ylabel('Voltage across capacitor')
text(0.03, 5.0, '+ for R = 1 Kilohms')
text(0.03, 6.0, 'o for R = 10 Kilohms')
text(0.03, 7.0, '* for R = 0.1 Kilohms')
```

Figure 5.3 shows the charging curves.

From Figure 5.3, it can be seen that as the time constant is small, it takes a short time for the capacitor to charge up.

Example 5.2 Charging and Discharging of a Capacitor

For Figure 5.2, the input voltage is a rectangular pulse with an amplitude of 5 V and a width of 0.5 sec. If $C = 10\ \mu\text{F}$, plot the output voltage, $v_0(t)$, for resistance R equal to (a) 1000 Ω and (b) 10,000 Ω. The plots should start from 0 sec and end at 1.5 sec.

Solution

MATLAB script

```
% The problem will be solved using function program
% rceval
% The output is obtained for the various resistances
c = 10.0e-6;
r1 = 2500;
[v1,t1] = rceval(r1,c);
r2 = 10000;
[v2,t2] = rceval(r2,c);
% plot the voltages
plot(t1, v1, t1,v1,'*b', t2, v2, t2, v2,'ob')
axis([0 1 0 6])
title('Response of RC circuit to a Pulse input')
xlabel('Time, s')
ylabel('Voltage, V')
text(0.55,5.5,'* is for 2500 Ohms')
text(0.55,5.0, 'o is for 10000 Ohms')

%  The problem will be solved using a function program
rceval
function [v, t] = rceval(r, c)
% rceval is a function program for calculating
%       the output voltage given the values of
%       resistance and capacitance.
% usage [v, t] = rceval(r, c)
%      r is the resistance value(ohms)
%      c is the capacitance value(Farads)
%      v is the output voltage
%      t is the time corresponding to voltage v
tau  = r*c;
for i=1:50
    t(i) = i/100;
    v(i) = 5*(1-exp(-t(i)/tau));
end
vmax = v(50);
for i = 51:100
    t(i) = i/100;
    v(i) = vmax*exp(-t(i-50)/tau);
end
```

Figure 5.4 shows the charging and discharging curves.

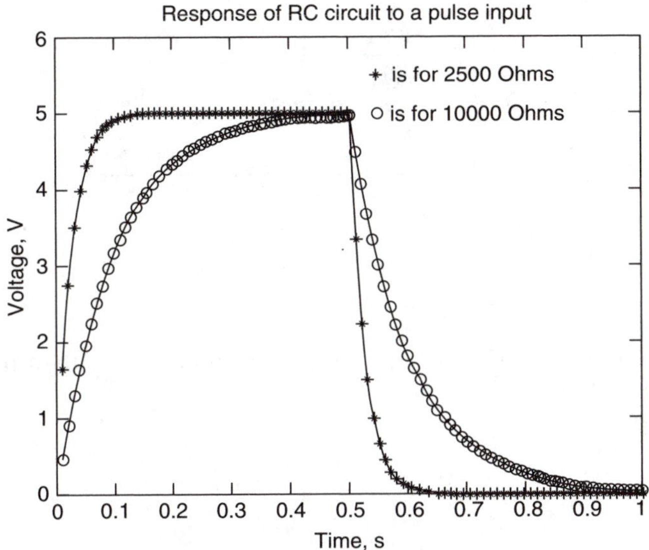

FIGURE 5.4
Charging and discharging of a capacitor with different time constants.

5.2 RL Network

Consider the RL circuit shown in Figure 5.5.

Using Kirchhoff's voltage law (KVL), we get

$$L\frac{di(t)}{dt} + Ri(t) = 0 \tag{5.5}$$

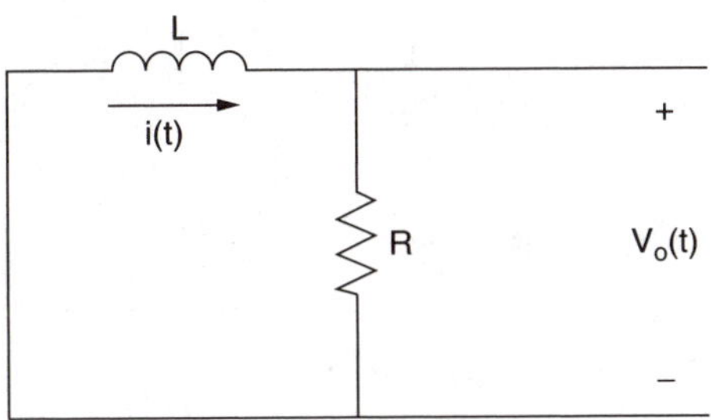

FIGURE 5.5
Source-free RL circuit.

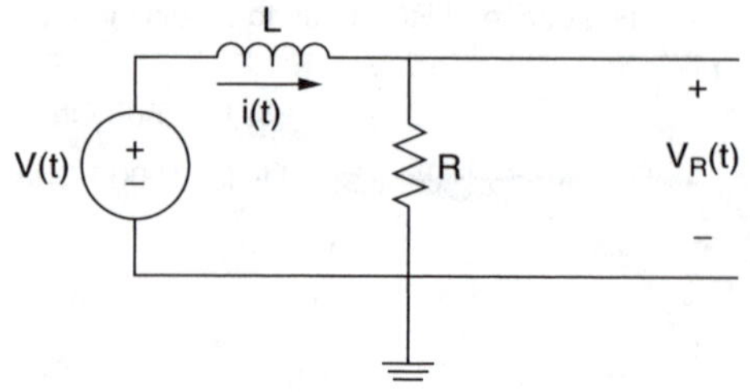

FIGURE 5.6
RL circuit with a voltage source.

If the initial current flowing through the inductor is I_m, then the solution to Equation (5.5) is

$$i(t) = I_m e^{-\left(\frac{t}{\tau}\right)} \tag{5.6}$$

where

$$\tau = L/R \tag{5.7}$$

Equation (5.6) represents the current response of a source-free RL circuit with initial current I_m, and it represents the natural response of an RL circuit.

Figure 5.6 is an RL circuit with source voltage $v(t) = V_S$.

Using KVL, we get

$$L\frac{di(t)}{dt} + Ri(t) = V_S \tag{5.8}$$

If the initial current flowing through the series circuit is zero, the solution of Equation (5.8) is

$$i(t) = \frac{V_S}{R}\left(1 - e^{-\left(\frac{Rt}{L}\right)}\right) \tag{5.9}$$

The voltage across the resistor is

$$\begin{aligned} v_R(t) &= Ri(t) \\ &= V_S\left(1 - e^{-\left(\frac{Rt}{L}\right)}\right) \end{aligned} \tag{5.10}$$

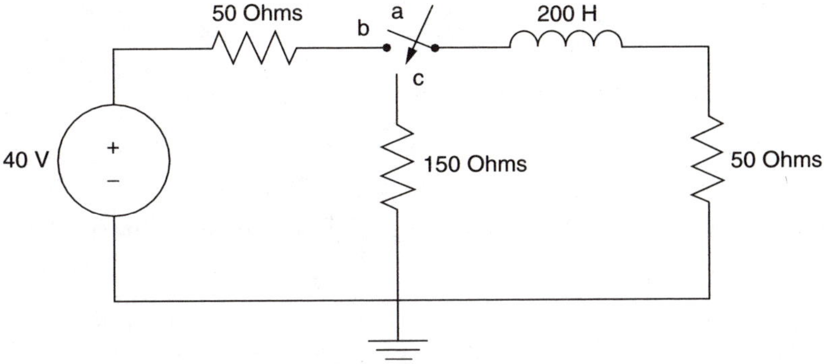

FIGURE 5.7
RL circuit for Example 5.3.

The voltage across the inductor is

$$\begin{aligned} v_L(t) &= V_S - v_R(t) \\ &= V_S e^{-\left(\frac{Rt}{L}\right)} \end{aligned} \tag{5.11}$$

The following example illustrates the use of MATLAB for solving RL circuit problems.

Example 5.3 Current in an RL Circuit

For the sequential circuit shown in Figure 5.7, the current flowing through the inductor is zero. At $t = 0$, the switch moved from position a to b, where it remained for 1 sec. After the 1-sec delay, the switch moved from position b to position c, where it remained indefinitely. Sketch the current flowing through the inductor vs. time.

Solution

For $0 < t < 1$ sec, we can use Equation (5.9) to find the current.

$$i(t) = 0.4\left(1 - e^{-\left(\frac{t}{\tau_1}\right)}\right) \tag{5.12}$$

where

$$\tau_1 = L/R = 200/100 = 2 \text{ sec}$$

At $t = 1$ sec

$$\begin{aligned} i(t) &= 0.4\left(1 - e^{-0.5}\right) \\ &= I_{max} \end{aligned} \tag{5.13}$$

For $t > 1$ sec, we can use Equation (5.6) to obtain the current,

$$i(t) = I_{max} e^{-\left(\frac{t-0.5}{\tau_2}\right)} \tag{5.14}$$

where

$$\tau_2 = L/R_{eq2} = 200/200 = 1 \text{ sec}$$

The MATLAB program for plotting $i(t)$ is shown below.

MATLAB script

```
% tau1 is time constant when switch is at b
% tau2 is the time constant when the switch is in
position c
%
tau1 = 200/100;
for k=1:20
  t(k) = k/20;
  i(k) = 0.4*(1-exp(-t(k)/tau1));
end
imax = i(20);
tau2 = 200/200;
for k = 21:120
  t(k) = k/20;
  i(k) = imax*exp(-t(k-20)/tau2);
end
% plot the current
plot(t, i)
axis([0 6 0 0.18])
title('Current of  RL Circuit')
xlabel('Time, s')
ylabel('Current, A')
```

Figure 5.8 shows the current $i(t)$.

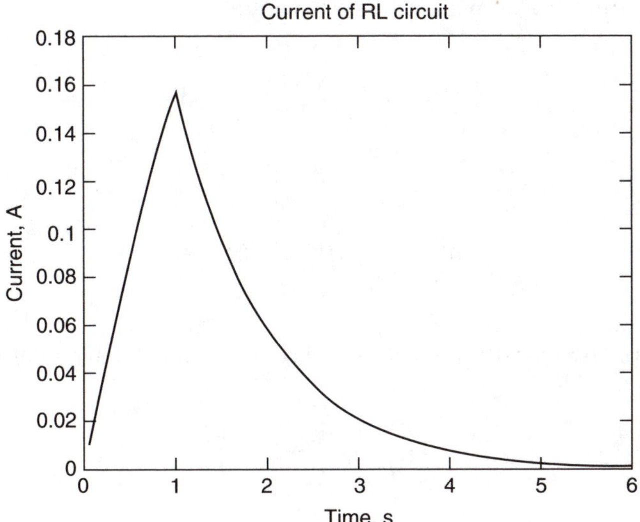

FIGURE 5.8
Current flowing through inductor.

5.3 RLC Circuit

For the series RLC circuit shown in Figure 5.9, we can use KVL to obtain Equation (5.15).

$$v_S(t) = L\frac{di(t)}{dt} + \frac{1}{C}\int_{-\infty}^{t} i(\tau)d\tau + Ri(t) \tag{5.15}$$

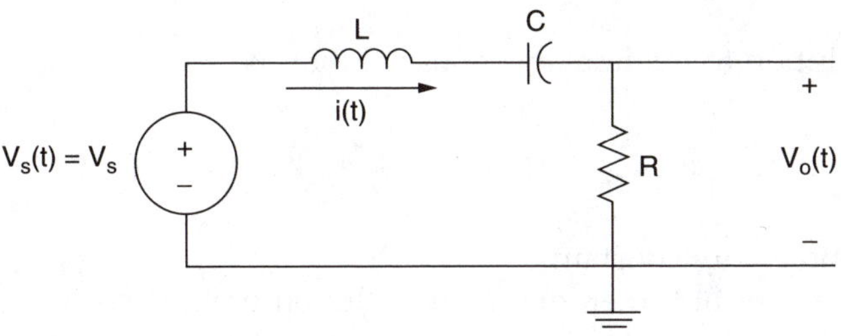

FIGURE 5.9
Series RLC circuit.

Differentiating the above expression, we get

$$\frac{dv_S(t)}{dt} = L\frac{d^2i(t)}{dt^2} + R\frac{di(t)}{dt} + \frac{i(t)}{C}$$

i.e.,

$$\frac{1}{L}\frac{dv_S(t)}{dt} = \frac{d^2i(t)}{dt^2} + \frac{R}{L}\frac{di(t)}{dt} + \frac{i(t)}{LC} \tag{5.16}$$

The homogeneous solution can be found by making $v_S(t)$ = constant, thus

$$0 = \frac{d^2i(t)}{dt^2} + \frac{R}{L}\frac{di(t)}{dt} + \frac{i(t)}{LC} \tag{5.17}$$

The characteristic equation is

$$0 = \lambda^2 + a\lambda + b \tag{5.18}$$

where

$$a = R/L$$

and

$$b = 1/LC$$

The roots of the characteristic equation can be determined. If we assume that the roots are

$$\lambda = \alpha_1,\ \alpha_2$$

then the solution to the homogeneous solution is

$$i_h(t) = A_1e^{\alpha_1 t} + A_2e^{\alpha_2 t} \tag{5.19}$$

where A_1 and A_2 are constants.

If $v_S(t)$ is a constant, then the forced solution will also be a constant and be given as

$$i_f(t) = A_3 \tag{5.20}$$

The total solution is given as

$$i(t) = A_1 e^{\alpha_1 t} + A_2 e^{\alpha_2 t} + A_3 \tag{5.21}$$

where A_1, A_2, and A_3 are obtained from initial conditions.

Example 5.4 illustrates the use of MATLAB for finding the roots of characteristic equations. The MATLAB function **roots,** described in Section 6.3.1, is used to obtain the roots of characteristic equations.

Example 5.4 Current Flowing through a Series RLC Circuit

For the series RLC circuit shown in Figure 5.9, if $L = 10$ H, $R = 400\ \Omega$, and $C = 100\ \mu$F, $v_S(t) = 0$, $i(0) = 4$ A, and

$$\frac{di(0)}{dt} = 15 \text{ A/sec}$$

find $i(t)$.

Solution

Since $v_S(t) = 0$, we use Equation (5.17) to get

$$0 = \frac{d^2 i(t)}{dt^2} + \frac{400}{10}\frac{di(t)}{dt} + 1000 i(t)$$

The characteristic equation is

$$0 = \lambda^2 + 40\lambda + 1000$$

The MATLAB function **roots** is used to obtain the roots of the characteristics equation.

MATLAB script

```
% roots function is used to find the
% roots of the characteristic equation
p = [1 40 1000];
lambda = roots(p)
```

The roots are

```
lambda =
              -20.0000 +24.4949i
              -20.0000 -24.4949i
```

Using the roots obtained from MATLAB, $i(t)$ is given as

$$i(t) = e^{-20t}(A_1 \cos(24.4949t) + A_2 \sin(24.4949t)$$

$$i(0) = e^{-0}(A_1 + A_2(0)) \qquad \Rightarrow A_1 = 4$$

$$\frac{di(t)}{dt} = -20e^{-20t}\left[A_1 \cos(24.4949t) + A_2 \sin(24.4949t)\right] +$$

$$e^{-20t}\left[-24.4949A_1 \sin(24.4949t) + 24.4949A_2 \cos(24.4949t)\right]$$

$$\frac{di(0)}{dt} = 24.4949A_2 - 20A_1 = 15$$

Since $A_1 = 4$ and $A_2 = 3.8784$,

$$i(t) = e^{-20t}\left[4\cos(24.4949t) + 3.8784\sin(24.4949t)\right]$$

Perhaps the simplest way to obtain voltages and currents in an RLC circuit is to use the Laplace transform. Table 5.1 shows Laplace transform pairs that are useful for solving RLC circuit problems.

From the RLC circuit, we write differential equations by using network analysis tools. The differential equations are converted into algebraic equations using the Laplace transform. The unknown current or voltage is then solved in the s-domain. Using an inverse Laplace transform, the solution can be expressed in the time domain. We will illustrate this method using Example 5.5.

Example 5.5 Voltage across a Parallel RLC Circuit

The switch in Figure 5.10 has been opened for a long time. If the switch opens at t = 0, find the voltage $v(t)$. Assume that $R = 10\ \Omega$, $L = 1/32$ H, $C =$ 50 µF, and $I_S = 2\ A$.

At t < 0, the voltage across the capacitor is

$$v_C(0) = (2)(10) = 20 \text{ V}$$

In addition, the current flowing through the inductor

$$i_L(0) = 0$$

At t > 0, the switch closes and all four elements of Figure 5.10 remain in parallel. Using KCL, we get

TABLE 5.1
Laplace Transform Pairs

	f(t)	f(s)
1	1	$\frac{1}{s}$; s > 0
2	t	$\frac{1}{s^2}$; s > 0
3	t^n	$\frac{n!}{s^{n+1}}$; s > 0
4	e^{-at}	$\frac{1}{s+a}$; s > a
5	te^{-at}	$\frac{1}{(s+a)^2}$; s > a
6	$\sin(wt)$	$\frac{w}{s^2+w^2}$; s > 0
7	$\cos(wt)$	$\frac{s}{s^2+w^2}$; s > 0
8	$e^{at}\sin(wt)$	$\frac{w}{(s+a)^2+w^2}$
9	$e^{at}\cos(wt)$	$\frac{s+a}{(s+a)^2+w^2}$
10	$\frac{df}{dt}$	$sF(s)-f(0^+)$
11	$\int_0^t f(\tau)d\tau$	$\frac{F(s)}{s}$
12	$f(t-t_1)$	$e^{-t_1 s}F(s)$

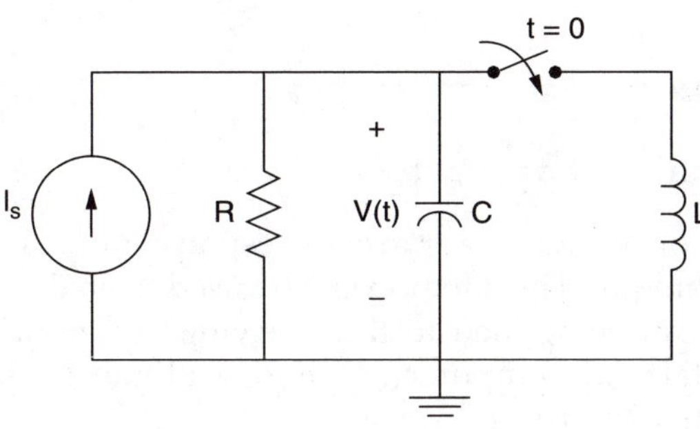

FIGURE 5.10
Circuit for Example 5.5.

$$I_S = \frac{v(t)}{R} + C\frac{dv(t)}{dt} + \frac{1}{L}\int_0^t v(\tau)d\tau + i_L(0)$$

Taking the Laplace transform of the above expression, we get

$$\frac{I_S}{s} = \frac{V(s)}{R} + C[sV(s) - V_C(0)] + \frac{V(s)}{sL} + \frac{i_L(0)}{s}$$

Simplifying the above expression, we get

$$V(s) = \frac{1/C\left[I_S - I_L(0) + sCV_C(0)\right]}{s^2 + \frac{s}{CR} + \frac{1}{LC}}$$

For $I_S = 2$ A, $C = 50$ μF, $R = 10$ Ω, $L = 1/32$ H, $V(s)$ becomes

$$V(s) = \frac{40000 + 20s}{s^2 + 2000s + 64 * 10^4}$$

$$V(s) = \frac{40000 + 20s}{(s+1600)(s+400)} = \frac{A}{(s+1600)} + \frac{B}{(s+400)}$$

$$A = \lim_{s \to -1600} V(s)(s+1600) = -6.67$$

$$B = \lim_{s \to -400} V(s)(s+400) = 26.67$$

$$v(t) = -6.67e^{-1600t} + 26.67e^{-400t}$$

The plot of $v(t)$ is shown in Figure 5.13.

5.4 State Variable Approach

Another method of finding the transient response of an RLC circuit is the state variable technique. This method can be used to analyze and synthesize control systems, can be applied to time-varying and nonlinear systems, is suitable for digital and computer solution, and can be used to develop general system characteristics.

A state of a system is a minimal set of variables chosen such that if their values are known at the time t, and all inputs are known for times greater than t_1, one can calculate the output of the system for times greater than t_1. In general, if we designate x as the state variable, u as the input, and y as the output of a system, we can express the input u and output y as

$$\dot{x}(t) = Ax(t) + Bu(t) \tag{5.22}$$

$$y(t) \quad = Cx(t) + Du(t) \tag{5.23}$$

where

$$u(t) = \begin{bmatrix} u_1(t) \\ u_2(t) \\ . \\ . \\ u_n(t) \end{bmatrix} \qquad x(t) = \begin{bmatrix} x_1(t) \\ x_2(t) \\ . \\ . \\ x_n(t) \end{bmatrix} \qquad y(t) = \begin{bmatrix} y_1(t) \\ y_2(t) \\ . \\ . \\ y_n(t) \end{bmatrix}$$

and A, B, C, and D are matrices determined by constants of a system.

For example, consider a single-input and a single-output system described by the differential equation

$$\frac{d^4y(t)}{dt^4} + 3\frac{d^3y(t)}{dt^3} + 4\frac{d^2y(t)}{dt^2} + 8\frac{dy(t)}{dt} + 2y(t) \quad = \quad 6u(t) \tag{5.24}$$

We define the components of the state vector as

$$x_1(t) = y(t)$$

$$x_2(t) = \frac{dy(t)}{dt} = \frac{dx_1(t)}{dt} = \dot{x}_1(t)$$

$$x_3(t) = \frac{d^2y(t)}{dt^2} = \frac{dx_2(t)}{dt} = \dot{x}_2(t)$$

$$x_4(t) = \frac{d^3y(t)}{dt^3} = \frac{dx_3(t)}{dt} = \dot{x}_3(t)$$

$$x_5(t) = \frac{d^4y(t)}{dt^4} = \frac{dx_4(t)}{dt} = \dot{x}_4(t) \tag{5.25}$$

Using Equation (5.24) and Equation (5.25), we get

$$\dot{x}_4(t) = 6u(t) - 3x_4(t) - 4x_3(t) - 8x_2(t) - 2x_1(t) \tag{5.26}$$

From Equation (5.25) and Equation (5.26), we get

$$\begin{bmatrix} \dot{x}_1(t) \\ \dot{x}_2(t) \\ \dot{x}_3(t) \\ \dot{x}_4(t) \end{bmatrix} = \begin{bmatrix} 0 & 1 & 0 & 0 \\ 0 & 0 & 1 & 0 \\ 0 & 0 & 0 & 1 \\ -2 & -8 & -4 & -3 \end{bmatrix} \begin{bmatrix} x_1(t) \\ x_2(t) \\ x_3(t) \\ x_4(t) \end{bmatrix} + \begin{bmatrix} 0 \\ 0 \\ 0 \\ 6 \end{bmatrix} u(t) \tag{5.27}$$

or

$$\dot{x}(t) = Ax(t) + Bu(t) \tag{5.28}$$

where

$$\dot{x} = \begin{bmatrix} \dot{x}_1(t) \\ \dot{x}_2(t) \\ \dot{x}_3(t) \\ \dot{x}_4(t) \end{bmatrix}; \quad A = \begin{bmatrix} 0 & 1 & 0 & 0 \\ 0 & 0 & 1 & 0 \\ 0 & 0 & 0 & 1 \\ -2 & -8 & -4 & -3 \end{bmatrix}; B = \begin{bmatrix} 0 \\ 0 \\ 0 \\ 6 \end{bmatrix} \tag{5.29}$$

Since $y(t) = x_1(t)$, we can express the output $y(t)$ in terms of the state $x(t)$ and input $u(t)$ as

$$y(t) = Cx(t) + Du(t) \tag{5.30}$$

where

$$C = \begin{bmatrix} 1 & 0 & 0 & 0 \end{bmatrix} \text{ and } D = \begin{bmatrix} 0 \end{bmatrix} \tag{5.31}$$

In RLC circuits, if the voltage across a capacitor and the current flowing in an inductor are known at some initial time t, then the capacitor voltage and inductor current will allow the description of system behavior for all subsequent times. This suggests the following guidelines for the selection of acceptable state variables for RLC circuits:

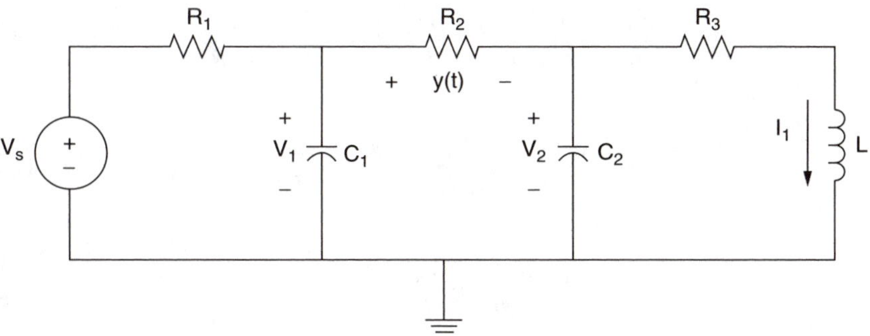

FIGURE 5.11
Circuit for state analysis.

1. Currents associated with inductors are state variables.
2. Voltages associated with capacitors are state variables.
3. Currents or voltages associated with resistors do not specify independent state variables.
4. When closed loops of capacitors or junctions of inductors exist in a circuit, the state variables chosen according to guidelines 1 and 2 are not independent.

Consider the circuit shown in Figure 5.11. Using the above guidelines, we select the state variables to be V_1, V_2, and i_1. Using nodal analysis, we have

$$C_1 \frac{dv_1(t)}{dt} + \frac{V_1 - V_s}{R_1} + \frac{V_1 - V_2}{R_2} = 0 \tag{5.32}$$

$$C_2 \frac{dv_2(t)}{dt} + \frac{V_2 - V_1}{R_2} + i_1 = 0 \tag{5.33}$$

Using loop analysis,

$$V_2 = i_1 R_3 + L \frac{di_1(t)}{dt} \tag{5.34}$$

The output $y(t)$ is given as

$$y(t) = v_1(t) - v_2(t) \tag{5.35}$$

Simplifying Equation (5.32) to Equation (5.34), we get

$$\frac{dv_1(t)}{dt} = -(\frac{1}{C_1R_1} + \frac{1}{C_1R_2})V_1 + \frac{V_2}{C_1R_2} + \frac{V_s}{C_1R_1} \tag{5.36}$$

$$\frac{dv_2(t)}{dt} = \frac{V_1}{C_2R_2} - \frac{V_2}{C_2R_2} - \frac{i_1}{C_2} \tag{5.37}$$

$$\frac{di_1(t)}{dt} = \frac{V_2}{L} - \frac{R_3}{L}i_1 \tag{5.38}$$

Expressing the equations in matrix form, we get

$$\begin{bmatrix} \dot{V}_1 \\ \dot{V}_2 \\ \dot{i}_1 \end{bmatrix} = \begin{bmatrix} -(\frac{1}{C_1R_1} + \frac{1}{C_1R_2}) & \frac{1}{C_1R_2} & 0 \\ \frac{1}{C_2R_2} & -\frac{1}{C_2R_2} & -\frac{1}{C_2} \\ 0 & \frac{1}{L} & -\frac{R_3}{L} \end{bmatrix} \begin{bmatrix} V_1 \\ V_2 \\ i_1 \end{bmatrix} + \begin{bmatrix} \frac{1}{C_1R_1} \\ 0 \\ 0 \end{bmatrix} V_s \tag{5.39}$$

and the output is

$$y = \begin{bmatrix} 1 & -1 & 0 \end{bmatrix} \begin{bmatrix} V_1 \\ V_2 \\ i_1 \end{bmatrix} \tag{5.40}$$

MATLAB functions for solving ordinary differential equations are ode functions. These are described in the following section.

5.4.1 MATLAB Ode Functions

MATLAB has two functions, **ode23** and **ode45**, for computing numerical solutions to ordinary differential equations. The **ode23** function integrates a system of ordinary differential equations using second- and third-order Runge-Kutta formulas; the **ode45** function uses fourth- and fifth-order Runge-Kutta integration equations.

The general forms of the ode functions are

[t, x] = ode23 (xprime, tspan, xo, tol)

or

[t, x] = ode45 (xprime, tspan, xo, tol)

where

xprime is the name (in quotation marks) of the MATLAB function or m-file that contains the differential equations to be integrated. The function will compute the state derivative vector $\dot{x}(t)$ given the current time t and state vector $x(t)$. The function must have two input arguments, scalar t (time) and column vector x (state), and the function returns the output argument *xdot*, $(\dot{x})$, a column vector of state derivatives

$$\dot{x}(t_1) = \frac{dx(t_1)}{dt}$$

tspan = [tstart tfinal]
tstart is the starting time for the integration
tfinal is the final time for the integration
xo is a column vector of initial conditions
tol is optional. It specifies the desired accuracy of the solution

Let us illustrate the use of MATLAB ode functions with the following two examples.

Example 5.6 State Variable Approach for RC Circuit Analysis

For Figure 5.2, $V_S = 10$ V, $R = 10{,}000\ \Omega$, and $C = 10\ \mu$F. Find the output voltage $v_0(t)$, between the interval 0 to 20 msec, assuming $v_0(0) = 0$ by (a) using a numerical solution to the differential equation and (b) using an analytical solution.

Solution

From Equation (5.3), we have

$$C\frac{dv_o(t)}{dt} + \frac{v_o(t) - V_s}{R} = 0$$

thus

$$\frac{dv_o(t)}{dt} = \frac{V_s}{CR} - \frac{v_o(t)}{CR} = 100 - 10v_0(t)$$

From Equation (5.4), the analytical solution is

$$v_0(t) = 10\left(1 - e^{-\left(\frac{t}{CR}\right)}\right)$$

MATLAB script

```
% Solution for first order differential equation
% the function diff1(t,y) is created to evaluate
% the differential equation
% Its m-file is diff1.m
%
% Transient analysis of RC circuit using ode
% function and analytical solution
% numerical solution using ode
t0 = 0;
tf = 20e-3;
tspan = [t0 tf];
xo = 0;  % initial conditions
[t, vo] = ode23('diff1',tspan, xo);

% the analytical solution given by Equation(5.4) is
vo_analy = 10*(1-exp(-10*t));

% plot two solutions
subplot(121)
plot(t,vo,'b')
title('State Variable Approach')
xlabel('Time, s'),ylabel('Capacitor Voltage, V'),grid
subplot(122)
plot(t,vo_analy,'b')
title('Analytical Approach')
xlabel('Time, s'),ylabel('Capacitor Voltage, V'),grid
function dy = diff1(t,y)
dy = 100 - 10*y;
```

Figure 5.12 shows the plot obtained using Equation (5.4) and that obtained from the MATLAB ode23 function. From the two plots, we can see that the two results are identical.

Example 5.7 State Variable Approach to RLC Circuit Analysis

For Figure 5.10, if $R = 10\ \Omega$, $L = 1/32$ H, and $C = 50\ \mu$F, use a numerical solution of the differential equation to solve $v(t)$. Compare the numerical solution to the analytical solution obtained from Example 5.5.

Solution

From Example 5.5, $v_C(0) = 20$ V, $i_L(0) = 0$, and

$$L\frac{di_L(t)}{dt} = v_C(t)$$

$$C\frac{dv_C(t)}{dt} + i_L + \frac{v_C(t)}{R} - I_S = 0$$

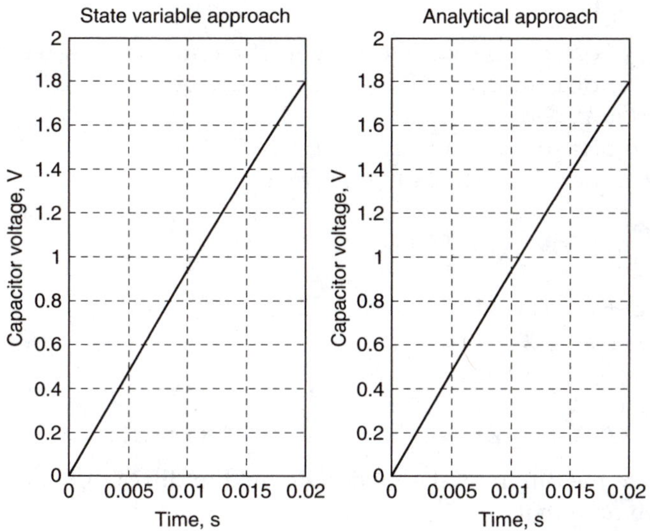

FIGURE 5.12
Output voltage $v_0(t)$ obtained from (left) the state variable approach and (right) an analytical method.

Simplifying, we get

$$\frac{di_L(t)}{dt} = \frac{v_C(t)}{L}$$

$$\frac{dv_C(t)}{dt} = \frac{I_S}{C} - \frac{i_L(t)}{C} - \frac{v_C(t)}{RC}$$

Assuming that

$$x_1(t) = i_L(t)$$

$$x_2(t) = v_C(t)$$

we get

$$\dot{x}_1(t) = \frac{1}{L} x_2(t)$$

$$\dot{x}_2(t) = \frac{I_S}{C} - \frac{1}{C} x_1(t) - \frac{1}{RC} x_2(t)$$

We create a function m-file containing the above differential equations.

MATLAB script

```
%  Solution of second-order differential equation
% The function diff2(x,y) is created to evaluate the
diff. equation
%  the name of the m-file is diff2.m
%  the function is defined as:
%
function xdot = diff2(t,x)
is = 2;
c = 50e-6;  L = 1/32;  r = 10;
k1 = 1/c ;   %  1/C
k2 =  1/L ;   %  1/L
k3 = 1/(r*c);   % 1/RC
xdot = [0 k2; -k1 -k3]*[x] + [0; k1*is];
```

To simulate the differential equation defined in diff2 in the interval $0 \leq t \leq$ 30 msec, we note that

$$x_1(0) = i_L(0) = 0 \quad \text{V}$$

$$x_2(0) = v_C(0) = 20$$

Using the MATLAB ode23 function, we get

```
% solution of second-order differential equation
% the function diff2(x,y) is created to evaluate
% the differential equation
% the name of m-file is diff2.m
%
% Transient analysis of RLC circuit using ode function
% numerical solution

t0 = 0;
tf = 30e-3;
tspan = [t0 tf];
x0 = [0 20]; % Initial conditions
[t,x] = ode23('diff2',tspan, x0);

% Second column of matrix x represents capacitor voltage
subplot(211), plot(t,x(:,2))
xlabel('Time, s'), ylabel('Capacitor voltage, V')
text(0.01, 7, 'State Variable Approach')

% Transient analysis of RLC circuit from Example 5.5
t2 =0:1e-3:30e-3;
vt = -6.667*exp(-1600*t2) + 26.667*exp(-400*t2);
subplot(212), plot(t2,vt)
xlabel('Time, s'), ylabel('Capacitor voltage, V')
text(0.01, 4.5, 'Results from Example 5.5')
```

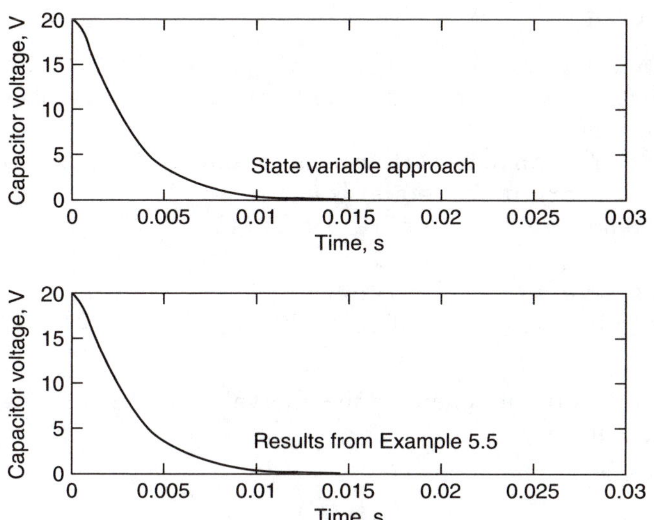

FIGURE 5.13
Capacitor voltage $v_0(t)$ obtained from both the state variable approach and the use of a Laplace transform.

The plot is shown in Figure 5.13. The results from the state variable approach and those obtained from Example 5.5 are identical.

Example 5.8 State Variable Analysis of a Network

For Figure 5.11, if

$$v_2 = i_1 R_3 + L\frac{di_1(t)}{dt}$$

where $u(t)$ is the unit step function and $R_1 = R_2 = R_3 = 10\ \text{K}\Omega$, $C_1 = C_2 = 5\ \mu\text{F}$, and $L = 10$ H, find and plot the voltage $y(t)$ within the intervals of 0 to 2 sec.

Solution

Using the element values and Equation (5.36) to Equation (5.38), we have

$$\frac{dv_1(t)}{dt} = -40v_1(t) + 20v_2(t) + 20V_s$$

$$\frac{dv_2(t)}{dt} = 20v_1(t) - 20v_2(t) - 200000i_1(t)$$

$$\frac{di_1(t)}{dt} = 0.1v_2(t) - 1000i_1(t)$$

We create an m-file containing the above differential equations.

MATLAB script

```
%
% solution of a set of first order differential
equations
% the function diff3(t,v) is created to evaluate
% the differential equation
% the name of the m-file is diff3.m
%
function vdot = diff3(t,v)
vdot =[-40 20 0; 20 -20 200000; 0 0.1 -1000]*[v]+ [100;
0; 0];
```

To obtain the output voltage in the interval of $0 \le t \le 5$ sec, we note that the output voltage

$$y(t) = v_1(t) - v_2(t)$$

Note that at $t < 0$, the step signal is zero, so

$$v_1(0) = v_2(0) = i_1(0) = 0$$

Using ode23 we get

```
% solution of a set of first-order differential
equations
% the function diff3(t,v) is created to evaluate
% the differential equation
% the name of the m-file is diff3.m
%
% Transient analysis of RLC circuit using state
% variable approach
t0 = 0;
tf = 2;
tspan = [t0 tf];
x0 = [0 0 0]; % initial conditions
[t,x] = ode23('diff3', tspan, x0);
tt = length(t);
for i = 1:tt
   vo(i) = x(i,1) - x(i,2);
end
plot(t, vo)
title('Transient Analysis of an RLC Circuit')
xlabel('Time, s'), ylabel('Output voltage, V')
```

```
% solution of a set of first-order differential
equations
% the function diff3(t,v) is created to evaluate
% the differential equation
% the name of the m-file is diff3.m
%
% Transient analysis of RLC circuit using state
% variable approach
t0 = 0;
tf = 2;
tspan = [t0 tf];
x0 = [0 0 0]; % initial conditions
[t,x] = ode23('diff3', tspan, x0);
tt = length(t);
for i = 1:tt
   vo(i) = x(i,1) - x(i,2);
end
plot(t, vo)
title('Transient analysis of RLC')
xlabel('Time, s'), ylabel('Output voltage')
```

The plot of the output voltage $y(t)$ of Figure 5.11 is shown in Figure 5.14.

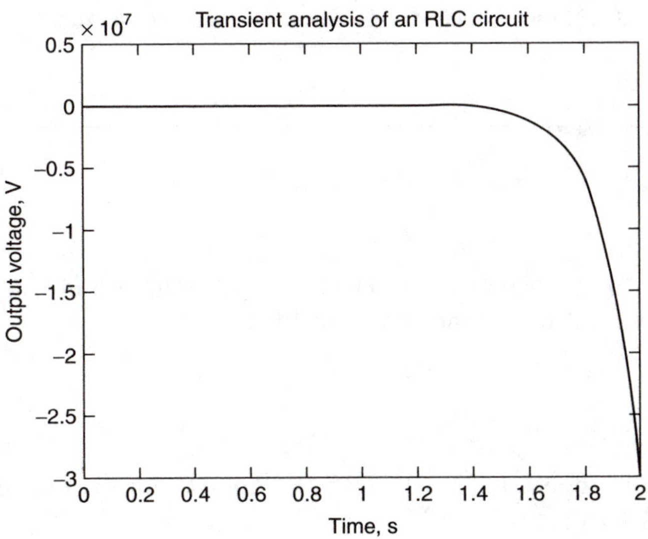

FIGURE 5.14
Output voltage $y(t)$ of Figure 5.11.

Bibliography

1. Alexander, C.K. and Sadiku, M.N.O., *Fundamentals of Electric Circuits*, 2nd ed., McGraw-Hill, New York, 2004.
2. Attia, J.O., *PSPICE and MATLAB for Electronics: An Integrated Approach*, CRC Press, Boca Raton, FL, 2002.
3. Biran, A. and Breiner, M., *MATLAB for Engineers*, Addison-Wesley, Reading, MA, 1995
4. Chapman, S.J., *MATLAB Programming for Engineers*, Brook, Cole Thompson Learning, Pacific Grove, CA, 2000.
5. Dorf, R.C. and Svoboda, J.A., *Introduction to Electric Circuits*, 3rd ed., John Wiley & Sons, New York, 1996.
6. Etter, D.M., *Engineering Problem Solving with MATLAB*, 2nd ed., Prentice Hall, Upper Saddle River, NJ, 1997.
7. Etter, D.M., Kuncicky, D.C., and Hull, D., *Introduction to MATLAB 6*, Prentice Hall, Upper Saddle River, NJ, 2002.
8. Gottling, J.G., *Matrix Analysis of Circuits Using MATLAB*, Prentice Hall, Englewood Cliffs, NJ, 1995.
9. Johnson, D.E., Johnson, J.R., and Hilburn, J.L., *Electric Circuit Analysis*, 3rd ed., Prentice Hall, Englewood Cliffs, NJ, 1997.
10. Meader, D.A., *Laplace Circuit Analysis and Active Filters*, Prentice Hall, Englewood Cliffs, NJ, 1991.
11. Sigmor, K., *MATLAB Primer*, 4th ed., CRC Press, Boca Raton, FL, 1998.
12. *Using MATLAB, The Language of Technical Computing, Computation, Visualization, Programming*, Version 6, MathWorks, Inc., Natick, MA, 2000.
13. Vlach, J.O., Network theory and CAD, *IEEE Trans. on Education*, 36, 23–27, 1993.

Problems

Problem 5.1

In Figure 5.10, $R = 800\ \Omega$, $L = 10$ H, $C = 2.5\ \mu$F and $I_S = 5$ mA. If the switch has been opened for a long time and the switch closes at $t = 0$, find the voltage $v(t)$.

Problem 5.2

If the switch is opened at $t = 0$, find $v_0(t)$. Plot $v_0(t)$ for the time interval $0 \le t \le 5$ sec. Refer to Figure P5.2.

Problem 5.3

The switch is closed at $t = 0$; find $i(t)$ for the interval 0 to 10 msec. The resistance values are in ohms. Refer to Figure P5.3. Plot $i(t)$.

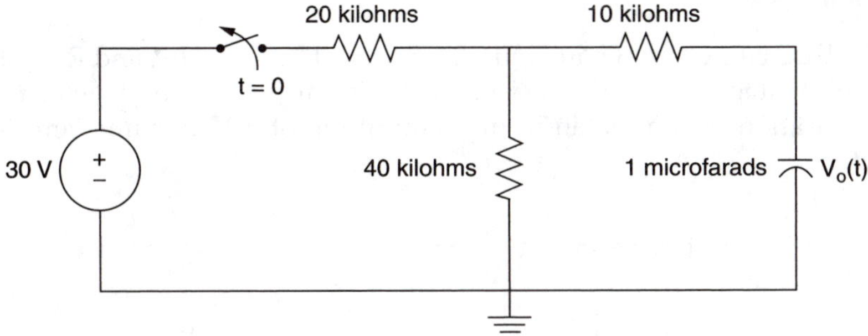

FIGURE P5.2
Figure for Problem 5.2.

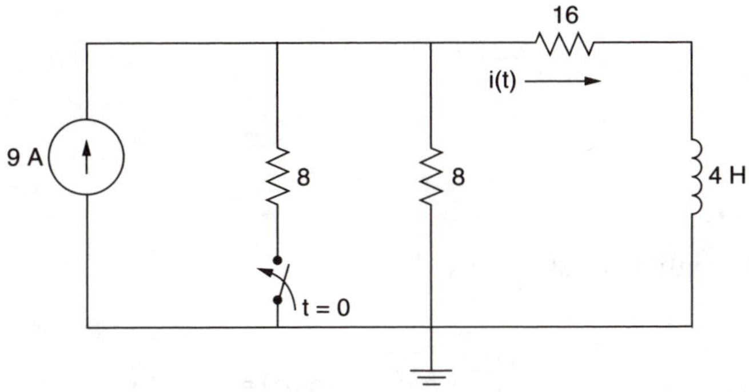

FIGURE P5.3
Figure for Problem 5.3.

Problem 5.4

For the series RLC circuit, the switch is closed at $t = 0$. The initial energy in the storage elements is zero. Use MATLAB to find $v_0(t)$. Refer to Figure P5.4.

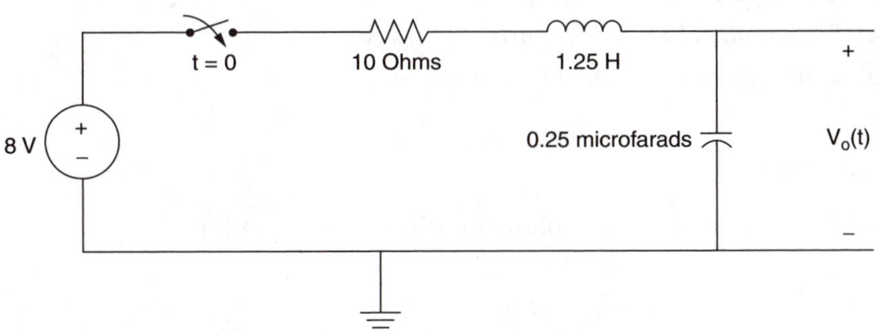

FIGURE P5.4
Circuit for Problem 5.4

Problem 5.5

For the RLC circuit shown in Figure P5.5, L = 4 H, C = 2.5 μF, and R = 1000 Ω. Find the voltage across the resistor if the input is a pulse waveform with pulse duration of 5 msec and pulse amplitude of 5 V. Assume zero initial conditions.

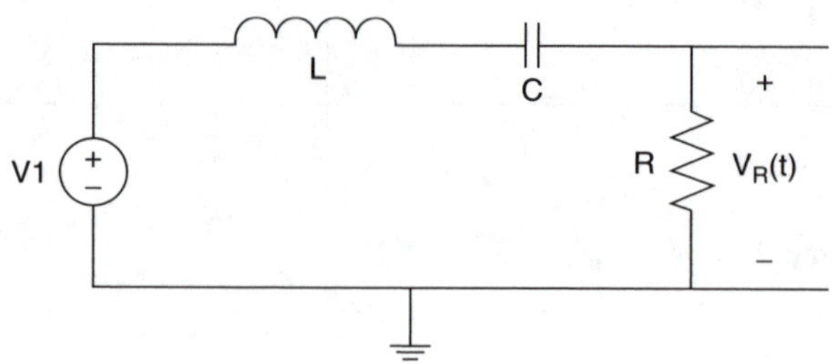

FIGURE P5.5
RLC circuit.

Problem 5.6

For the differential equation

$$\frac{d^2y(t)}{dt^2} + 5\frac{dy(t)}{dt} + 6y(t) = 20$$

with initial conditions $y(0) = 4$ and

$$\frac{dy(0)}{dt} = -1$$

(a) Determine $y(t)$ using Laplace transforms.
(b) Use MATLAB to determine $y(t)$.
(c) Sketch $y(t)$ obtained in parts (a) and (b).
(d) Compare the results obtained in part (c).

Problem 5.7

Use MATLAB to solve the following differential equation:

$$\frac{d^3y(t)}{dt^3} + 7\frac{d^2y(t)}{dt^2} + 14\frac{dy(t)}{dt} + 12y(t) = 10$$

with initial conditions

$$y(0) = 1\ , \quad \frac{dy(0)}{dt} = 2\ , \quad \frac{d^2y(0)}{dt^2} = 5$$

Plot y(t) within the interval of 0 to 10 sec.

Problem 5.8

For Figure P5.8, if $V_S = 5\ u(t)$, determine the voltages $V_1(t)$ and $V_4(t)$ in the interval 0 to 20 sec. Assume that the initial voltage across each capacitor is zero.

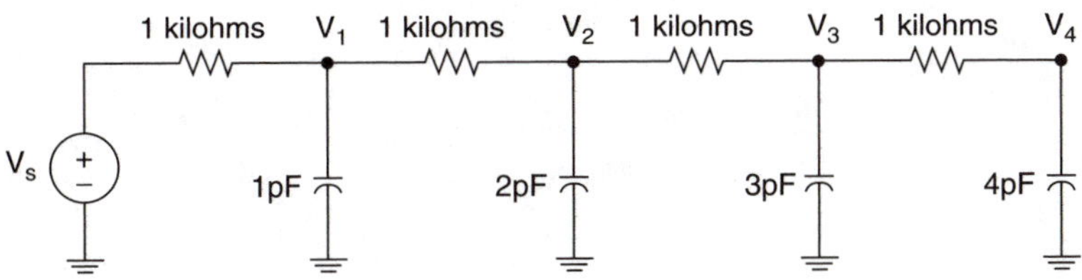

FIGURE P5.8
RC network of Problem 5.8.

Problem 5.9

For the multi-stage RC circuit shown in Figure P5.9, C1 = C2 = C3 = 1 μF and R1 = R2 = R3 = 1 kΩ. If the input signal $V_s(t) = 10u(t)$ V, determine the voltage $V_o(t)$.

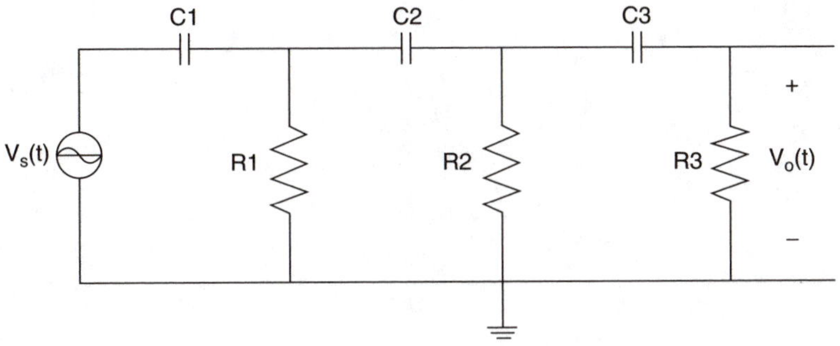

FIGURE P5.9
Multi-stage RC network.

with initial conditions

[illegible]

Plot y(t) using the [illegible] from t = 0 to t = [illegible]

[illegible]

[illegible]

[illegible]

6

AC Analysis and Network Functions

This chapter discusses sinusoidal steady-state power calculations. Numerical integration is used to obtain the root-mean-squared (rms) value, average power, and quadrature power. Three-phase circuits are analyzed by converting the circuits into the frequency domain and by using the Kirchhoff voltage and current laws. The unknown voltages and currents are solved using matrix techniques.

Given a network function or transfer function, MATLAB has functions that can be used to obtain the poles and zeros, perform partial fraction expansion, and evaluate the transfer function at specific frequencies. Furthermore, the frequency response of networks can be obtained using a MATLAB function. These features of MATLAB are applied in this chapter.

6.1 Steady-State AC Power

Figure 6.1 shows an impedance with voltage across it given by $v(t)$ and current through it by $i(t)$.

The instantaneous power $p(t)$ is

$$p(t) = v(t)i(t) \tag{6.1}$$

If $v(t)$ and $i(t)$ are periodic with period T, the rms or effective values of the voltage and current are

$$V_{rms} = \sqrt{\frac{1}{T}\int_0^T v^2(t)dt} \tag{6.2}$$

$$I_{rms} = \sqrt{\frac{1}{T}\int_0^T i^2(t)dt} \tag{6.3}$$

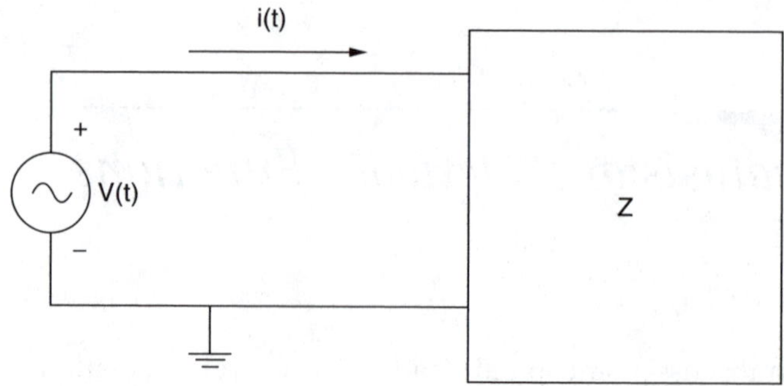

FIGURE 6.1
One-port network with impedance Z.

where V_{rms} is the rms value of $v(t)$ and I_{rms} is the rms value of $i(t)$.

The average power dissipated by the one-port network is

$$P = \frac{1}{T}\int_{0}^{T} v(t)i(t)dt \tag{6.4}$$

The power factor, *pf*, is given as

$$pf = \frac{P}{V_{rms} I_{rms}} \tag{6.5}$$

For the special case where both the current $i(t)$ and voltage $v(t)$ are both sinusoidal, that is,

$$v(t) = V_m \cos(wt + \theta_V) \tag{6.6}$$

and

$$i(t) = I_m \cos(wt + \theta_I) \tag{6.7}$$

the rms value of the voltage $v(t)$ is

$$V_{rms} = \frac{V_m}{\sqrt{2}} \tag{6.8}$$

and that of the current is

$$I_{rms} = \frac{I_m}{\sqrt{2}} \tag{6.9}$$

The average power P is

$$P = V_{rms} I_{rms} \cos(\theta_V - \theta_I) \tag{6.10}$$

The power factor, pf, is

$$pf = \cos(\theta_V - \theta_I) \tag{6.11}$$

The reactive power Q is

$$Q = V_{rms} I_{rms} \sin(\theta_V - \theta_I) \tag{6.12}$$

and the complex power, S, is

$$S = P + jQ \tag{6.13}$$

$$S = V_{rms} I_{rms} \left[\cos(\theta_V - \theta_I) + j\sin(\theta_V - \theta_I)\right] \tag{6.14}$$

Equation (6.2) to Equation (6.4) involve the use of integration in the determination of the rms value and the average power. MATLAB has two functions, quad and quad8, for performing numerical function integration.

6.1.1 MATLAB Functions quad and quad8

The **quad** function uses an adaptive, recursive Simpson's rule. The **quad8** function uses an adaptive, recursive Newton Cutes eight-panel rule. The quad8 function is better than the quad at handling functions with "soft" singularities such as $\int \sqrt{x}dx$. Suppose we want to find q given as

$$q = \int_a^b funct(x)dx$$

The general forms of the quad and quad8 functions that can be used to find q are

$$quad('funct', a, b, tol, trace)$$

$$quad8('funct', a, b, tol, trace)$$

where *funct* is a MATLAB function name (in quotes) that returns a vector of values of $f(x)$ for a given vector of input values x; a is the lower limit of integration; *b* is the upper limit of integration; and *tol* is the tolerance limit set for stopping the iteration of the numerical integration. The iteration continues until the relative error is less than *tol*. The default value is 1.0e-3. The trace allows the plot of a graph showing the process of the numerical integration. If the trace is nonzero, a graph is plotted. The default value is zero.

Example 6.1 shows the use of the quad function to perform alternating current power calculations.

Example 6.1 AC Power Calculations

For Figure 6.1, if $v(t) = 10\cos(120\pi t + 30^0)$ and $i(t) = 6\cos(120\pi t + 60^0)$, determine the average power, rms value of $v(t)$, and the power factor using (a) analytical solution and (b) numerical solution.

Solution

MATLAB script

```
% This program computes the average power, rms value and
% power factor using quad function. The analytical and
% numerical results are compared.
% numerical calculations

T = 2*pi/(120*pi); % period of the sin wave
a = 0; % lower limit of integration
b = T; % upper limit of integration
x = 0:0.02:1;
t = x.*b;
v_int = quad('voltage1', a, b);
v_rms = sqrt(v_int/b);  % rms of voltage
i_int = quad('current1',a,b);
i_rms = sqrt(i_int/b);  % rms of current

p_int = quad('inst_pr', a, b);
p_ave = p_int/b;     % average power
pf = p_ave/(i_rms*v_rms); % power factor
%
% analytical solution
%
p_ave_an = (60/2)*cos(30*pi/180);  % average power
v_rms_an = 10.0/sqrt(2);
pf_an = cos(30*pi/180);

% results are printed
fprintf('Average power, analytical %f \n Average power,
numerical: %f \n', p_ave_an,p_ave)
fprintf('rms voltage, analytical: %f \n rms voltage,
numerical: %f \n', v_rms_an, v_rms)
```

```
fprintf('power factor, analytical: %f \n power factor,
numerical: %f \n', pf_an, pf)
```

The following functions are used in the above m-file:

```
function vsq = voltage1(t)
% voltage1  This function is used to
%           define the voltage function
vsq = (10*cos(120*pi*t + 60*pi/180)).^2;

function isq = current1(t)
% current1  This function is to define the current
%
isq = (6*cos(120*pi*t + 30.0*pi/180)).^2;

function pt = inst_pr(t)
% inst_pr   This function is used to define
%          instantaneous power obtained by multiplying
%           sinusoidal voltage and current
it = 6*cos(120*pi*t + 30.0*pi/180);
vt = 10*cos(120*pi*t + 60*pi/180);
pt = it.*vt;
```

The results obtained are

```
Average power, analytical 25.980762
Average power, numerical: 25.980762
rms voltage, analytical: 7.071068
rms voltage, numerical: 7.071076
power factor, analytical: 0.866025
power factor, numerical: 0.866023
```

From the results, it can be seen that the two techniques give almost the same answers.

6.2 Single- and Three-Phase AC Circuits

Voltages and currents of a network can be obtained in the time domain. This normally involves solving differential equations. By transforming the differential equations into algebraic equations using phasors or complex frequency representation, the analysis can be simplified. For a voltage given by

$$v(t) = V_m e^{\sigma t} \cos(wt + \theta)$$

$$v(t) = \mathrm{Re}\left[V_m e^{\sigma t} \cos(wt + \theta)\right] \tag{6.15}$$

the phasor is

$$V = V_m e^{j\theta} = V_m \angle \theta \tag{6.16}$$

and the complex frequency s is

$$s = \sigma + jw \tag{6.17}$$

When the voltage is purely sinusoidal, that is,

$$v_2(t) = V_{m2} \cos(wt + \theta_2) \tag{6.18}$$

then the phasor

$$V_2 = V_{m2} e^{j\theta_2} = V_{m2} \angle \theta_2 \tag{6.19}$$

and complex frequency is purely imaginary, that is,

$$s = jw \tag{6.20}$$

To analyze circuits with sinusoidal excitations, we convert the circuits into the s-domain with $s = jw$. Network analysis laws, theorems, and rules are used to solve for unknown currents and voltages in the frequency domain. The solution is then converted into the time domain using inverse phasor transformation. For example, Figure 6.2 shows an RLC circuit in both the time and frequency domains.

If the values of R_1, R_2, R_3, L_1, L_2, and C_1 are known, the voltage V_3 can be obtained using circuit analysis tools. Suppose V_3 is

$$V_3 = V_{m3} \angle \theta_3$$

Then the time domain voltage $V_3(t)$ is

$$v_3(t) = V_{m3} \cos(wt + \theta_3)$$

The following two examples illustrate the use of MATLAB for solving one-phase circuits.

Example 6.2 AC Voltage Calculations Using Nodal Analysis

In Figure 6.2, if $R_1 = 20\ \Omega$, $R_2 = 100\ \Omega$, $R_3 = 50\ \Omega$, $L_1 = 4$ H, $L_2 = 8$ H, and $C_1 = 250\ \mu F$, find $v_3(t)$ when $w = 10$ rad/sec.

(a)

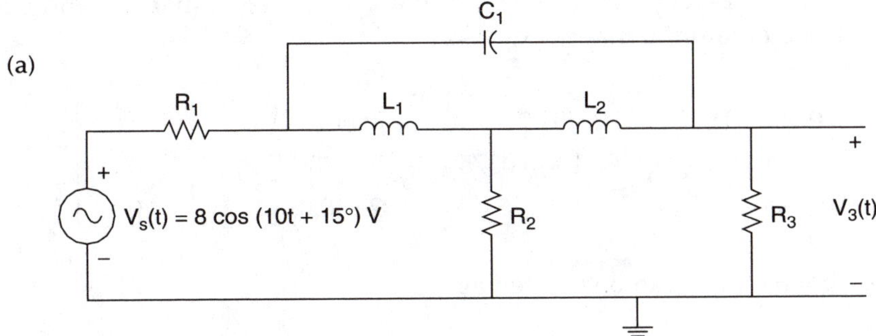

(b)

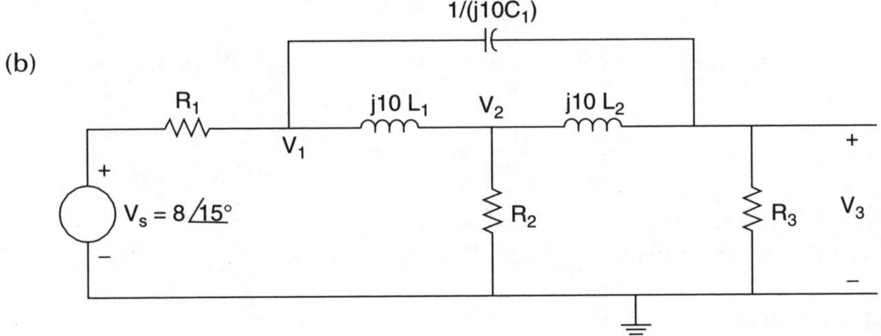

FIGURE 6.2
RLC circuit with sinusoidal excitation: (a) time domain; (b) frequency domain equivalent.

Solution

Using nodal analysis, we obtain the following equations.

At node 1,

$$\frac{V_1 - V_s}{R_1} + \frac{V_1 - V_2}{j10L_1} + \frac{V_1 - V_3}{1/(j10C_1)} = 0 \tag{6.21}$$

At node 2,

$$\frac{V_2 - V_1}{j10L_1} + \frac{V_2}{R_2} + \frac{V_2 - V_3}{j10L_2} = 0 \tag{6.22}$$

At node 3,

$$\frac{V_3}{R_3} + \frac{V_3 - V_2}{j10L_2} + \frac{V_3 - V_1}{1/(j10C_1)} = 0 \tag{6.23}$$

Substituting the element values into the above three equations and simplifying, we obtain the matrix equation

$$\begin{bmatrix} 0.05-j0.0225 & j0.025 & -j0.0025 \\ j0.025 & 0.01-j0.0375 & j0.0125 \\ -j0.0025 & j0.0125 & 0.02-j0.01 \end{bmatrix} \begin{bmatrix} V_1 \\ V_2 \\ V_3 \end{bmatrix} = \begin{bmatrix} 0.4\angle 15^0 \\ 0 \\ 0 \end{bmatrix}$$

The above matrix can be written as

$$[Y][V]=[I]$$

We can compute the vector [v] using the MATLAB command

$$V = inv(Y) * I$$

where $inv(Y)$ is the inverse of the matrix $[Y]$.

A MATLAB program for solving V_3 is as follows.

MATLAB script

```
% This program computes the nodal voltage v3 of circuit
Figure 6.2
% Y is the admittance matrix; % I is the current matrix
% V is the voltage vector

Y = [0.05-0.0225*j    0.025*j           -0.0025*j;
      0.025*j          0.01-0.0375*j     0.0125*j;
     -0.0025*j         0.0125*j          0.02-0.01*j];

c1 = 0.4*exp(pi*15*j/180);
I = [c1
     0
     0];   % current vector entered as column vector

V = inv(Y)*I;   % solve for nodal voltages
v3_abs = abs(V(3));
v3_ang = angle(V(3))*180/pi;
fprintf('voltage V3, magnitude: %f \n voltage V3, angle
in degree: %f', v3_abs, v3_ang)
```

The following results are obtained:

```
voltage V3, magnitude: 1.850409
voltage V3, angle in degree: -72.453299
```

From the MATLAB results, the time domain voltage $v_3(t)$ is

$$v_3(t) = 1.85\cos(10t - 72.45^0) \quad \text{V}$$

Example 6.3 AC Circuit Analysis Using Loop Analysis

For the circuit shown in Figure 6.3, find the current $i_1(t)$ and the voltage $v_C(t)$.

Solution

Figure 6.3 is transformed into the frequency domain. The resulting circuit is shown in Figure 6.4. The impedances are in ohms.

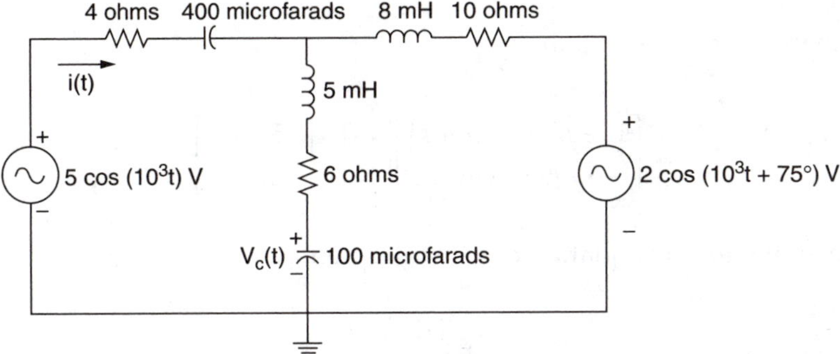

FIGURE 6.3
Circuit with two sources.

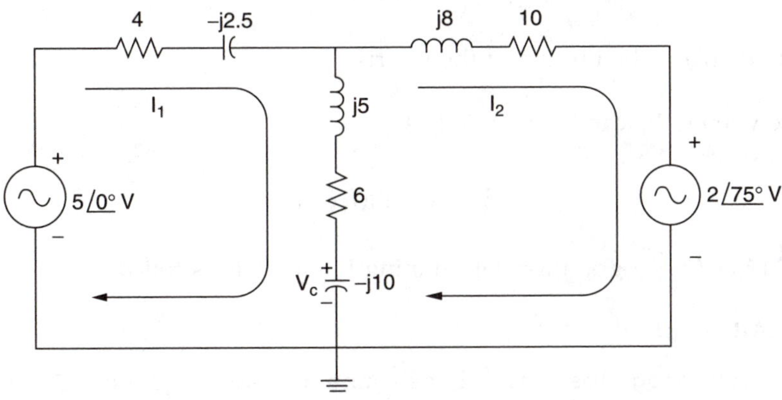

FIGURE 6.4
Frequency domain equivalent of Figure 6.3.

Using loop analysis, we have

$$-5\angle 0^\circ + (4 - j2.5)I_1 + (6 + j5 - j10)(I_1 - I_2) = 0 \tag{6.24}$$

$$(10 + j8)I_2 + 2\angle 75^\circ + (6 + j5 - j10)(I_2 - I_1) = 0 \tag{6.25}$$

Simplifying, we have

$$(10 - j7.5)I_1 - (6 - j5)I_2 = 5\angle 0^\circ$$

$$-(6 - j5)I_1 + (16 + j3)I_2 = -2\angle 75^\circ$$

In matrix form, we obtain

$$\begin{bmatrix} 10 - j7.5 & -6 + j5 \\ -6 + j5 & 16 + j3 \end{bmatrix} \begin{bmatrix} I_1 \\ I_2 \end{bmatrix} = \begin{bmatrix} 5\angle 0^\circ \\ -2\angle 75^\circ \end{bmatrix}$$

The above matrix equation can be rewritten as

$$[Z][I] = [V]$$

We obtain the current vector $[I]$ using the MATLAB command

$$I = inv(Z) * V$$

where $inv(Z)$ is the inverse of the matrix [Z].

The voltage V_C can be obtained as

$$V_C = (-j10)(I_1 - I_2)$$

A MATLAB program for determining I_1 and V_a is as follows.

MATLAB script

```
% This programs calculates the phasor current I1 and
% phasor voltage Va.
% Z is impedance matrix
% V is voltage vector
% I is current vector
```

```
Z = [10-7.5*j    -6+5*j;
      -6+5*j    16+3*j];

b = -2*exp(j*pi*75/180);
V = [5
        b];  % voltage vector in column form

I = inv(Z)*V; % solve for loop currents
i1 = I(1);
i2 = I(2);

Vc = -10*j*(i1 - i2);
i1_abs = abs(I(1));
i1_ang = angle(I(1))*180/pi;
Vc_abs = abs(Vc);
Vc_ang = angle(Vc)*180/pi;
%results are printed
fprintf('phasor current i1, magnitude: %f \n phasor
current i1, angle in degree: %f \n', i1_abs,i1_ang)
fprintf('phasor voltage Vc, magnitude: %f \n phasor
voltage Vc, angle in degree: %f \n',Vc_abs,Vc_ang)
```

The following results were obtained:

```
phasor current i1, magnitude: 0.387710
phasor current i1, angle in degree: 15.019255
phasor voltage Vc, magnitude: 4.218263
phasor voltage Vc, angle in degree: -40.861691
```

The current $i_1(t)$ is

$$i_1(t) = 0.388\cos(10^3 t + 15.02^0) \text{ A}$$

and the voltage $v_C(t)$ is

$$v_C(t) = 4.21\cos(10^3 t - 40.86^0) \text{ V}$$

Power utility companies use three-phase circuits for the generation, transmission, and distribution of large blocks of electrical power. The basic structure of a three-phase system consists of a three-phase voltage source connected to a three-phase load through transformers and transmission lines. The three-phase voltage source can be wye or delta connected. Also the three-phase load can be delta or wye connected. Figure 6.5 shows a three-phase system with wye-connected source and wye-connected load.

For a balanced abc system, the voltages V_{an}, V_{bn}, and V_{cn} have the same magnitude, and they are out of phase by 120°. Specifically, for a balanced abc system, we have

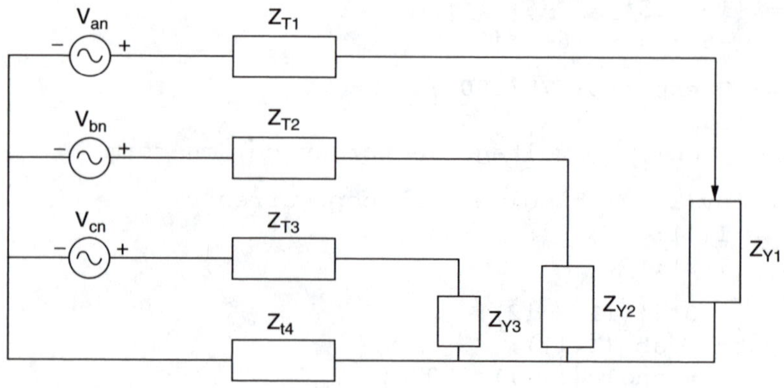

FIGURE 6.5
Three-phase system, wye-connected source, and wye-connected load.

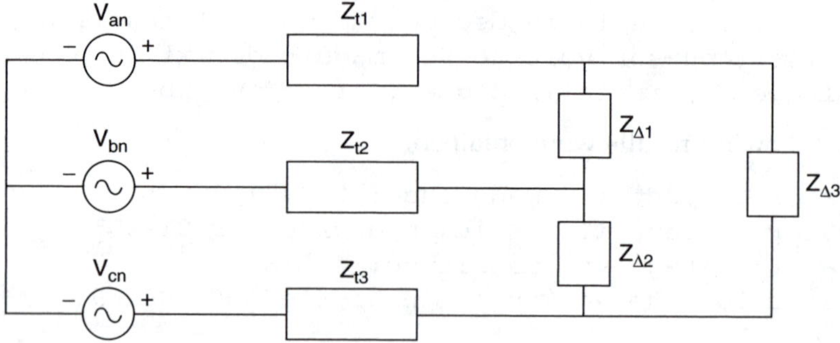

FIGURE 6.6
Three-phase system, wye-connected source, and delta-connected load.

$$\begin{aligned} V_{an} &= V_P \angle 0^0 \\ V_{bn} &= V_P \angle -120^0 \\ V_{cn} &= V_P \angle 120^0 \end{aligned} \tag{6.26}$$

For a cba system

$$\begin{aligned} V_{an} &= V_P \angle 0^0 \\ V_{bn} &= V_P \angle 120^0 \\ V_{cn} &= V_P \angle -120^0 \end{aligned} \tag{6.27}$$

The wye-connected load is balanced if

$$Z_{Y1} = Z_{Y2} = Z_{Y3} \tag{6.28}$$

Similarly, the delta-connected load is balanced if

$$Z_{\Delta 1} = Z_{\Delta 2} = Z_{\Delta 3} \tag{6.29}$$

We have a balanced three-phase system of Equation (6.26) to Equation (6.29), which is satisfied with the additional condition

$$Z_{T1} = Z_{T2} = Z_{T3} \tag{6.30}$$

Analysis of balanced three-phase systems can easily be performed by converting the three-phase system into an equivalent one-phase system and performing simple hand calculations. The method of symmetrical components can be used to analyze unbalanced three-phase systems. Another way to analyze three-phase systems is to use KVL and KCL. The unknown voltage or currents are solved using MATLAB. This is illustrated by the following example.

Example 6.4 Unbalanced Wye-Wye Three-Phase System

In Figure 6.7, showing an unbalanced wye-wye system, find the phase voltages V_{AN}, V_{BN}, and V_{CN}.

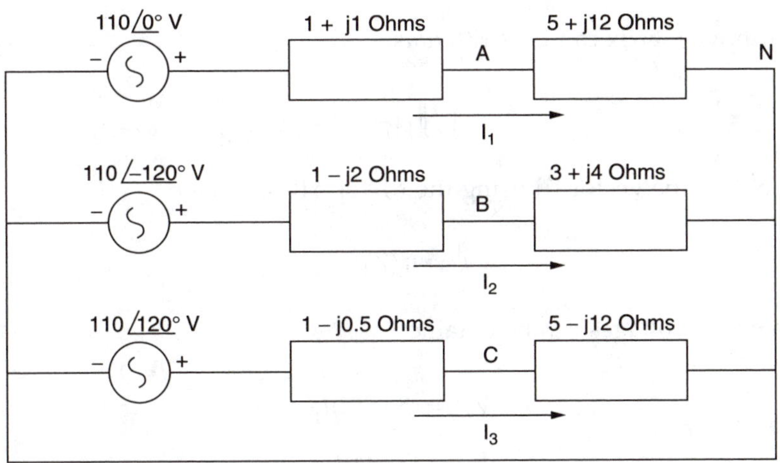

FIGURE 6.7
Unbalanced three-phase system.

Solution

Using KVL, we can solve for I_1, I_2, and I_3. From the figure, we have

$$110\angle 0° = (1 + j1)I_1 + (5 + j12)I_1 \tag{6.31}$$

$$110\angle -120° = (1 - j2)I_2 + (3 + j4)I_2 \tag{6.32}$$

$$110\angle 120° = (1 - j0.5)I_3 + (5 - j12)I_3 \tag{6.33}$$

Simplifying Equation (6.31), Equation (6.32), and Equation (6.33), we have

$$110\angle 0° = (6 + j13)I_1 \tag{6.34}$$

$$110\angle -120° = (4 + j2)I_2 \tag{6.35}$$

$$110\angle 120° = (6 - j12.5)I_3 \tag{6.36}$$

and expressing the above three equations in matrix form, we have

$$\begin{bmatrix} 6 + j13 & 0 & 0 \\ 0 & 4 + j2 & 0 \\ 0 & 0 & 6 - j12.5 \end{bmatrix} \begin{bmatrix} I_1 \\ I_2 \\ I_3 \end{bmatrix} = \begin{bmatrix} 110\angle 0° \\ 110\angle -120° \\ 110\angle 120° \end{bmatrix}$$

The above matrix can be written as

$$[Z][I] = [V]$$

We obtain the vector [*I*] using the MATLAB command

$$I = inv(Z) * V$$

The phase voltages can be obtained as

$$V_{AN} = (5 + j12)I_1$$

$$V_{BN} = (3 + j4)I_2$$

$$V_{CN} = (5 - j12)(I_3)$$

The MATLAB program for obtaining the phase voltages is:

MATLAB script

```
% This program calculates the phase voltage of an
% unbalanced three-phase system
% Z is impedance matrix
% V is voltage vector and
% I is current vector
Z = [6-13*j    0        0;
     0         4+2*j    0;
     0         0        6-12.5*j];
c2 = 110*exp(j*pi*(-120/180));
c3 = 110*exp(j*pi*(120/180));

V = [110; c2; c3]; % column voltage vector
I = inv(Z)*V;  % solve for loop currents
% calculate the phase voltages
%
Van = (5+12*j)*I(1);
Vbn = (3+4*j)*I(2);
Vcn = (5-12*j)*I(3);
Van_abs = abs(Van);
Van_ang = angle(Van)*180/pi;
Vbn_abs = abs(Vbn);
Vbn_ang = angle(Vbn)*180/pi;
Vcn_abs = abs(Vcn);
Vcn_ang = angle(Vcn)*180/pi;

% print out results
fprintf('phase voltage Van,magnitude: %f \n phase
voltage Van, angle in degree: %f \n', Van_abs, Van_ang)
fprintf('phase voltage Vbn,magnitude: %f \n phase
voltage Vbn, angle in degree: %f \n', Vbn_abs, Vbn_ang)
fprintf('phase voltage Vcn,magnitude: %f \n phase
voltage Vcn, angle in degree: %f \n', Vcn_abs, Vcn_ang)
```

The following results are obtained:

```
phase voltage Van,magnitude: 99.875532
phase voltage Van, angle in degree: 132.604994
phase voltage Vbn,magnitude: 122.983739
phase voltage Vbn, angle in degree: -93.434949
phase voltage Vcn,magnitude: 103.134238
phase voltage Vcn, angle in degree: 116.978859
```

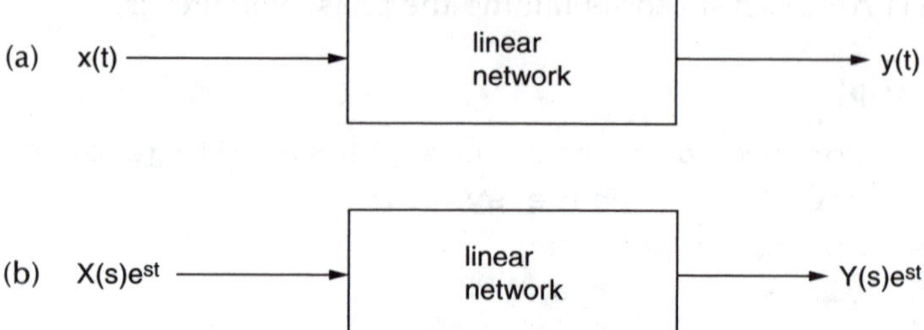

FIGURE 6.8
Linear network representation: (a) time domain; (b) s-domain.

6.3 Network Characteristics

Figure 6.8 shows a linear network with input $x(t)$ and output $y(t)$. Its complex frequency representation is also shown.

In general, the input $x(t)$ and output $y(t)$ are related by the differential equation

$$\begin{aligned} a_n \frac{d^n y(t)}{dt^n} + a_{n-1} \frac{d^{n-1} y(t)}{dt^{n-1}} + \mathrm{K} + a_1 \frac{dy(t)}{dt} + a_0 y(t) = \\ b_m \frac{d^m x(t)}{dt^m} + b_{m-1} \frac{d^{m-1} x(t)}{dt^{m-1}} + \Lambda b_1 \frac{dx(t)}{dt} + b_0 x(t) \end{aligned} \tag{6.37}$$

where a_n, a_{n-1}, …, a_0, b_m, b_{m-1}, …, b_0 are real constants.

If $x(t) = X(s)e^{st}$, then the output must have the form $y(t) = Y(s)e^{st}$, where $X(s)$ and $Y(s)$ are phasor representations of $x(t)$ and $y(t)$. From Equation (6.37), we have

$$\begin{aligned} (a_n s^n + a_{n-1} s^{n-1} + \Lambda + a_1 s + a_0) Y(s) e^{st} = \\ (b_m s^m + b_{m-1} s^{m-1} + \Lambda + b_1 s + b_0) X(s) e^{st} \end{aligned} \tag{6.38}$$

and the network function

$$H(s) = \frac{Y(s)}{X(s)} = \frac{b_m s^m + b_{m-1} s^{m-1} + \Lambda\, b_1 s + b_0}{a_n s^n + a_{n-1} s^{n-1} + \Lambda\, a_1 s + a_0} \tag{6.39}$$

The network function can be rewritten in factored form

$$H(s) = \frac{k(s - z_1)(s - z_2)\,\Lambda\,(s - z_m)}{(s - p_1)(s - p_2)\,\Lambda\,(s - p_n)} \tag{6.40}$$

where k is a constant; $z_1, z_2, \ldots, z_m$ are zeros of the network function; and p_1, $p_2, \ldots, p_n$ are poles of the network function.

The network function can also be expanded using partial fractions as

$$H(s) = \frac{r_1}{s - p_1} + \frac{r_2}{s - p_2} + \quad .... \quad + \frac{r_n}{s - p_n} + k(s) \tag{6.41}$$

6.3.1 MATLAB Functions roots, residue, and polyval

MATLAB has the function **roots**, which can be used to obtain the poles and zeros of a network function. The MATLAB function **residue** can be used for partial fraction expansion. Furthermore, the MATLAB function **polyval** can be used to evaluate the network function.

The MATLAB function **roots** determines the roots of a polynomial. The general form of the roots function is

$$r = roots(p) \tag{6.42}$$

where p is a vector containing the coefficients of the polynomial in descending order and r is a column vector containing the roots of the polynomials.

For example, given the polynomial

$$f(x) = x^3 + 9x^2 + 23x + 15$$

the commands to compute and print out the roots of $f(x)$ are

```
p  = [1   9   23   15]
r  = roots (p)
```

and the values printed are

```
r  =
          -1.0000
          -3.0000
          -5.0000
```

Given the roots of a polynomial, we can obtain the coefficients of the polynomial by using the MATLAB function **poly**.

Thus

$$S = \text{poly}\ ([\ -1 \quad -3 \quad -5\ ]^1) \tag{6.43}$$

will give a row vector s given as

```
S =
     1.0000    9.0000           23.0000          15.0000
```

The coefficients of S are the same as those of p.

The MATLAB function **polyval** is used for polynomial evaluation. The general form of polyval is

$$polyval(p, x) \tag{6.44}$$

where p is a vector whose elements are the coefficients of a polynomial in descending powers and $polyval(p, x)$ is the value of the polynomial evaluated at x.

For example, to evaluate the polynomial

$$f(x) = x^3 - 3x^2 - 4x + 15$$

at $x = 2$, we use the command

```
p  = [1   -3      -4     15];
polyval(p, 2)
```

Then we get

```
ans =
          3
```

The MATLAB function **residue** can be used to perform partial fraction expansion. Assuming $H(s)$ is the network function, since $H(s)$ may represent an improper fraction, we may express $H(s)$ as a mixed fraction:

$$H(s) = \frac{B(s)}{A(s)} \tag{6.45}$$

$$H(s) = \sum_{n=0}^{N} k_n s^n + \frac{N(s)}{D(s)} \tag{6.46}$$

where

$$\frac{N(s)}{D(s)}$$

is a proper fraction.

From Equation (6.41) and Equation (6.46), we get

$$H(s) = \frac{r_1}{s-p_1} + \frac{r_2}{s-p_2} + \quad .... \quad + \frac{r_n}{s-p_n} + \sum_{n=0}^{N} k_n s^n \tag{6.47}$$

Given the coefficients of the numerator and denominator polynomials, the MATLAB **residue** function provides the values of $r_1, r_2, \ldots, r_n, p_1, p_2, \ldots, p_n$, and $k_1, k_2, \ldots, k_n$. The general form of the residue function is

$$[r, p, k] = residue(num, den) \tag{6.48}$$

where *num* is a row vector whose entries are the coefficients of the numerator polynomial in descending order, *den* is a row vector whose entries are the coefficient of the denominator polynomial in descending order, *r* is returned as a column vector, *p* (pole locations) is returned as a column vector, and *k* (direct term) is returned as a row vector.

The command

$$[num, den] = residue(r, p, k) \tag{6.49}$$

converts the partial fraction expansion back to the polynomial ratio

$$H(s) = \frac{B(s)}{A(s)}$$

The following example illustrates the use of the residue function to perform partial fraction expansion.

Example 6.5 Poles and Zeros of a Network Function

Given a network function,

$$H(s) = \frac{4s^4 + 3s^3 + 6s^2 + 10s + 20}{s^4 + 2s^3 + 5s^2 + 2s + 8} \tag{6.50}$$

determine its partial fraction expansion.

Solution

The following commands will perform partial fraction expansion:

```
%
num = [4 3 6 10 20];
den = [1 2 5 2 8];
% poles and zeros are determined
[r, p, k] = residue(num, den)
```

and we shall get the following results:

```
r =
        -1.6970 + 3.0171i
        -1.6970 - 3.0171i
        -0.8030 - 0.9906i
        -0.8030 + 0.9906i

p =
        -1.2629 + 1.7284i
        -1.2629 - 1.7284i
         0.2629 + 1.2949i
         0.2629 - 1.2949i

k =
        4
```

The results shown above and Equation (6.47) can be used to obtain the partial fraction expansion of Equation (6.50)

Example 6.6 Output Voltage of a Network

For the circuit shown below:

(a) Find the network function

$$H(s) = \frac{V_o(s)}{V_S(s)}$$

(b) Find the poles and zeros of $H(s)$.

(c) If $v_S(t) = 10e^{-3t}\cos(2t + 40^0)$, find $v_0(t)$.

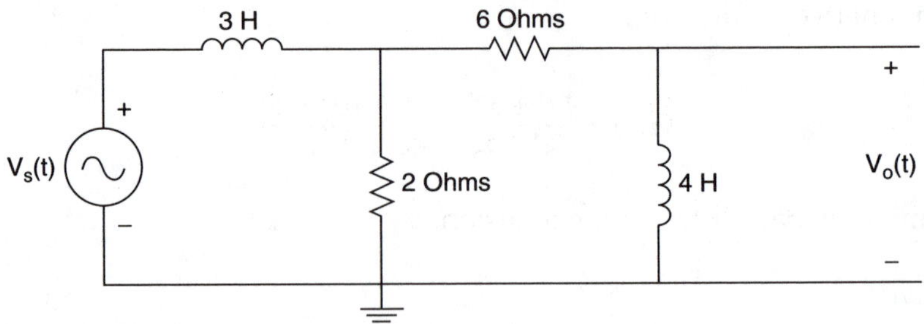

FIGURE 6.9
Circuit for Example 6.6.

Solution

In the s-domain, the above figure becomes

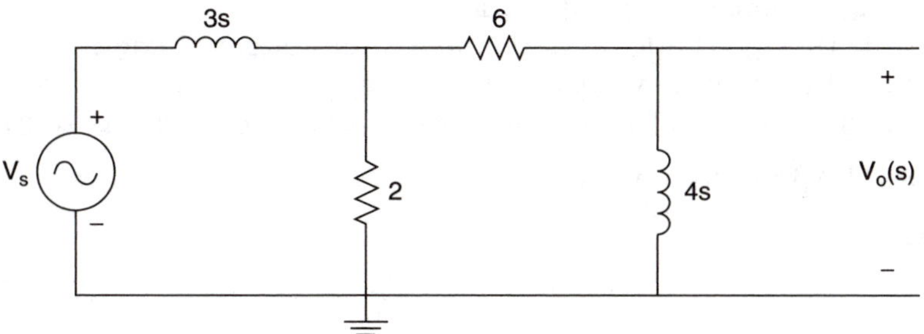

FIGURE 6.10
S-domain equivalent circuit of Figure 6.9.

$$\frac{V_0(s)}{V_S(s)} = \frac{V_0(s)}{V_X(s)}\frac{V_X(s)}{V_S(s)} = \frac{4s}{(6+4s)}\frac{\left[2\|(6+4s)\right]}{\left[\left(2\|(6+4s)\right)+3s\right]} \tag{6.51}$$

Simplifying, we get

$$\frac{V_0(s)}{V_S(s)} = \frac{4s^2+6s}{6s^3+25s^2+30s+9} \tag{6.52}$$

The phasor voltage $V_S = 10\angle 40^o$; $s = -3+j2$

$$V_0(s) = (10\angle 40^o)H(s)\Big|_{s=-3+j2}$$

(b, c) MATLAB is used to find the poles, zeros, and $v_0(t)$.

MATLAB script

```
% Poles and zeros determination
num = [4  6  0];
den = [6  25  30  9];
disp('the zeros are')
z = roots(num)
disp('the poles are')
p = roots(den)
% program to evaluate transfer function and
% find the output voltage
s1 = -3+2*j;
n1 = polyval(num,s1);
d1 = polyval(den,s1);
vo = 10.0*exp(j*pi*(40/180))*n1/d1;
vo_abs = abs(vo);
```

```
vo_ang = angle(vo)*180/pi;
% print magnitude and phase of output voltage
fprintf('phasor voltage vo, magnitude: %f \n phasor
voltage vo, angle in degrees: %f', vo_abs, vo_ang)
```

The MATLAB results are

```
Zeros

z =

       0
   -1.5000

Poles

p =

   -2.2153
   -1.5000
   -0.4514

phasor voltage vo, magnitude: 3.453492
phasor voltage vo, angle in degrees: -66.990823
```

From the results, the output voltage is given as

$$v(t) = 3.45e^{-3t}\cos(2t - 66.99°)$$

Example 6.7 Inverse Laplace Transform

Find the inverse Laplace transform of

$$G(s) = \frac{10s^2 + 20s + 40}{s^3 + 12s^2 + 47s + 60}$$

Solution

MATLAB script

```
% MATLAB is used to do the partial fraction expansion
%
num = [10 20 40];
den = [1 12 47 60];
% we get the following results
[r, p, k] = residue(num,den)
```

The MATLAB results are

```
r =

    95.0000
  -120.0000
    35.0000
```

```
p =
        -5.0000
        -4.0000
        -3.0000

k =
        []
```

From the results, we get

$$G(s) = \frac{95}{s+5} - \frac{120}{s+4} + \frac{35}{s+3}$$

and the inverse Laplace transform is

$$g(t) = 35e^{-3t} - 120e^{-4t} + 95e^{-5t} \tag{6.53}$$

6.4 Frequency Response

The general form of a transfer function of an analog circuit is given in Equation (6.39). It is repeated here.

$$H(s) = \frac{Y(s)}{X(s)} = \frac{b_m s^m + b_{m-1}s^{m-1} + \Lambda\, b_1 s + b_0}{a_n s^n + a_{n-1}s^{n-1} + \Lambda\, a_1 s + a_0}$$

More specifically, for a second-order analog filter, the following transfer functions can be obtained:

1. Lowpass

$$H_{LP}(s) = \frac{k_1}{s^2 + Bs + w_0^2} \tag{6.54}$$

2. Highpass

$$H_{HP}(s) = \frac{k_2 s^2}{s^2 + Bs + w_0^2} \tag{6.55}$$

3. Bandpass

$$H_{BP}(s) = \frac{k_3 s}{s^2 + Bs + w_0^2} \tag{6.56}$$

4. Bandreject

$$H_{BR}(s) = \frac{k_4 s^2 + k_5}{s^2 + Bs + w_0^2} \tag{6.57}$$

where k_1, k_2, k_3, k_4, B, and w_0 are constants.

Figure 6.11 shows the circuit diagram of some filter sections.

Frequency response is the response of a network to sinusoidal input signal. If we substitute $s = jw$ in the general network function, $H(s)$ we get

$$H(s)\Big|_{s=jw} = M(w)\angle\theta(w) \tag{6.58}$$

where

$$M(w) = |H(jw)| \tag{6.59}$$

and

$$\theta(w) = \angle H(jw) \tag{6.60}$$

The plot of $M(\omega)$ vs. ω is the magnitude characteristics or response. Also, the plot of $\theta(w)$ vs. ω is the phase response. The magnitude and phase characteristics can be obtained using MATLAB function **freqs**.

6.4.1 MATLAB Function freqs

MATLAB function **freqs** is used to obtain the frequency response of transfer function $H(s)$. The general form of the frequency function is

$$hs = freqs(num, den, range) \tag{6.61}$$

where

$$H(s) = \frac{Y(s)}{X(s)} = \frac{b_m s^m + b_{m-1} s^{m-1} + \Lambda\, b_1 s + b_0}{a_n s^n + a_{n-1} s^{n-1} + \Lambda\, a_1 s + a_0} \tag{6.62}$$

$$num = [b_m \; b_{m-1} \; \ldots \; b_1 \; b_0] \tag{6.63}$$

$$den = [a_n \; a_{n-1} \; \ldots \; a_1 \; a_0] \tag{6.64}$$

range is range of frequencies for the case

hs is the frequency response (in complex number form)

(a)

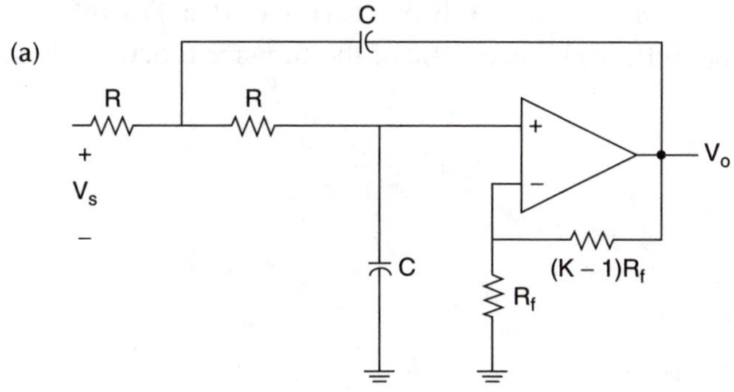

(b)

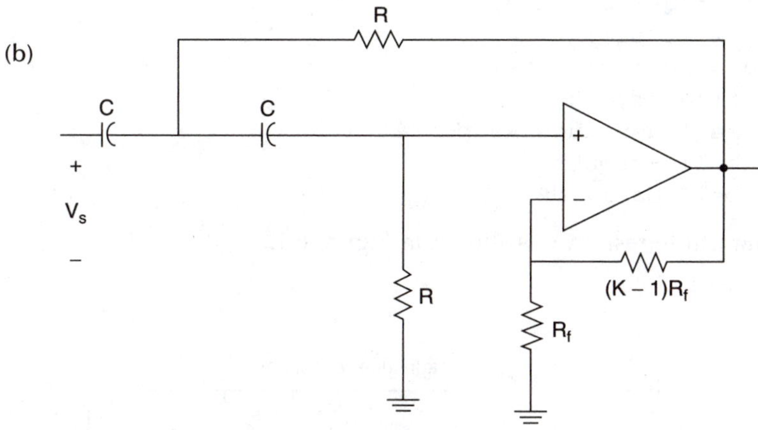

(c)

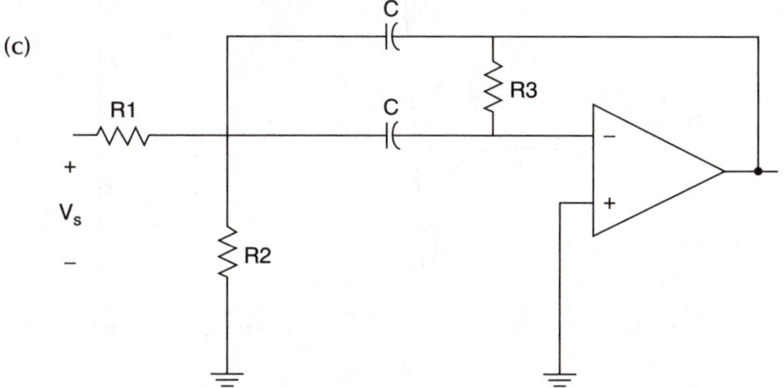

FIGURE 6.11
Active filters: (a) lowpass, (b) highpass, and (c) bandpass.

Example 6.8 Magnitude Characteristic of a Transfer Function

Obtain the magnitude characteristic of the transfer function given as

$$H(s) = \frac{2s^2 + 4}{s^2 + 4s + 16} \tag{6.65}$$

Solution

MATLAB script

```
num = [2 0 4];
den = [1 4 16];
w = logspace(-2, 4);
 h = freqs(num, den, w);
 f = w/(2*pi);
 mag = 20*log10(abs(h));
 semilogx(f, mag)
 title('Magnitude Response')
 xlabel('Frequency, Hz')
 ylabel('Gain, dB')
```

The magnitude response is shown in Figure 6.12.

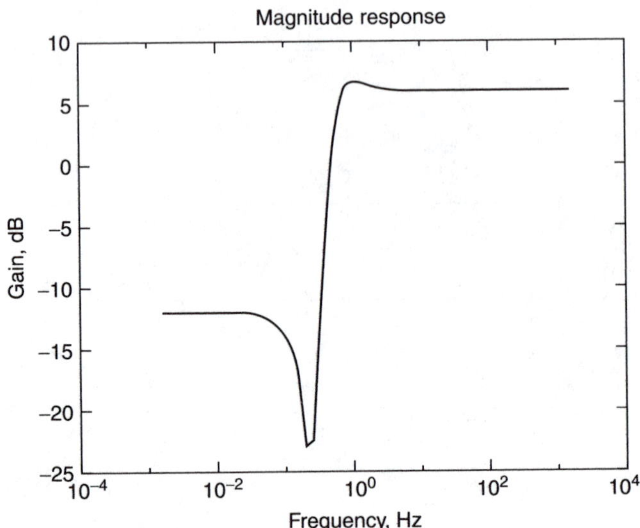

FIGURE 6.12
Magnitude response of Equation (6.65).

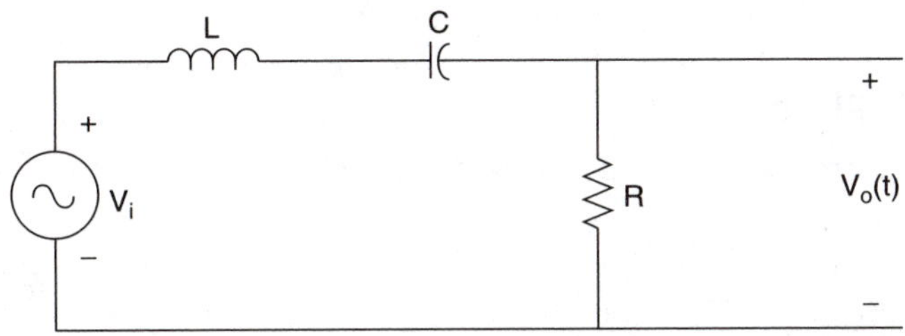

FIGURE 6.13
RLC circuit.

The following example shows how to obtain and plot the frequency response of an RLC circuit.

Example 6.9 Magnitude and Phase Response of an RLC Circuit

For the RLC circuit shown in Figure 6.13:

(a) Show that the transfer function is

$$H(s) = \frac{V_o(s)}{V_i(s)} = \frac{s\frac{R}{L}}{s^2 + s\frac{R}{L} + \frac{1}{LC}} \tag{6.66}$$

(b) If $L = 5$ H, $C = 1.12$ μF, and $R = 10{,}000\ \Omega$, plot the frequency response.

(c) What happens when $R = 100\ \Omega$, but L and C remain unchanged?

Solution

(a) In the frequency domain,

$$H(s) = \frac{V_0(s)}{V_i(s)} = \frac{R}{R + sL + \frac{1}{sC}} = \frac{sCR}{s^2LC + sCR + 1} \tag{6.67}$$

which is

$$H(s) = \frac{V_0(s)}{V_i(s)} = \frac{s\frac{R}{L}}{s^2 + s\frac{R}{L} + \frac{1}{LC}}$$

Parts (b) and (c) are solved using MATLAB.

MATLAB script

```
% Frequency response of RLC circuit
%
l = 5;
c = 1.25e-6;
r1 = 10000;
r2 = 100;

num1 = [r1/l 0];
den1 = [1 r1/l  1/(l*c)];

w = logspace(1,4);
h1 = freqs(num1,den1,w);
f = w/(2*pi);
mag1 = abs(h1);
phase1 = angle(h1)*180/pi;
num2 = [r2/l 0];
den2 = [1  r2/l  1/(l*c)];
h2 = freqs(num2,den2,w);
mag2 = abs(h2);
phase2 = angle(h2)*180/pi;
% Plot the response
subplot(221), loglog(f, mag1,'.')
title('Magnitude Response R=10K')
ylabel('Magnitude')
subplot(222), loglog(f,mag2,'.')
title('Magnitude Response R=0.1K')
ylabel('Magnitude')
subplot(223), semilogx(f, phase1,'.')
title('Phase Response R=10K'),...
xlabel('Frequency, Hz'), ylabel('Angle in degrees')
subplot(224), semilogx(f, phase2,'.')
title('Phase Response R=0.1K'),...
xlabel('Frequency, Hz'), ylabel('Angle in degrees')
```

The plots are shown in Figure 6.14. As the resistance is decreased from 10,000 to 100 Ω, the bandwidth of the frequency response decreases and the quality factor of the circuit increases.

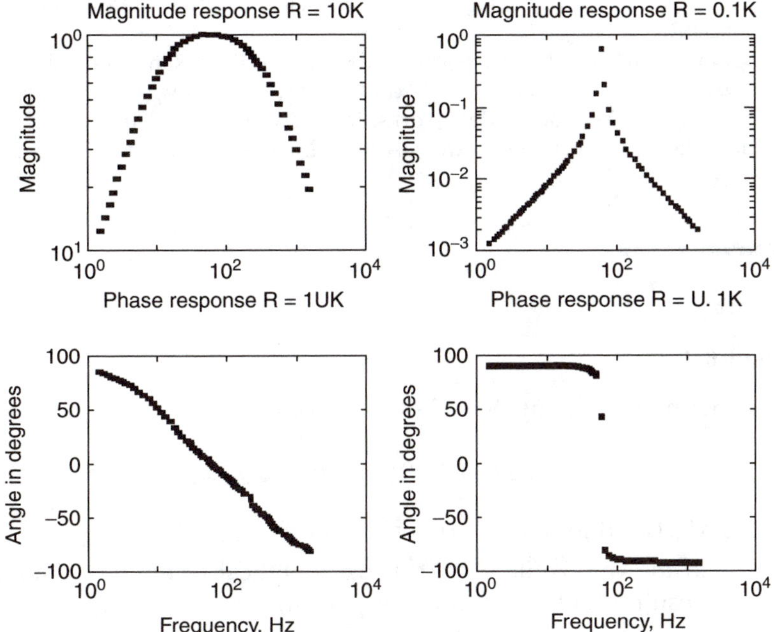

FIGURE 6.14
Frequency response of an RLC circuit.

Bibliography

1. Alexander, C.K. and Sadiku, M.N.O., *Fundamentals of Electric Circuits*, 2nd ed., McGraw-Hill, New York, 2004.
2. Attia, J.O., *PSPICE and MATLAB for Electronics: An Integrated Approach*, CRC Press, Boca Raton, FL, 2002.
3. Biran, A. and Breiner, M., *MATLAB for Engineers*, Addison-Wesley, Reading, MA, 1995.
4. Chapman, S.J., *MATLAB Programming for Engineers*, Brook, Cole Thompson Learning, Pacific Grove, CA, 2000.
5. Dorf, R.C. and Svoboda, J.A., *Introduction to Electric Circuits*, 3rd ed., John Wiley & Sons, New York, 1996.
6. Etter, D.M., *Engineering Problem Solving with MATLAB*, 2nd ed., Prentice Hall, Upper Saddle River, NJ, 1997.
7. Etter, D.M., Kuncicky, D.C., and Hull, D., *Introduction to MATLAB 6*, Prentice Hall, Upper Saddle River, NJ, 2002.
8. Gottling, J.G., *Matrix Analysis of Circuits Using MATLAB*, Prentice Hall, Englewood Cliffs, NJ, 1995.
9. Johnson, D.E., Johnson, J.R., and Hilburn, J.L., *Electric Circuit Analysis*, 3rd ed., Prentice Hall, Upper Saddle River, NJ, 1997.

10. Meader, D.A., *Laplace Circuit Analysis and Active Filters*, Prentice Hall, Upper Saddle River, NJ, 1991.
11. Sigmor, K., *MATLAB Primer*, 4th ed., CRC Press, Boca Raton, FL, 1998.
12. *Using MATLAB, The Language of Technical Computing, Computation, Visualization, Programming*, Version 6, MathWorks, Inc., Natick, MA, 2000
13. Vlach, J.O., Network theory and CAD, *IEEE Trans. on Education*, 36, 23–27, 1993.

Problems

Problem 6.1

If $v(t)$ is periodic with one period of $v(t)$ given as

$$v(t) = 16(1 - e^{-6t}) \quad \text{V}; \quad 0 \le t < 2 \ \text{sec}$$

(a) Use MATLAB to find the rms value of $v(t)$.

(b) Obtain the rms value of $v(t)$ using analytical technique. Compare your result with that obtained in part (a).

(c) Find the power dissipated in the 4-Ω resistor when the voltage $v(t)$ is applied across the 4-Ω resistor. Refer to Figure P6.1.

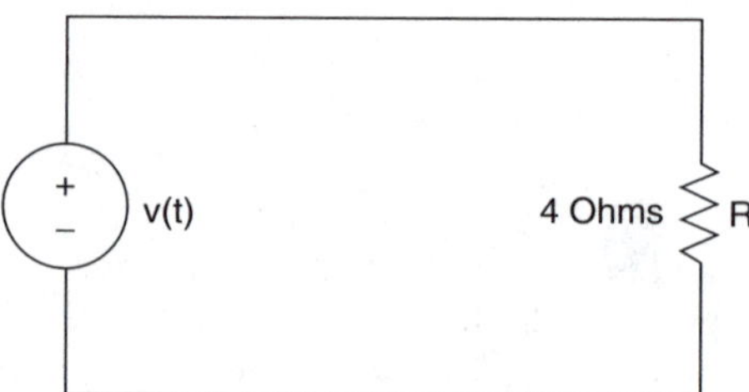

FIGURE P6.1
Resistive circuit for part (c) of Exercise 6.1.

Problem 6.2

A balanced Y-Y positive sequence system has phase voltage of the source $V_{an} = 120\angle 0^0$ rms if the load impedance per phase is $(11 + j4.5)$ Ω, and the transmission line has an impedance per phase of $(1 + j0.5)$ Ω.

(a) Use analytical techniques to find the magnitude of the line current and the power delivered to the load.

(b) Use MATLAB to solve for the line current and the power delivered to the load.

(c) Compare the results of parts (a) and (b).

Problem 6.3

For the unbalanced three-phase system shown in Figure P6.3, find the currents I_1, I_2, I_3 and hence I_{bB}. Assume that $Z_A = 10 + j5\ \Omega$, $Z_B = 15 + j7\ \Omega$, and $Z_C = 12 - j3\ \Omega$.

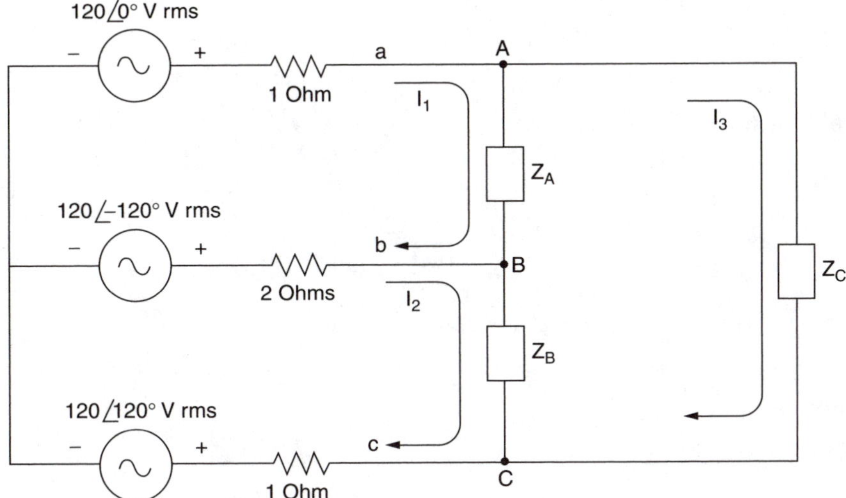

FIGURE P6.3
Unbalanced three-phase system.

Problem 6.4

For the system with network function

$$H(s) = \frac{s^3 + 4s^2 + 16s + 4}{s^4 + 20s^3 + 12s^2 + s + 10}$$

find the poles and zeros of $H(s)$.

Problem 6.5

For the system with network function

$$H(s) = \frac{s^2 + 14s + 20}{s^3 + 12s^2 + 41s + 30}$$

find the poles and zeros of $H(s)$.

Problem 6.6

Use MATLAB to determine the roots of the following polynomials. Plot the polynomial over the appropriate interval to verify the roots location:

(a) $f_1(x) = x^2 + 4x + 3$
(b) $f_2(x) = x^3 + 5x^2 + 9x + 5$
(c) $f_3(x) = 2x^5 - 4x^4 - 12x^3 + 27x^2 + 8x - 16$

Problem 6.7

If

$$\frac{V_o(s)}{V_i(s)} = \frac{20s}{15s^2 + 23s + 16}$$

find $v_0(t)$ given that $v_i(t) = 2.3e^{-2t}\cos(5t + 30^0)$.

Problem 6.8

If

$$\frac{V_o(s)}{V_i(s)} = \frac{s+4}{s^3 + 11s^2 + 36s + 36}$$

find $v_0(t)$ given that $v_i(t) = 5\cos(8t + 30^0)$.

Problem 6.9

For the circuit of Figure P6.9:

(a) Find the transfer function

$$\frac{V_o(s)}{V_i(s)}$$

(b) If $v_i(t) = 10e^{-5t}\cos(t + 10^0)$, find $v_0(t)$.

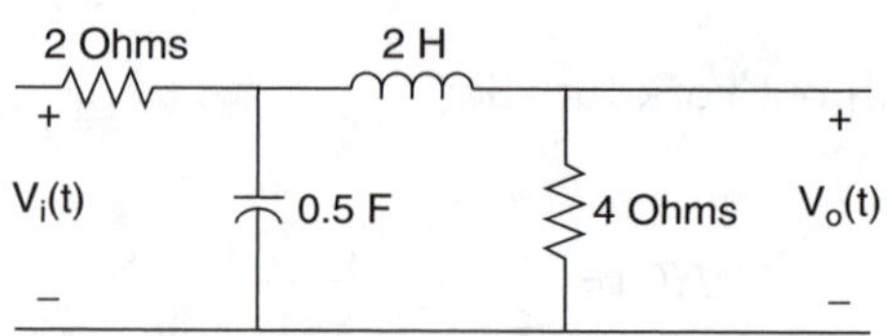

FIGURE P6.9
RLC circuit of Exercise 6.9.

Problem 6.10

For Figure P6.10:

(a) Find the transfer function

$$H(s) = \frac{V_o(s)}{V_i(s)}$$

(b) Use MATLAB to plot the magnitude characteristics.

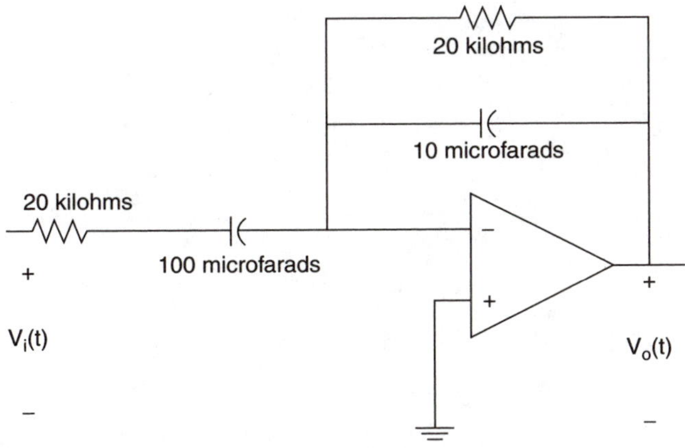

FIGURE P6.10
Simple active filter.

Problem 6.11

In the bandpass filter shown in Figure 6.11(c), R1 = 1 kΩ, R2 = 250 Ω, R3 = 10 kΩ, and C = 0.1 μF.

(a) Find the transfer function

$$H(s) = \frac{V_o(s)}{V_s(s)}$$

(b) Use MATLAB to plot the magnitude characteristics.
(c) Determine the center frequency.

7

Two-Port Networks

This chapter discusses the application of MATLAB for analysis of two-port networks. The describing equations for the various two-port network representations are given. The use of MATLAB for solving problems involving parallel, series, and cascaded two-port networks is shown. Example problems involving both passive and active circuits will be solved using MATLAB.

7.1 Two-Port Network Representations

A general two-port network is shown in Figure 7.1. I_1 and V_1 are input current and voltage, respectively. Also, I_2 and V_2 are output current and voltage, respectively. It is assumed that the linear two-port circuit contains no independent sources of energy and that the circuit is initially at rest (no stored energy). Furthermore, any controlled sources within the linear two-port circuit cannot depend on variables that are outside the circuit.

7.1.1 z-Parameters

A two-port network can be described by z-parameters as

$$V_1 = z_{11} I_1 + z_{12} I_2 \tag{7.1}$$

$$V_2 = z_{21} I_1 + z_{22} I_2 \tag{7.2}$$

In matrix form, the above equation can be rewritten as

$$\begin{bmatrix} V_1 \\ V_2 \end{bmatrix} = \begin{bmatrix} z_{11} & z_{12} \\ z_{21} & z_{22} \end{bmatrix} \begin{bmatrix} I_1 \\ I_2 \end{bmatrix} \tag{7.3}$$

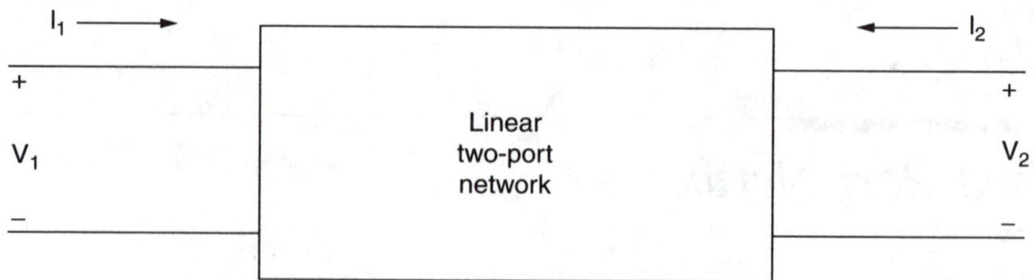

FIGURE 7.1
General two-port network.

The z-parameter can be found as follows:

$$z_{11} = \frac{V_1}{I_1}\bigg|_{I_2=0} \tag{7.4}$$

$$z_{12} = \frac{V_1}{I_2}\bigg|_{I_1=0} \tag{7.5}$$

$$z_{21} = \frac{V_2}{I_1}\bigg|_{I_2=0} \tag{7.6}$$

$$z_{22} = \frac{V_2}{I_2}\bigg|_{I_1=0} \tag{7.7}$$

The z-parameters are also called open-circuit impedance parameters, since they are obtained as a ratio of voltage and current by open-circuiting port 2 ($I_2 = 0$) or port 1 ($I_1 = 0$). The following example shows a technique for finding the z-parameters of a simple circuit.

Example 7.1 z-Parameters of a T-Network

For the T-network shown in Figure 7.2, find the z-parameters.

Solution

Using KVL,

$$V_1 = Z_1 I_1 + Z_3(I_1 + I_2) = (Z_1 + Z_3)I_1 + Z_3 I_2 \tag{7.8}$$

$$V_2 = Z_2 I_2 + Z_3(I_1 + I_2) = (Z_3)I_1 + (Z_2 + Z_3)I_2 \tag{7.9}$$

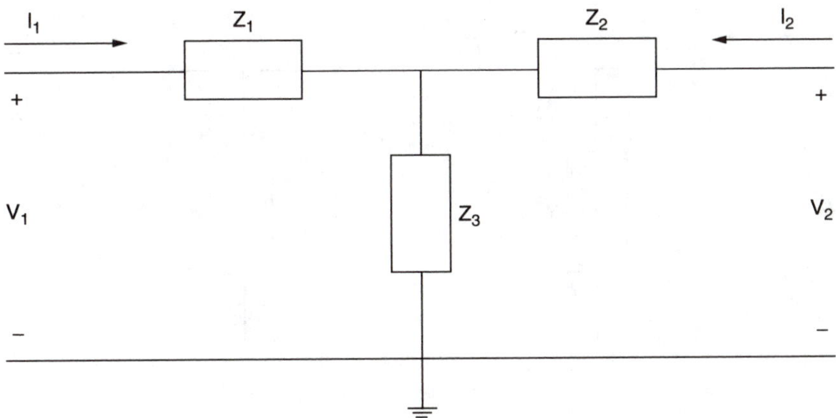

FIGURE 7.2
T-network.

Thus

$$\begin{bmatrix} V_1 \\ V_2 \end{bmatrix} = \begin{bmatrix} Z_1 + Z_3 & Z_3 \\ Z_3 & Z_2 + Z_3 \end{bmatrix} \begin{bmatrix} I_1 \\ I_2 \end{bmatrix} \tag{7.10}$$

and the z-parameters are

$$[Z] = \begin{bmatrix} Z_1 + Z_3 & Z_3 \\ Z_3 & Z_2 + Z_3 \end{bmatrix} \tag{7.11}$$

7.1.2 y-Parameters

A two-port network can also be represented using y-parameters. The describing equations are

$$I_1 = y_{11}V_1 + y_{12}V_2 \tag{7.12}$$

$$I_2 = y_{21}V_1 + y_{22}V_2 \tag{7.13}$$

where V_1 and V_2 are independent variables and I_1 and I_2 are dependent variables.

In matrix form, the above equations can be rewritten as

$$\begin{bmatrix} I_1 \\ I_2 \end{bmatrix} = \begin{bmatrix} y_{11} & y_{12} \\ y_{21} & y_{22} \end{bmatrix} \begin{bmatrix} V_1 \\ V_2 \end{bmatrix} \tag{7.14}$$

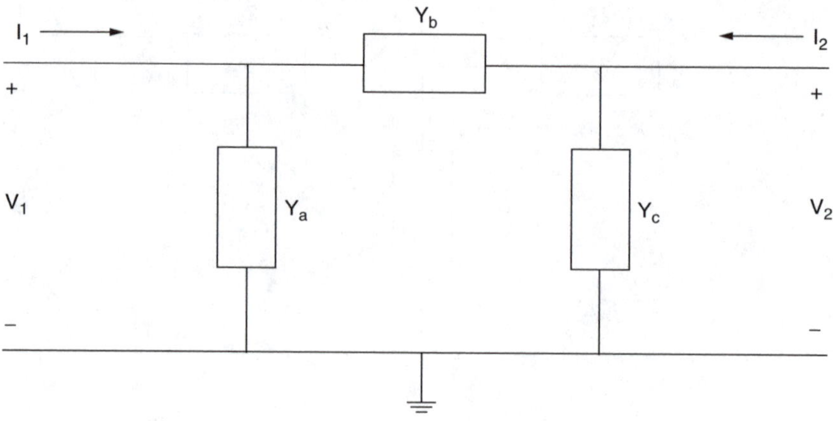

FIGURE 7.3
Pi-network.

The y-parameters can be found as follows:

$$y_{11} = \frac{I_1}{V_1}\bigg|_{V_2=0} \tag{7.15}$$

$$y_{12} = \frac{I_1}{V_2}\bigg|_{V_1=0} \tag{7.16}$$

$$y_{21} = \frac{I_2}{V_1}\bigg|_{V_2=0} \tag{7.17}$$

$$y_{22} = \frac{I_2}{V_2}\bigg|_{V_1=0} \tag{7.18}$$

The y-parameters are also called short-circuit admittance parameters. They are obtained as a ratio of current and voltage and found by short-circuiting port 2 ($V_2 = 0$) or port 1 ($V_1 = 0$). The following two examples show how to obtain the y-parameters of simple circuits.

Example 7.2 y-Parameters of a Pi-Network

Find the y-parameters of the pi (π) network shown in Figure 7.3.

Solution

Using KCL, we have

$$I_1 = V_1 Y_a + (V_1 - V_2) Y_b = V_1 (Y_a + Y_b) - V_2 Y_b \tag{7.19}$$

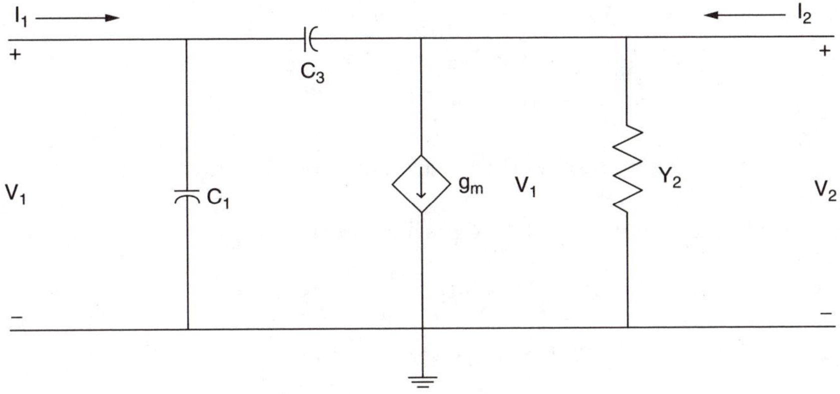

FIGURE 7.4
Simplified model of a field effect transistor.

$$I_2 = V_2Y_c + (V_2 - V_1)Y_b = -V_1Y_b + V_2(Y_b + Y_c) \tag{7.20}$$

Comparing Equation (7.19) and Equation (7.20) to Equation (7.12) and Equation (7.13), the y-parameters are

$$[Y] = \begin{bmatrix} Y_a + Y_b & -Y_b \\ -Y_b & Y_b + Y_c \end{bmatrix} \tag{7.21}$$

Example 7.3 y-Parameters of a Field Effect Transistor

Figure 7.4 shows the simplified model of a field effect transistor. Find its y-parameters.

Using KCL,

$$I_1 = V_1sC_1 + (V_1 - V_2)sC_3 = V_1(sC_1 + sC_3) + V_2(-sC_3) \tag{7.22}$$

$$I_2 = V_2Y_2 + g_mV_1 + (V_2 - V_1)sC_3 = V_1(g_m - sC_3) + V_2(Y_2 + sC_3) \tag{7.23}$$

Comparing the above two equations to Equation (7.12) and Equation (7.13), the y-parameters are

$$[Y] = \begin{bmatrix} sC_1 + sC_3 & -sC_3 \\ g_m - sC_3 & Y_2 + sC_3 \end{bmatrix} \tag{7.24}$$

7.1.3 h-Parameters

A two-port network can be represented using the h-parameters. The describing equations for the h-parameters are

$$V_1 = h_{11}I_1 + h_{12}V_2 \tag{7.25}$$

$$I_2 = h_{21}I_1 + h_{22}V_2 \tag{7.26}$$

where I_1 and V_2 are independent variables and V_1 and I_2 are dependent variables.

In matrix form, the above two equations become

$$\begin{bmatrix} V_1 \\ I_2 \end{bmatrix} = \begin{bmatrix} h_{11} & h_{12} \\ h_{21} & h_{22} \end{bmatrix} \begin{bmatrix} I_1 \\ V_2 \end{bmatrix} \tag{7.27}$$

The h-parameters can be found as follows:

$$h_{11} = \left. \frac{V_1}{I_1} \right|_{V_2=0} \tag{7.28}$$

$$h_{12} = \left. \frac{V_1}{V_2} \right|_{I_1=0} \tag{7.29}$$

$$h_{21} = \left. \frac{I_2}{I_1} \right|_{V_2=0} \tag{7.30}$$

$$h_{22} = \left. \frac{I_2}{V_2} \right|_{I_1=0} \tag{7.31}$$

The h-parameters are also called hybrid parameters since they contain both open-circuit parameters ($I_1 = 0$) and short-circuit parameters ($V_2 = 0$). The h-parameters of a bipolar junction transistor are determined in the following example.

Example 7.4 h-Parameters of a Bipolar Junction Transistor Equivalent Circuit

A simplified equivalent circuit of a bipolar junction transistor is shown in Figure 7.5. Find its h-parameters.

Solution

Using KCL for port 1,

$$V_1 = I_1 Z_1 + V_2 \tag{7.32}$$

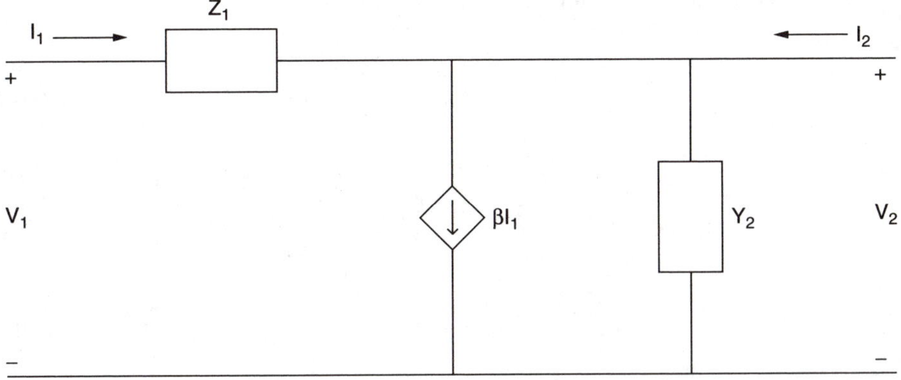

FIGURE 7.5
Simplified equivalent circuit of a bipolar junction transistor.

Using KCL at port 2, we get

$$I_2 = \beta I_1 + Y_2 V_2 - I_1 = (\beta - 1) I_1 + Y_2 V_2 \tag{7.33}$$

Comparing the above two equations to Equation (7.25) and Equation (7.26), we get the h-parameters.

$$[h] = \begin{bmatrix} Z_1 & 1 \\ \beta - 1 & Y_2 \end{bmatrix} \tag{7.34}$$

7.1.4 Transmission Parameters

A two-port network can be described by transmission parameters. The describing equations are

$$V_1 = a_{11} V_2 - a_{12} I_2 \tag{7.35}$$

$$I_1 = a_{21} V_2 - a_{22} I_2 \tag{7.36}$$

where V_2 and I_2 are independent variables and V_1 and I_1 are dependent variables.

In matrix form, the above two equations can be rewritten as

$$\begin{bmatrix} V_1 \\ I_1 \end{bmatrix} = \begin{bmatrix} a_{11} & a_{12} \\ a_{21} & a_{22} \end{bmatrix} \begin{bmatrix} V_2 \\ -I_2 \end{bmatrix} \tag{7.37}$$

The transmission parameters can be found as

$$a_{11} = \left.\frac{V_1}{V_2}\right|_{I_2=0} \tag{7.38}$$

$$a_{12} = -\left.\frac{V_1}{I_2}\right|_{V_2=0} \tag{7.39}$$

$$a_{21} = \left.\frac{I_1}{V_2}\right|_{I_2=0} \tag{7.40}$$

$$a_{22} = -\left.\frac{I_1}{I_2}\right|_{V_2=0} \tag{7.41}$$

The transmission parameters express the primary (sending end) variables V_1 and I_1 in terms of the secondary (receiving end) variables V_2 and $-I_2$. The negative of I_2 is used to allow the current to enter the load at the receiving end. Example 7.5 and Example 7.6 show some techniques for obtaining the transmission parameters of impedance and admittance networks.

Example 7.5 Transmission Parameters of a Simple Impedance Network

Find the transmission parameters of Figure 7.6.

Solution

By inspection,

$$I_1 = -I_2 \tag{7.42}$$

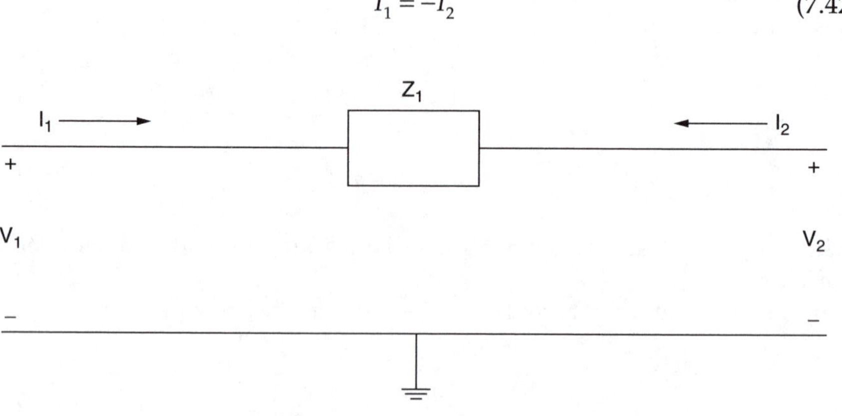

FIGURE 7.6
Simple impedance network.

Using KVL,

$$V_1 = V_2 + Z_1 I_1 \tag{7.43}$$

Since $I_1 = -I_2$, Equation (7.43) becomes

$$V_1 = V_2 - Z_1 I_2 \tag{7.44}$$

Comparing Equation (7.42) and Equation (7.44) to Equation (7.35) and Equation (7.36), we have

$$\begin{aligned} a_{11} &= 1 \qquad a_{12} = Z_1 \\ a_{21} &= 0 \qquad a_{22} = 1 \end{aligned} \tag{7.45}$$

Example 7.6 Transmission Parameters of a Simple Admittance Network

Find the transmission parameters for the network shown in Figure 7.7.

Solution

By inspection,

$$V_1 = V_2 \tag{7.46}$$

Using KCL, we have

$$I_1 = V_2 Y_2 - I_2 \tag{7.47}$$

Comparing Equation (7.46) and Equation (7.47) to Equation (7.35) and Equation (7.36), we have

$$\begin{aligned} a_{11} &= 1 \qquad a_{12} = 0 \\ a_{21} &= Y_2 \qquad a_{22} = 1 \end{aligned} \tag{7.48}$$

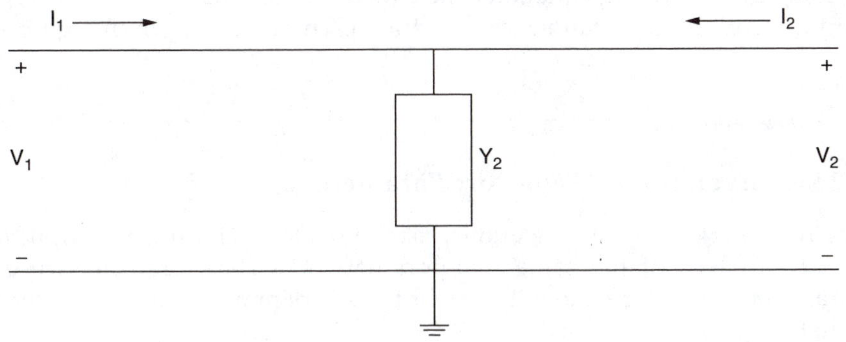

FIGURE 7.7
Simple admittance network.

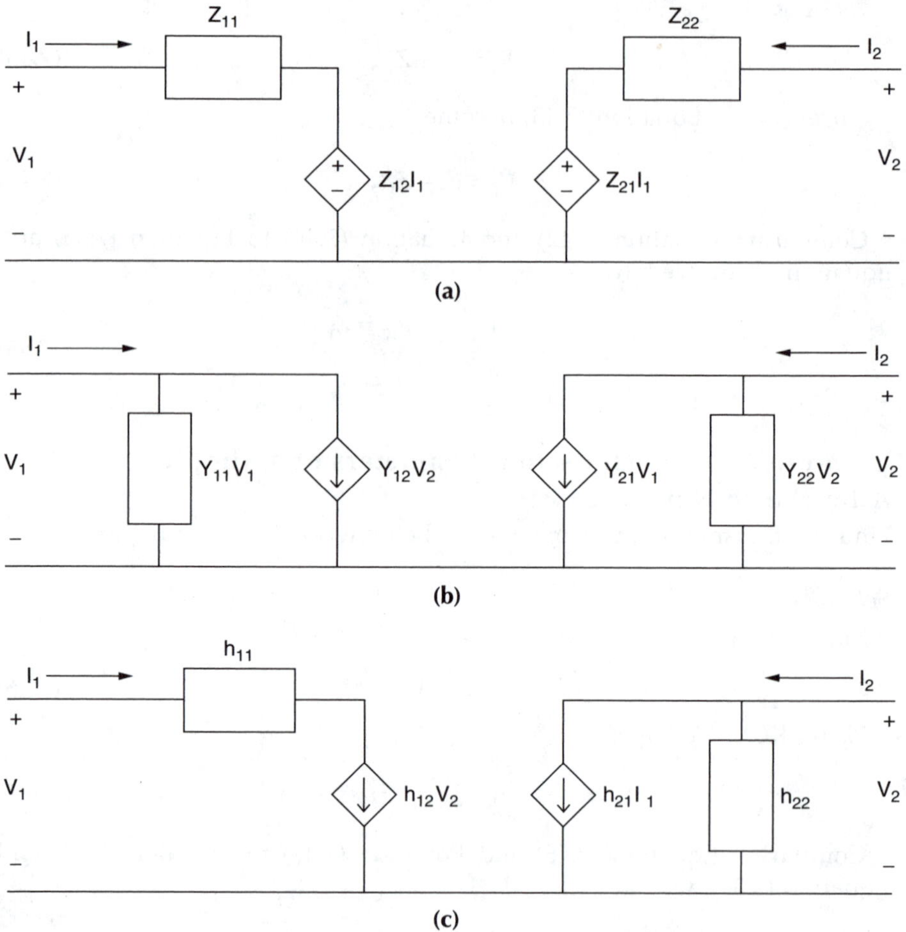

FIGURE 7.8
Equivalent circuit of two-port networks: (a) z-parameters, (b) y-parameters, and (c) h-parameters.

Using the describing equations, the equivalent circuits of the various two-port network representations can be drawn. These are shown in Figure 7.8.

7.2 Conversion of Two-Port Parameters

The two-port parameters mentioned above relate the input and output terminal variables of the same two-port network. Therefore, the two-port parameters are interrelated. To convert the y-parameters to z-parameters, Equation (7.49) can be used.

$$z_{11} = \frac{y_{22}}{y_{11}y_{22} - y_{12}y_{21}}$$

$$z_{12} = \frac{-y_{12}}{y_{11}y_{22} - y_{12}y_{21}}$$

$$z_{21} = \frac{-y_{21}}{y_{11}y_{22} - y_{12}y_{21}} \tag{7.49}$$

$$z_{22} = -\frac{y_{11}}{y_{11}y_{22} - y_{12}y_{21}}$$

To convert z-parameters to y-parameters, we can use Equation (7.50),

$$y_{11} = \frac{z_{22}}{z_{11}z_{22} - z_{12}z_{21}}$$

$$y_{12} = \frac{-z_{12}}{z_{11}z_{22} - z_{12}z_{21}}$$

$$y_{21} = \frac{-z_{21}}{z_{11}z_{22} - z_{12}z_{21}} \tag{7.50}$$

$$y_{22} = -\frac{z_{11}}{z_{11}z_{22} - z_{12}z_{21}}$$

The y-parameters can be converted to h-parameters using Equation (7.51),

$$h_{11} = \frac{1}{y_{11}}$$

$$h_{12} = \frac{-y_{12}}{y_{11}}$$

$$h_{21} = \frac{y_{21}}{y_{11}} \tag{7.51}$$

$$h_{22} = -\frac{y_{11}y_{22} - y_{12}y_{21}}{y_{11}}$$

To convert h-parameters to y-parameters, we can use Equation (7.52),

$$
\begin{aligned}
y_{11} &= \frac{1}{h_{11}} \\
y_{12} &= \frac{-h_{12}}{h_{11}} \\
y_{21} &= \frac{-h_{21}}{h_{11}} \\
y_{22} &= -\frac{h_{11}h_{22} - h_{12}h_{21}}{h_{11}}
\end{aligned}
\tag{7.52}
$$

The z-parameters can be converted into the h-parameters through the use of Equation (7.53),

$$
\begin{aligned}
h_{11} &= \frac{z_{11}z_{22} - z_{12}z_{21}}{z_{22}} \\
h_{12} &= \frac{z_{12}}{z_{22}} \\
h_{21} &= \frac{-z_{21}}{z_{22}} \\
h_{22} &= \frac{1}{z_{22}}
\end{aligned}
\tag{7.53}
$$

To convert h-parameters to z-parameters, we can use Equation (7.54),

$$
\begin{aligned}
z_{11} &= \frac{h_{11}h_{22} - h_{12}h_{21}}{h_{22}} \\
z_{12} &= \frac{h_{12}}{h_{22}} \\
z_{21} &= \frac{-h_{21}}{h_{22}} \\
z_{22} &= \frac{1}{h_{22}}
\end{aligned}
\tag{7.54}
$$

Other two-port parameter conversion equations can be found in references 1 and 9 of this chapter.

Example 7.7 Two-Port Parameters Conversion

The z-parameter of a circuit is given as

$$[Z] = \begin{bmatrix} 14+j9 & 4+j3 \\ 4+j3 & 12+j2 \end{bmatrix} \tag{7.55}$$

Determine the equivalent h-parameter of the circuit.

Solution

MATLAB script

```
%  Two-port parameter conversion
% z-parameters
z11 = 14+j*9;
z12 = 4+j*3;
z21 = 4+j*3;
z22 = 12+j*2;
% Conversion equations
h11 =  (z11*z22 - z12*z21)/z22
h12 = z12/z22
h21 = -z21/z22
h22 = 1/z22
```

The results obtained from MATLAB are

```
h11 = 13.1081 + 7.1486i
h12 = 0.3649 + 0.1892i
h21 = -0.3649 - 0.1892i
h22 =  0.0811 - 0.0135i
```

7.3 Interconnection of Two-Port Networks

Two-port networks can be connected in series, parallel, or cascade. Figure 7.9 shows the various two-port interconnections.

It can be shown that if two-port networks with z-parameters $[Z]_1$, $[Z]_2$, $[Z]_3$, …, $[Z]_n$ are connected in series, then the equivalent two-port z-parameters are given as

$$[Z]_{eq} = [Z]_1 + [Z]_2 + [Z]_3 + \dots + [Z]_n \tag{7.56}$$

If two-port networks with y-parameters $[Y]_1$, $[Y]_2$, $[Y]_3$, …, $[Y]_n$ are connected in parallel, then the equivalent two-port y-parameters are given as

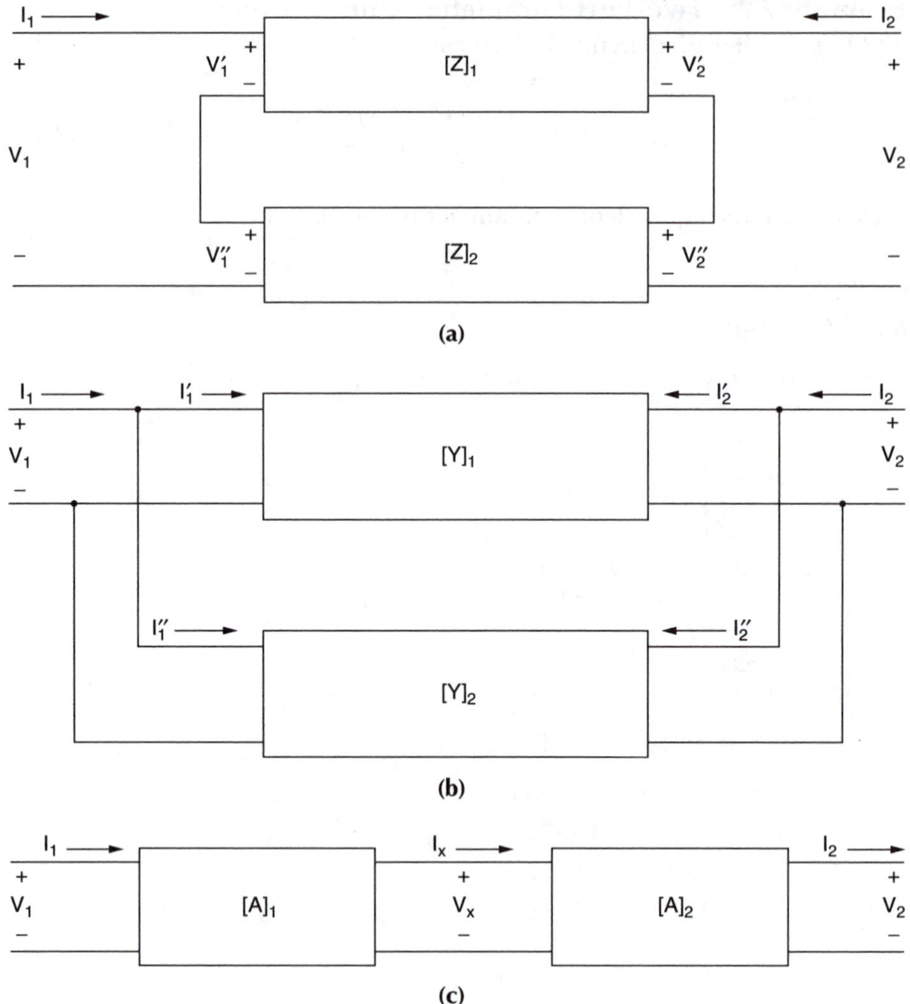

FIGURE 7.9
Interconnection of two-port networks: (a) series, (b) parallel, and (c) cascade.

$$[Y]_{eq} = [Y]_1 + [Y]_2 + [Y]_3 + \ldots + [Y]_n \tag{7.57}$$

When several two-port networks are connected in cascade, and the individual networks have transmission parameters $[A]_1$, $[A]_2$, $[A]_3$, …, $[A]_n$, then the equivalent two-port parameter will have a transmission parameter given as

$$[A]_{eq} = [A]_1 * [A]_2 * [A]_3 * \ldots * [A]_n \tag{7.58}$$

The following three examples illustrate the use of MATLAB for determining the equivalent parameters of interconnected two-port networks.

Example 7.8 y-Parameters of a Bridge-T Network

Find the equivalent y-parameters for the bridge-T network shown in Figure 7.10.

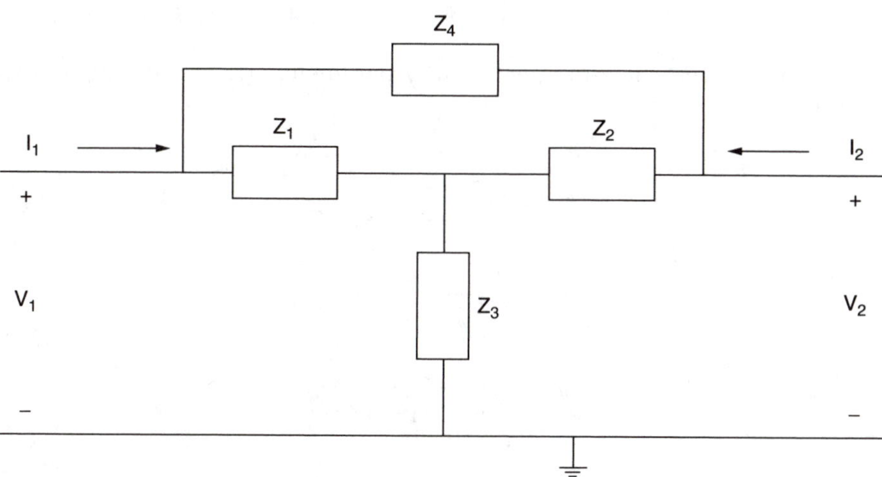

FIGURE 7.10
Bridge-T network.

Solution

The bridge-T network can be redrawn as

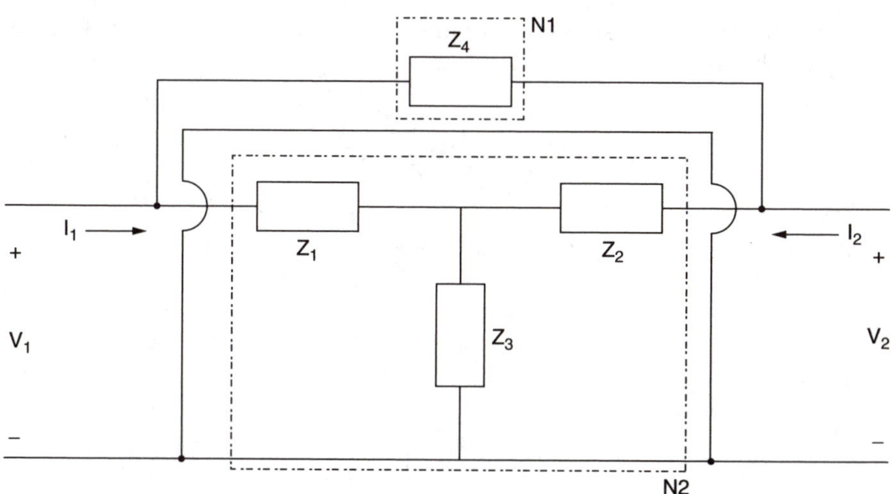

FIGURE 7.11
An alternative representation of bridge-T network.

From Example 7.1, the z-parameters of network N2 are

$$[Z] = \begin{bmatrix} Z_1 + Z_3 & Z_3 \\ Z_3 & Z_2 + Z_3 \end{bmatrix}$$

We can convert the z-parameters to y-parameters (refs. 1 and 9), and we obtain

$$\begin{aligned}
y_{11} &= \frac{Z_2 + Z_3}{Z_1Z_2 + Z_1Z_3 + Z_2Z_3} \\
y_{12} &= \frac{-Z_3}{Z_1Z_2 + Z_1Z_3 + Z_2Z_3} \\
y_{21} &= \frac{-Z_3}{Z_1Z_2 + Z_1Z_3 + Z_2Z_3} \\
y_{22} &= -\frac{Z_1 + Z_3}{Z_1Z_2 + Z_1Z_3 + Z_2Z_3}
\end{aligned} \tag{7.59}$$

From Example 7.5, the transmission parameters of network N1 are

$$a_{11} = 1 \qquad a_{12} = Z_4$$

$$a_{21} = 0 \qquad a_{22} = 1$$

We convert the transmission parameters to y-parameters (refs. 1 and 9) and we obtain

$$\begin{aligned}
y_{11} &= \frac{1}{Z_4} \\
y_{12} &= -\frac{1}{Z_4} \\
y_{21} &= -\frac{1}{Z_4} \\
y_{22} &= \frac{1}{Z_4}
\end{aligned} \tag{7.60}$$

Using Equation (7.57), the equivalent y-parameters of the bridge-T network are

$$
\begin{aligned}
y_{11eq} &= \frac{1}{Z_4} + \frac{Z_2 + Z_3}{Z_1Z_2 + Z_1Z_3 + Z_2Z_3} \\
y_{12eq} &= -\frac{1}{Z_4} - \frac{Z_3}{Z_1Z_2 + Z_1Z_3 + Z_2Z_3} \\
y_{21eq} &= -\frac{1}{Z_4} - \frac{Z_3}{Z_1Z_2 + Z_1Z_3 + Z_2Z_3} \\
y_{22eq} &= \frac{1}{Z_4} + \frac{Z_1 + Z_3}{Z_1Z_2 + Z_1Z_3 + Z_2Z_3}
\end{aligned}
\tag{7.61}
$$

Example 7.9 Transmission Parameters of a Simple Cascaded Network

Find the transmission parameters of Figure 7.12.

Solution

Figure 7.12 can be redrawn as shown in Figure 7.13.

From Example 7.5, the transmission parameters of network N1 are

$$
\begin{aligned}
a_{11} &= 1 \qquad a_{12} = Z_1 \\
a_{21} &= 0 \qquad a_{22} = 1
\end{aligned}
$$

From Example 7.6, the transmission parameters of network N2 are

$$
\begin{aligned}
a_{11} &= 1 \qquad a_{12} = 0 \\
a_{21} &= Y_2 \qquad a_{22} = 1
\end{aligned}
$$

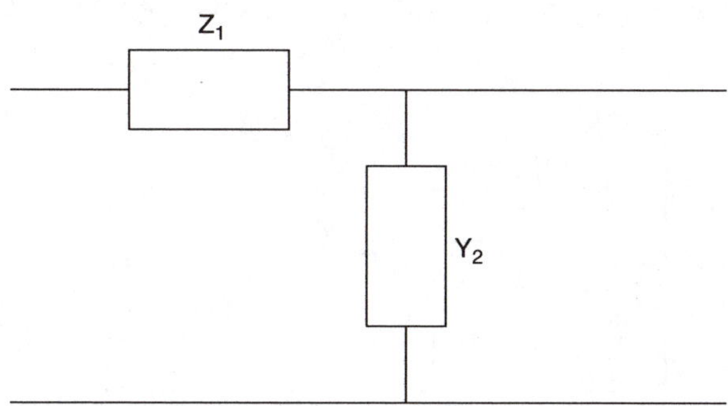

FIGURE 7.12
Simple cascaded network.

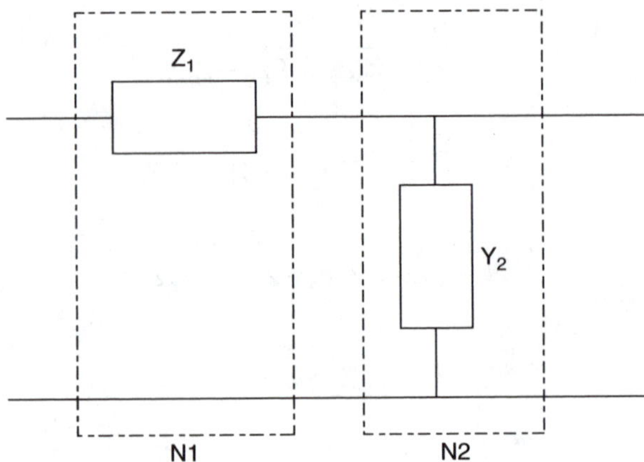

FIGURE 7.13
Cascade of two networks N1 and N2.

From Equation (7.58), the transmission parameters of Figure 7.13 are

$$\begin{bmatrix} a_{11} & a_{12} \\ a_{21} & a_{22} \end{bmatrix}_{eq} = \begin{bmatrix} 1 & Z_1 \\ 0 & 1 \end{bmatrix} \begin{bmatrix} 1 & 0 \\ Y_2 & 1 \end{bmatrix} = \begin{bmatrix} 1+Z_1Y_2 & Z_1 \\ Y_2 & 1 \end{bmatrix} \tag{7.62}$$

Example 7.10 Transmission Parameters of a Cascaded Resistive Network

Find the transmission parameters for the cascaded system shown in Figure 7.14. The resistance values are in ohms.

Solution

Figure 7.14 can be considered as four networks, N1, N2, N3, and N4, connected in cascade. From Example 7.8, the transmission parameters of Figure 7.12 are

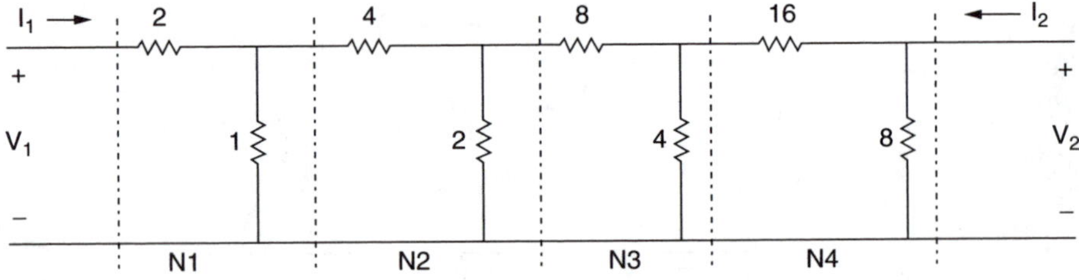

FIGURE 7.14
Cascaded resistive network.

$$[a]_{N1} = \begin{bmatrix} 3 & 2 \\ 1 & 1 \end{bmatrix}$$

$$[a]_{N2} = \begin{bmatrix} 3 & 4 \\ 0.5 & 1 \end{bmatrix}$$

$$[a]_{N3} = \begin{bmatrix} 3 & 8 \\ 0.25 & 1 \end{bmatrix}$$

$$[a]_{N4} = \begin{bmatrix} 3 & 16 \\ 0.125 & 1 \end{bmatrix}$$

The transmission parameters of Figure 7.14 can be obtained using the following MATLAB program.

MATLAB script

```
% Transmission parameters of cascaded resistive network
a1 = [3 2; 1 1];
a2 = [3 4; 0.5 1];
a3 = [3 8; 0.25 1];
a4 = [3 16; 0.125 1];
% equivalent transmission parameters
a = a1*(a2*(a3*a4))
```

The value of matrix a is

```
a =
      112.2500    630.0000
       39.3750    221.0000
```

7.4 Terminated Two-Port Networks

In normal applications, two-port networks are usually terminated. A terminated two-port network is shown in Figure 7.4.

In Figure 7.15, V_g and Z_g are the source generator voltage and impedance, respectively. Z_L is the load impedance. If we use z-parameter representation for the two-port network, the voltage transfer function can be shown to be

$$\frac{V_2}{V_g} = \frac{z_{21}Z_L}{(z_{11}+Z_g)(z_{22}+Z_L)-z_{12}z_{21}} \tag{7.63}$$

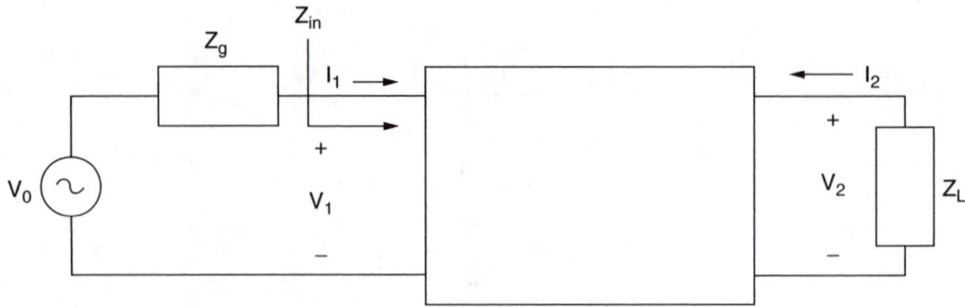

FIGURE 7.15
Terminated two-port network.

and the input impedance,

$$Z_{in} = z_{11} - \frac{z_{12}z_{21}}{z_{22} + Z_L} \tag{7.64}$$

and the current transfer function,

$$\frac{I_2}{I_1} = -\frac{z_{21}}{z_{22} + Z_L} \tag{7.65}$$

A terminated two-port network, represented using the y-parameters, is shown in Figure 7.16.

It can be shown that the input admittance, Y_{in}, is

$$Y_{in} = y_{11} - \frac{y_{12}y_{21}}{y_{22} + Y_L} \tag{7.66}$$

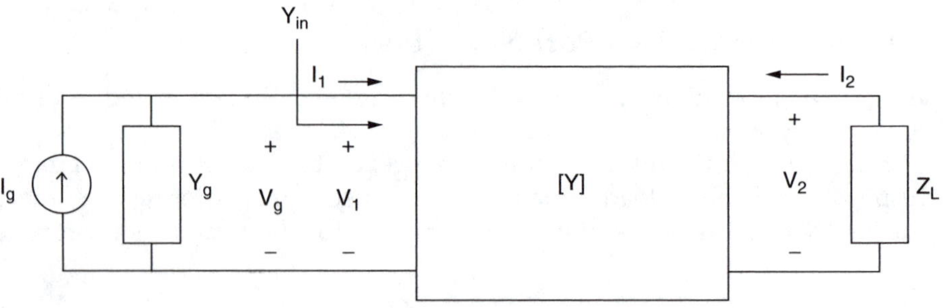

FIGURE 7.16
A terminated two-port network with y-parameters representation.

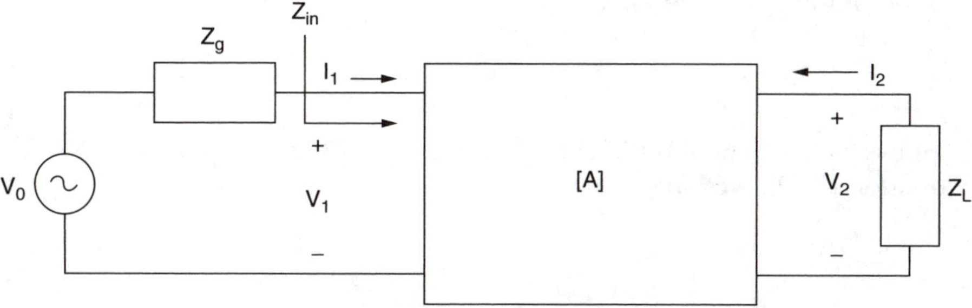

FIGURE 7.17
A terminated two-port network with transmission parameters representation.

and the current transfer function is given as

$$\frac{I_2}{I_g} = \frac{y_{21}Y_L}{(y_{11}+Y_g)(y_{22}+Y_L) - y_{12}y_{21}} \tag{7.67}$$

and the voltage transfer function

$$\frac{V_2}{V_g} = -\frac{y_{21}}{y_{22}+Y_L} \tag{7.68}$$

A doubly terminated two-port network, represented by transmission parameters, is shown in Figure 7.17. The voltage transfer function and the input impedance of the transmission parameters can be obtained as follows. From the transmission parameters, we have

$$V_1 = a_{11}V_2 - a_{12}I_2 \tag{7.69}$$

$$I_1 = a_{21}V_2 - a_{22}I_2 \tag{7.70}$$

From Figure 7.6,

$$V_2 = -I_2 Z_L \tag{7.71}$$

Substituting Equation (7.64) into Equation (7.62) and Equation (7.63), we get the input impedance,

$$Z_{in} = \frac{a_{11}Z_L + a_{12}}{a_{21}Z_L + a_{22}} \tag{7.72}$$

From Figure 7.17, we have

$$V_1 = V_g - I_1 Z_g \tag{7.73}$$

Substituting Equation (7.71) and Equation (7.73) into Equation (7.69) and Equation (7.70), we have

$$V_g - I_1 Z_g = V_2 \left[a_{11} + \frac{a_{12}}{Z_L} \right] \tag{7.74}$$

$$I_1 = V_2 \left[a_{21} + \frac{a_{22}}{Z_L} \right] \tag{7.75}$$

Substituting Equation (7.75) into Equation (7.74), we get

$$V_g - V_2 Z_g \left[a_{21} + \frac{a_{22}}{Z_L} \right] = V_2 \left[a_{11} + \frac{a_{12}}{Z_L} \right] \tag{7.76}$$

Simplifying Equation (7.76), we get the voltage transfer function

$$\frac{V_2}{V_g} = \frac{Z_L}{(a_{11} + a_{21} Z_g) Z_L + a_{12} + a_{22} Z_g} \tag{7.77}$$

The following examples illustrate the use of MATLAB for solving terminated two-port network problems.

Example 7.11 z-Parameters and Magnitude Response of an Active Lowpass Filter

Assuming that the operational amplifier of Figure 7.18 is ideal:

(a) Find the z-parameters of Figure 7.18.

(b) If the network is connected by a voltage source with source resistance of 50 Ω and a load resistance of 1 kΩ, find the voltage gain.

(c) Use MATLAB to plot the magnitude response.

Solution

Using KVL,

$$V_1 = R_1 I_1 + \frac{I_1}{sC} \tag{7.78}$$

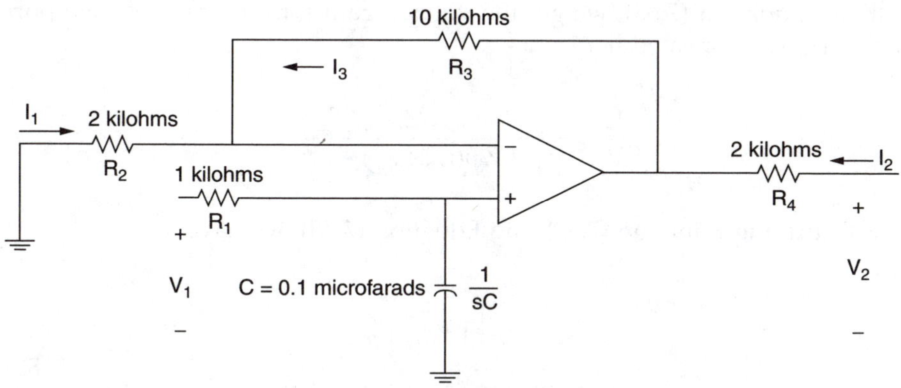

FIGURE 7.18
An active lowpass filter.

$$V_2 = R_4 I_2 + R_3 I_3 + R_2 I_3 \tag{7.79}$$

From the concept of virtual circuit discussed in Chapter 11,

$$R_2 I_3 = \frac{I_1}{sC} \tag{7.80}$$

Substituting Equation (7.80) into Equation (7.79), we get

$$V_2 = \frac{(R_2 + R_3) I_1}{sCR_2} + R_4 I_2 \tag{7.81}$$

Comparing Equation (7.79) and Equation (7.81) to Equation (7.1) and Equation (7.2), we have

$$\begin{aligned} z_{11} &= R_1 + \frac{1}{sC} \\ z_{12} &= 0 \\ z_{21} &= \left(1 + \frac{R_3}{R_2}\right)\left(\frac{1}{sC}\right) \\ z_{22} &= R_4 \end{aligned} \tag{7.82}$$

From Equation (7.63), we get the voltage gain for a terminated two-port network. It is repeated here.

$$\frac{V_2}{V_g} = \frac{z_{21}Z_L}{(z_{11}+Z_g)(z_{22}+Z_L)-z_{12}z_{21}}$$

Substituting Equation (7.82) into Equation (7.75), we have

$$\frac{V_2}{V_g} = \frac{\left(1+\frac{R_3}{R_2}\right)Z_L}{(R_4+Z_L)\left[1+sC(R_1+Z_g)\right]} \tag{7.83}$$

For $Z_g = 50\ \Omega$, $Z_L = 1\ \text{k}\omega$, $R_1 = 1\ \text{k}\Omega$, $R_2 = 12\ \text{k}\Omega$, $R_3 = 10\ \text{k}\Omega$, $R_4 = 2\ \text{k}\Omega$, and $C = 0.1\ \mu\text{F}$, Equation (7.83) becomes

$$\frac{V_2}{V_g} = \frac{2}{[1+1.05*10^{-4}s]} \tag{7.84}$$

The MATLAB script is

```
%
num = [2];
den = [1.05e-4 1];
w = logspace(1,5);
h = freqs(num,den,w);
f = w/(2*pi);
mag = 20*log10(abs(h));   % magnitude in dB
semilogx(f,mag)
title('Lowpass Filter Response')
xlabel('Frequency, Hz')
ylabel('Gain in dB')
```

The frequency response is shown in Figure 7.19.

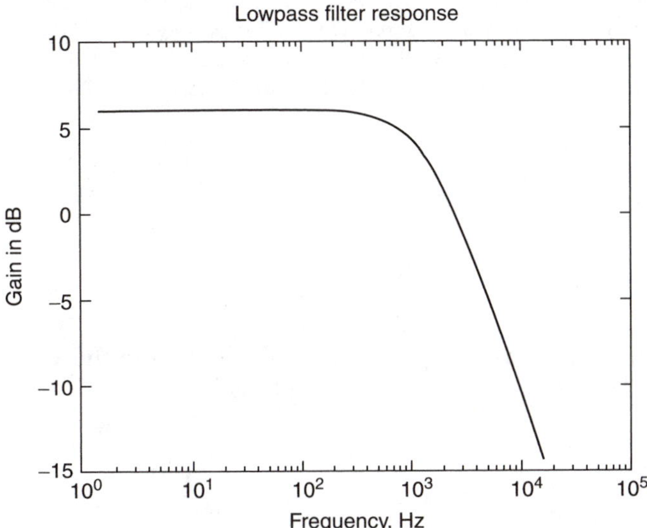

FIGURE 7.19
Magnitude response of an active lowpass filter.

Bibliography

1. Alexander, C.K. and Sadiku, M.N.O., *Fundamentals of Electric Circuits*, 2nd ed., McGraw-Hill, New York, 2004.
2. Attia, J.O., *PSPICE and MATLAB for Electronics: An Integrated Approach*, CRC Press, Boca Raton, FL, 2002.
3. Biran, A. and Breiner, M., *MATLAB for Engineers*, Addison-Wesley, Reading, MA, 1995.
4. Chapman, S.J., *MATLAB Programming for Engineers*, Brook, Cole Thompson Learning, Pacific Grove, CA, 2000.
5. Dorf, R.C. and Svoboda, J.A., *Introduction to Electric Circuits*, 3rd ed., John Wiley & Sons, New York, 1996.
6. Etter, D.M., *Engineering Problem Solving with MATLAB*, 2nd ed., Prentice Hall, Upper Saddle River, NJ, 1997.
7. Etter, D.M., Kuncicky, D.C., and Hull, D., *Introduction to MATLAB 6*, Prentice Hall, Upper Saddle River, NJ, 2002.
8. Gottling, J.G., *Matrix Analysis of Circuits Using MATLAB*, Prentice Hall, Englewood Cliffs, NJ, 1995.
9. Johnson, D.E., Johnson, J.R., and Hilburn, J.L., *Electric Circuit Analysis*, 3rd ed., Prentice Hall, Englewood Cliffs, NJ, 1997.
10. Meader, D.A., *Laplace Circuit Analysis and Active Filters*, Prentice Hall, Englewood Cliffs, NJ, 1991.

11. Sigmor, K., *MATLAB Primer*, 4th ed., CRC Press, Boca Raton, FL, 1998.
12. *Using MATLAB, The Language of Technical Computing, Computation, Visualization, Programming*, Version 6, MathWorks, Inc., Natick, MA, 2000
13. Vlach, J.O., Network theory and CAD, *IEEE Trans. on Education*, 36, 23–27, 1993.

Problems

Problem 7.1

(a) Find the transmission parameters of the circuit shown in Figure P7.1a. The resistance values are in ohms.

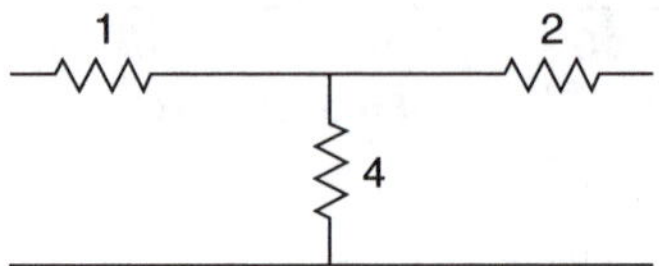

FIGURE P7.1a
Resistive T-network.

(b) From the result of part (a), use MATLAB to find the transmission parameters of Figure P7.2b. The resistance values are in ohms.

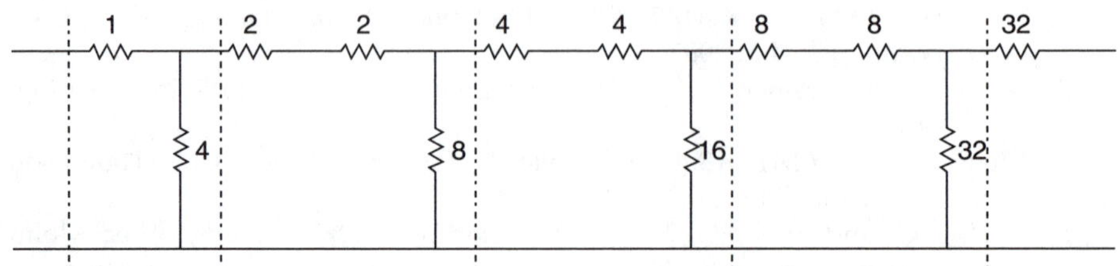

FIGURE P7.1b
Cascaded resistive network.

Problem 7.2

Find the y-parameters of the circuit with z-parameters given by

$$[Z] = \begin{bmatrix} 8+j6 & 3+j2 \\ 3+j2 & 16-j4 \end{bmatrix}$$

Problem 7.3

The y-parameters of a two-port network are

$$[Y] = \begin{bmatrix} 24 + \frac{1}{4s} & 3 + \frac{4}{s} \\ 3 + \frac{4}{s} & 16 + \frac{1}{2s} \end{bmatrix}$$

If $s = j2$, determine the equivalent z-parameters of the network.

Problem 7.4

Find the y-parameters of the circuit shown in Figure P7.4. The resistance values are in ohms.

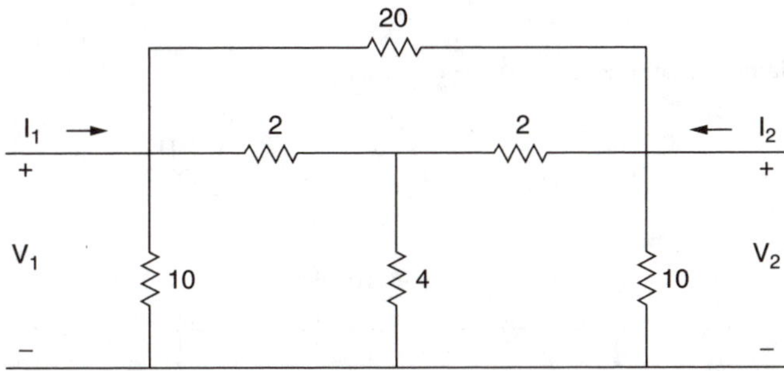

FIGURE P7.4
A resistive network.

Problem 7.5

(a) Show that for the symmetrical lattice structure shown in Figure P7.5,

$$z_{11} = z_{22} = 0.5(Z_c + Z_d)$$

$$z_{12} = z_{21} = 0.5(Z_c - Z_d)$$

(b) If $Z_c = 10\ \Omega$, $Z_d = 4\ \Omega$, find the equivalent y-parameters.

Problem 7.6

(a) Find the equivalent z-parameters of Figure P7.6.
(b) If the network is terminated by a load of 20 ohms and connected to a source of V_S with a source resistance of 4 ohms, use MATLAB to plot the frequency response of the circuit.

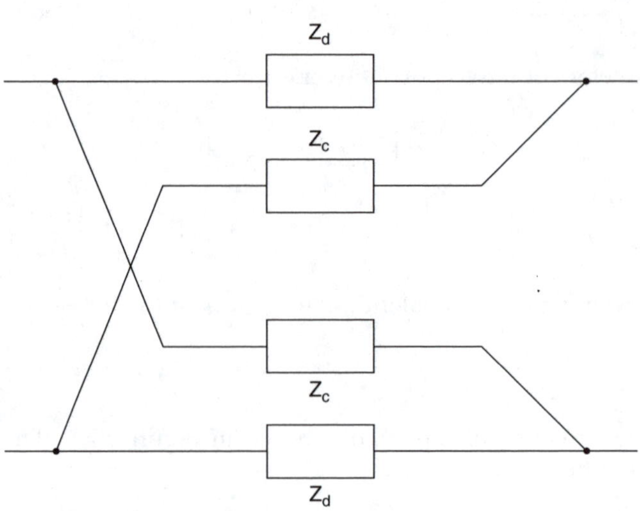

FIGURE P7.5
Symmetrical lattice structure.

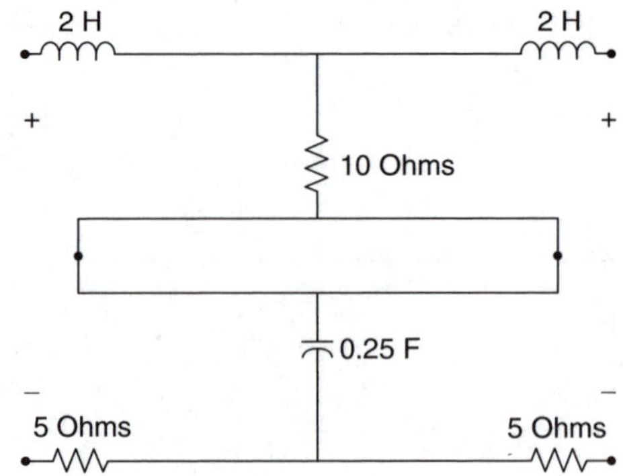

FIGURE P7.6
Circuit for Problem 7.6.

Problem 7.7

For Figure P7.7:

(a) Find the transmission parameters of the RC ladder network.
(b) Obtain the expression for

$$\frac{V_2}{V_1}$$

(c) Use MATLAB to plot the phase characteristics of

$$\frac{V_2}{V_1} \text{ if } C = 1\ \mu\text{F and R} = 10\ \text{k}\Omega$$

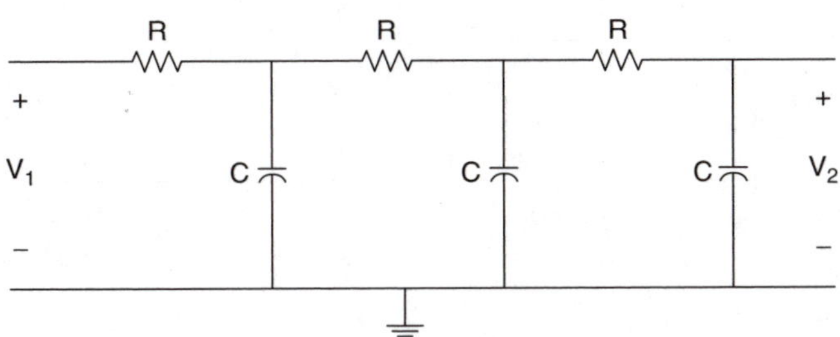

FIGURE P7.7
RC ladder network.

Problem 7.8

For the circuit shown in Figure P7.8, $R_1 = R_2 = R_3 = 5$ kΩ, L = 5H, and C = 10 μF.

(a) Find the y-parameters.
(b) Find the expression for the input admittance.
(c) Use MATLAB to plot the input admittance as a function of frequency.

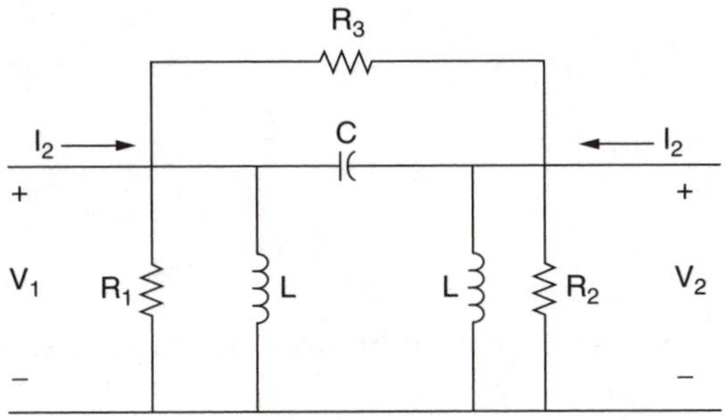

FIGURE P7.8
Circuit for Problem 7.8.

Problem 7.9

For the op amp circuit shown in Figure P7.9, find the y-parameters.

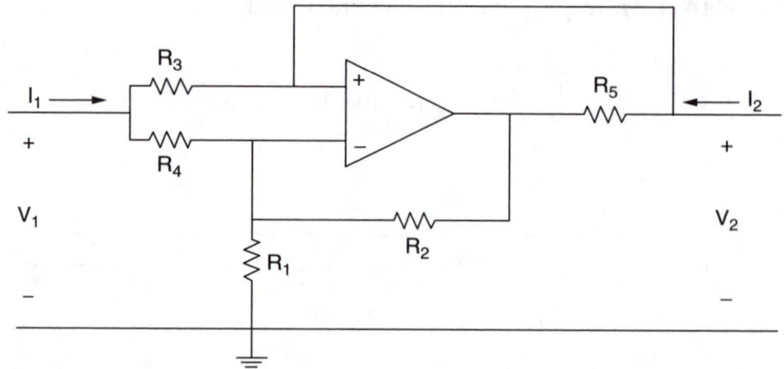

FIGURE P7.9
Op amp circuit.

Problem 7.10

For the two-port network shown in Figure 7.15, $Z_L = 10\ \text{k}\Omega$ and $Z_g = 100\ \Omega$. If the z-parameters are given by

$$[Z] = \begin{bmatrix} 500 + 10s & 0 \\ 200s & 1000 \end{bmatrix}$$

Plot the magnitude characteristics of the voltage gain

$$\frac{V_2}{V_g}$$

as a function of frequency.

Problem 7.11

For the two-port network shown in Figure 7.15, $Z_L = 2\ \text{k}\Omega$ and $Z_g = 50\ \Omega$. If the z-parameters are given by

$$[Z] = \begin{bmatrix} 1000 + \dfrac{10^5}{s} & 0 \\ \dfrac{1.1 * 10^6}{s} & 2000 \end{bmatrix}$$

Plot the input impedance, Z_{in}, as a function of frequency.

Problem 7.12

For the two-port network shown in Figure 7.16, $Z_L = 4\ \text{k}\Omega$ and $Y_g = \frac{1}{100}$ S. If the y-parameters are given by

$$[Y] = \begin{bmatrix} s + \dfrac{1}{2000} & s \\ s + \dfrac{1}{1000} & \dfrac{s}{4000} \end{bmatrix}$$

Plot the magnitude characteristics of the current gain

$$\frac{I_2}{I_g}$$

as a function of frequency.

8

Fourier Analysis

In this chapter, Fourier analysis will be discussed. The topics covered are Fourier series expansion, Fourier transform, discrete Fourier transform, and fast Fourier transform. Some applications of Fourier analysis using MATLAB will also be discussed.

8.1 Fourier Series

If a function $g(t)$ is periodic with period T_p, i.e.,

$$g(t) = g(t \pm T_p) \tag{8.1}$$

and in any finite interval $g(t)$ has at most a finite number of discontinuities and a finite number of maxima and minima (Dirichlet conditions), and in addition,

$$\int_0^{T_p} g(t)dt < \infty \tag{8.2}$$

then $g(t)$ can be expressed with a series of sinusoids. That is,

$$g(t) = \frac{a_0}{2} + \sum_{n=1}^{\infty} a_n \cos(nw_0 t) + b_n \sin(nw_0 t) \tag{8.3}$$

where

$$w_0 = \frac{2\pi}{T_p} \tag{8.4}$$

and the Fourier coefficients a_n and b_n are determined by the following equations:

$$a_n = \frac{2}{T_p} \int_{t_o}^{t_o+T_p} g(t)\cos(nw_0 t)dt \qquad n = 0, 1, 2, \ldots \tag{8.5}$$

$$b_n = \frac{2}{T_p} \int_{t_o}^{t_o+T_p} g(t)\sin(nw_0 t)dt \qquad n = 0, 1, 2 \ldots \tag{8.6}$$

Equation (8.3) is called the trigonometric Fourier series. The term

$$\frac{a_0}{2}$$

in Equation (8.3) is the dc component of the series and is the average value of $g(t)$ over a period. The term $a_n \cos(nw_0 t) + b_n \sin(nw_0 t)$ is called the n-*th* harmonic. The first harmonic is obtained when $n = 1$. The latter is also called the fundamental with the fundamental frequency of ω_0. When n = 2, we have the second harmonic and so on.

Equation (8.3) can be rewritten as

$$g(t) = \frac{a_0}{2} + \sum_{n=1}^{\infty} A_n \cos(nw_0 t + \Theta_n) \tag{8.7}$$

where

$$A_n = \sqrt{a_n^2 + b_n^2} \tag{8.8}$$

and

$$\Theta_n = -\tan^{-1}\left(\frac{b_n}{a_n}\right) \tag{8.9}$$

The total normalized power in $g(t)$ is given by Parseval's equation:

$$P = \frac{1}{T_p} \int_{t_o}^{t_o+T_p} g^2(t)dt = A_{dc}^2 + \sum_{n=1}^{\infty} \frac{A_n^2}{2} \tag{8.10}$$

where

$$A_{dc}^2 = \left(\frac{a_0}{2}\right)^2 \tag{8.11}$$

The following example shows the synthesis of a square wave using Fourier series expansion.

Example 8.1 Fourier Series Expansion of a Square Wave

Using Fourier series expansion, a square wave with a period of 2 msec, peak-to-peak value of 2 V, and average value of 0 V can be expressed as

$$g(t) = \frac{4}{\pi}\sum_{n=1}^{\infty}\frac{1}{(2n-1)}\sin[(2n-1)2\pi f_0 t] \tag{8.12}$$

where

$$f_0 = 500 \text{ Hz}$$

if $a(t)$ is given as

$$a(t) = \frac{4}{\pi}\sum_{n=1}^{12}\frac{1}{(2n-1)}\sin[(2n-1)2\pi f_0 t] \tag{8.13}$$

Write a MATLAB program to plot $a(t)$ from 0 to 4 msec at intervals of 0.05 msec and to show that $a(t)$ is a good approximation of $g(t)$.

Solution

MATLAB script

```
% Fourier series expansion
f = 500; c = 4/pi; dt = 5.0e-05;
tpts = (4.0e-3/5.0e-5) + 1;
for n = 1: 12
for m = 1: tpts
s1(n,m) = (4/pi)*(1/(2*n - 1))*sin((2*n -
1)*2*pi*f*dt*(m-1));
end
end
for m = 1:tpts
    s2 = s1(:,m);
    s3(m) = sum(s2);
```

```
end
f1 = s3';
t = 0.0:5.0e-5:4.0e-3;
plot(t,f1)
xlabel('Time, s')
ylabel('Amplitude, V')
title('Fourier Series Expansion')
```

Figure 8.1 shows the plot of $a(t)$.

By using the Euler's identity, the cosine and sine functions of Equation (8.3) can be replaced by exponential equivalents, yielding the expression

$$g(t) = \sum_{n=-\infty}^{\infty} c_n \exp(jnw_0 t) \tag{8.14}$$

where

$$c_n = \frac{1}{T_p} \int_{-t_p/2}^{T_p/2} g(t)\exp(-jnw_0 t)dt \tag{8.15}$$

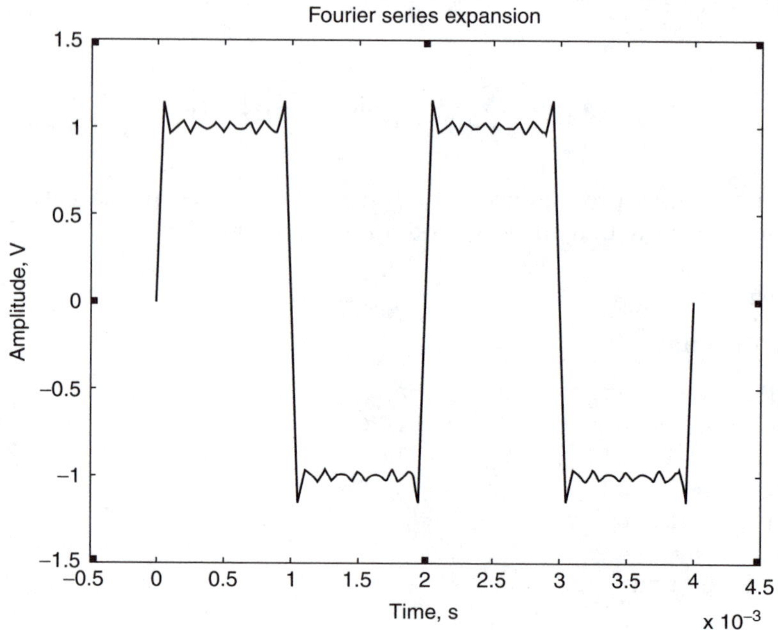

FIGURE 8.1
Approximation to square wave.

and

$$w_0 = \frac{2\pi}{T_p}$$

Equation (8.14) is termed the exponential Fourier series expansion. The coefficient c_n is related to the coefficients a_n and b_n of Equation (8.5) and Equation (8.6) by the expression

$$c_n = \frac{1}{2}\sqrt{a_n^2 + b_n^2}\angle - \tan^{-1}\left(\frac{b_n}{a_n}\right) \tag{8.16}$$

In addition, c_n relates to A_n and ϕ_n of Equation (8.8) and Equation (8.9) by the relation

$$c_n = \frac{A_n}{2}\angle\Theta_n \tag{8.17}$$

The plot of $|c_n|$ vs. frequency is termed the discrete amplitude spectrum or the line spectrum. It provides information on the amplitude spectral components of $g(t)$. A similar plot of $\angle c_n$ vs. frequency is called the discrete phase spectrum, and the latter gives information on the phase components with respect to the frequency of $g(t)$.

If an input signal $x_n(t)$

$$x_n(t) = c_n \exp(jnw_o t) \tag{8.18}$$

passes through a system with transfer function $H(w)$, then the output of the system $y_n(t)$ is

$$y_n(t) = H(jnw_o)c_n \exp(jnw_o t) \tag{8.19}$$

The block diagram of the input/output relation is shown in Figure 8.2.

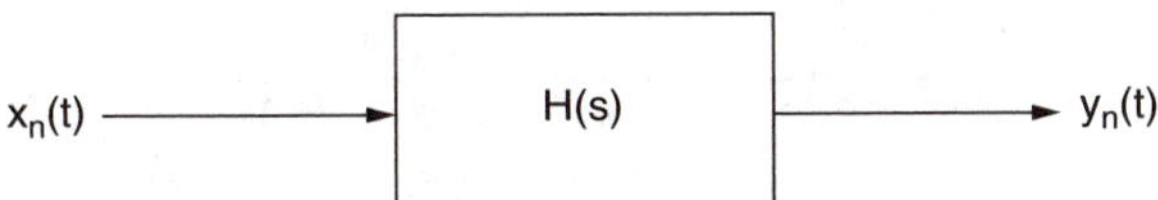

FIGURE 8.2
Input/output relationship.

However, with an input $x(t)$ consisting of a linear combination of complex excitations,

$$x_n(t) = \sum_{n=-\infty}^{\infty} c_n \exp(jnw_o t) \tag{8.20}$$

the response at the output of the system is

$$y_n(t) = \sum_{n=-\infty}^{\infty} H(jnw_o) c_n \exp(jnw_o t) \tag{8.21}$$

The following two examples show how to use MATLAB to obtain the coefficients of Fourier series expansion.

Example 8.2 Amplitude and Phase Spectrum of a Full-Wave Rectifier Waveform

For the full-wave rectifier waveform shown in Figure 8.3, the period is 0.0333 sec and the amplitude is 169.71 V.

(a) Write a MATLAB program to obtain the exponential Fourier series coefficients c_n for $n = 0, 1, 2, \ldots, 19$.

(b) Find the dc value.

(c) Plot the amplitude and phase spectrum.

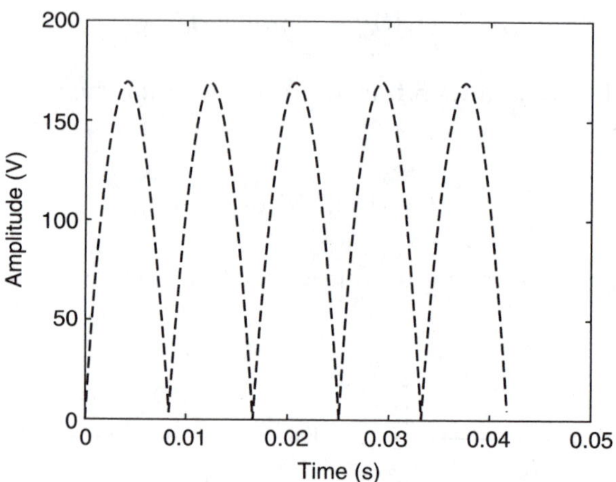

FIGURE 8.3
Full-wave rectifier waveform.

Solution

```
% generate the full-wave rectifier waveform
f1 = 60;
inv = 1/f1; inc = 1/(80*f1); tnum = 3*inv;
t = 0:inc:tnum;
g1 = 120*sqrt(2)*sin(2*pi*f1*t);
g = abs(g1);
N = length(g);
%
% obtain the exponential Fourier series coefficients

num = 20;
for i = 1:num
     for m = 1:N
      cint(m) = exp(-j*2*pi*(i-1)*m/N)*g(m);
     end
  c(i) = sum(cint)/N;
end
cmag = abs(c);
cphase = angle(c);

%print dc value
disp('dc value of g(t)'); cmag(1)
% plot the magnitude and phase spectrum

f = (0:num-1)*60;
subplot(121), stem(f(1:5),cmag(1:5))
title('Amplitude spectrum')
xlabel('Frequency, Hz')
subplot(122), stem(f(1:5),cphase(1:5))
title('Phase spectrum')
xlabel('Frequency, Hz')
% generate the full-wave rectifier waveform
f1 = 60;
inv = 1/f1; inc = 1/(80*f1); tnum = 3*inv;
t = 0:inc:tnum;
g1 = 120*sqrt(2)*sin(2*pi*f1*t);
g = abs(g1);
N = length(g);
%
% obtain the exponential Fourier series coefficients

num = 20;
for i = 1:num
     for m = 1:N
      cint(m) = exp(-j*2*pi*(i-1)*m/N)*g(m);
```

```
        end
    c(i) = sum(cint)/N;
  end
  cmag = abs(c);
  cphase = angle(c);

  %print dc value
  disp('dc value of g(t)'); cmag(1)
  % plot the magnitude and phase spectrum

  f = (0:num-1)*60;
  subplot(121), stem(f(1:5),cmag(1:5))
  title('Amplitude Spectrum')
  xlabel('Frequency, Hz')
  subplot(122), stem(f(1:5),cphase(1:5))
  title('Phase Spectrum')
  xlabel('Frequency, Hz')
```

The result obtained is

```
dc value of g(t)

ans =
         107.5344
```

Figure 8.4 shows the magnitude and phase spectra of Figure 8.3.

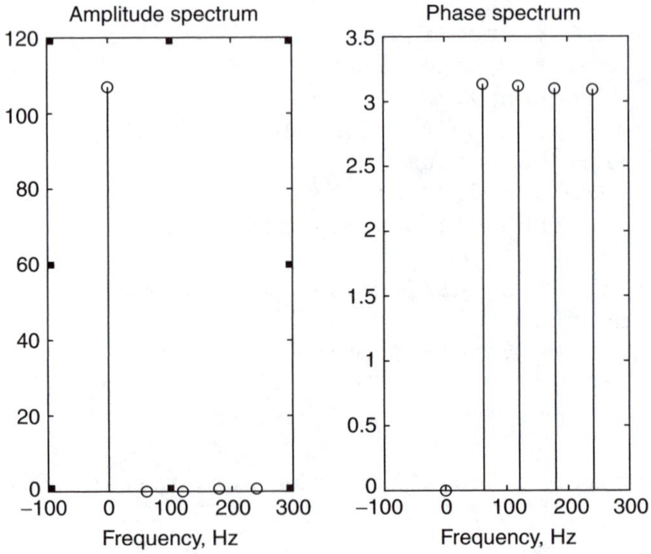

FIGURE 8.4
Magnitude and phase spectra of a full-wave rectification waveform.

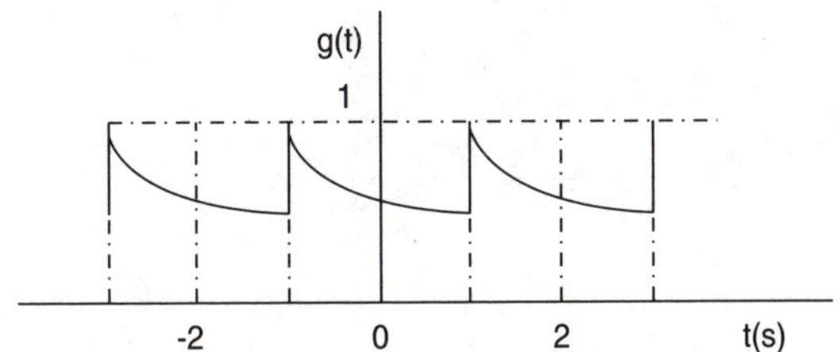

FIGURE 8.5
Periodic exponential signal.

Example 8.3 Synthesis of a Periodic Exponential Signal

The periodic signal shown in Figure 8.5 can be expressed as

$$g(t) = e^{-2t} \qquad -1 \le t < 1$$

$$g(t+2) = g(t)$$

(a) Show that its exponential Fourier series expansion can be expressed as

$$g(t) = \sum_{n=-\infty}^{\infty} \frac{(-1)^n (e^2 - e^{-2})}{2(2 + jn\pi)} \exp(jn\pi t) \tag{8.22}$$

(b) Using a MATLAB program, synthesize $g(t)$ using 20 terms, i.e.,

$$\hat{g}(t) = \sum_{n=-10}^{10} \frac{(-1)^n (e^2 - e^{-2})}{2(2 + jn\pi)} \exp(jn\pi t)$$

Solution

(a)

$$g(t) = \sum_{n=-\infty}^{\infty} c_n \exp(jnw_o t)$$

where

$$c_n = \frac{1}{T_p} \int_{-T_p/2}^{T_p/2} g(t) \exp(-jnw_o t) dt$$

and

$$w_o = \frac{2\pi}{T_p} = \frac{2\pi}{2} = \pi$$

$$c_n = \frac{1}{2}\int_{-1}^{1} \exp(-2t)\exp(-jn\pi t)dt$$

$$c_n = \frac{(-1)^n(e^2 - e^{-2})}{2(2+jn\pi)}$$

thus

$$g(t) = \sum_{n=-\infty}^{\infty} \frac{(-1)^n(e^2 - e^{-2})}{2(2+jn\pi)} \exp(jn\pi t)$$

(b) MATLAB script

```
% synthesis of g(t) using exponential Fourier series
expansion
dt = 0.05;
  tpts = 8.0/dt +1;
cst = exp(2) - exp(-2);
for n = -10:10
  for m = 1:tpts
    g1(n+11,m) = ((0.5*cst*((-1)^n))/
(2+j*n*pi))*(exp(j*n*pi*dt*(m-1)));
  end
end
for m = 1: tpts
 g2 = g1(:,m);
 g3(m) = sum(g2);
end
g = g3';
t = -4:0.05:4.0;
plot(t,abs(g))
xlabel('Time, s')
ylabel('Amplitude')
title('Approximation of g(t)')
```

Figure 8.6 shows the approximation of $g(t)$.

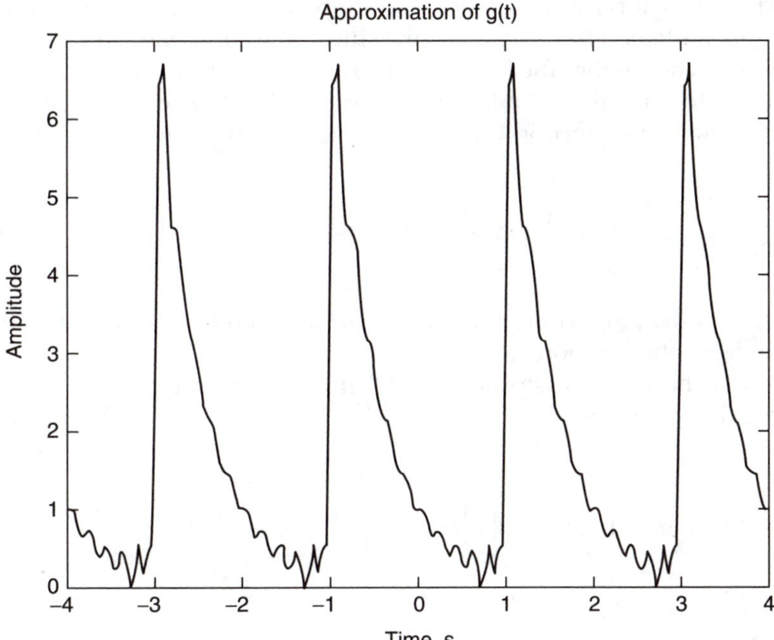

FIGURE 8.6
An approximation of $g(t)$.

8.2 Average Power and Harmonic Distortion

Since the Fourier series expansion decomposes a periodic signal into the sum of sinusoids, the root-mean-squared (rms) value of the periodic signal can be obtained by adding the rms value of each harmonic component vectorally, i.e.,

$$V_{rms} = \sqrt{V_{1,\,rms}^2 + V_{2,\,rms}^2 + V_{3,\,rms}^2 + \quad \ldots \quad + V_{n,\,rms}^2} \tag{8.23}$$

where V_{rms} is the rms value of the periodic signal and $V_{1,rms}$, $V_{2,rms}$, ..., $V_{n,rms}$, are rms values of the harmonic components.

From Equation (8.7) and Equation (8.23), the rms value of $v(t)$ is

$$g_{rms} = \sqrt{A_0^2 + \left(\frac{A_1}{\sqrt{2}}\right)^2 + \left(\frac{A_2}{\sqrt{2}}\right)^2 + \ldots. \quad \left(\frac{A_n}{\sqrt{2}}\right)^2} \tag{8.24}$$

Harmonic distortion can show the discrepancy between an approximation of a signal (obtained from synthesizing sinusoidal components) and its actual waveform. The smaller the harmonic distortion, the more nearly the approximation of the signal resembles the true signal. For the Fourier series expansion, the percent distortion for each individual component is given as

$$\textit{Percent distortion for n-th harmonic} = \frac{A_n}{A_1} * 100 \tag{8.25}$$

where A_n is the amplitude of the n-th harmonic and A_1 is the amplitude of the fundamental harmonic.

The total harmonic distortion (THD) involves all the frequency components and is given as

$$\text{Percent THD} = \sqrt{\left(\frac{A_2}{A_1}\right)^2 + \left(\frac{A_3}{A_1}\right)^2 + \Lambda \left(\frac{A_n}{A_1}\right)^2} * 100\% \tag{8.26}$$

or

$$\text{Percent THD} = \frac{\sqrt{A_2^2 + A_3^2 + \Lambda A_n^2}}{A_1} * 100\% \tag{8.27}$$

If a periodic voltage signal is applied to the input of a linear network and the voltage is given as

$$v(t) = V_0 + \sum_{n=1}^{\infty} V_n \cos(nw_0 t + \phi_V) \tag{8.28}$$

where V_0 is the average or dc value and V_n are the peak amplitudes.

The current flowing into the network may be represented as

$$i(t) = I_0 + \sum_{n=1}^{\infty} I_n \cos(nw_0 t + \phi_I) \tag{8.29}$$

where I_0 is the average or dc value and I_n's are the peak amplitudes.

The average power delivered to the network is given as

$$P_{av} = V_0 I_0 + \frac{1}{2} \sum_{n=1}^{\infty} V_n I_n \cos(\phi_V - \phi_I) \tag{8.30}$$

The following examples illustrate the calculation of total harmonic distortion and average power.

Example 8.4 Total Harmonic Distortion of a Waveform

A full-wave rectifier waveform can be expressed as:

$$g(t) = \frac{10}{\pi} - \frac{20}{\pi}\sum_{n=1}^{\infty}\frac{1}{(4n^2-1)}\cos\left[(2n\pi f_0 t)\right] \tag{8.31}$$

where $f_0 = 1000$ Hz if $g_{app}(t)$ is an approximation of $g(t)$ and

$$g_{app}(t) = \frac{10}{\pi} - \frac{20}{\pi}\sum_{n=1}^{15}\frac{1}{(4n^2-1)}\cos\left[(2n\pi f_0 t)\right] \tag{8.32}$$

Find the rms value and the total harmonic distortion of $g_{app}(t)$.

Solution

MATLAB script

```
% Harmonic Distortion and rms value
% calculate coefficients of harmonics
c0 = 10/pi;
for n = 1: 15
   c(n) = (20/pi)*(1/(4*n*n - 1));
end

% rms calculations
for n = 1: 15
   g1(n) = 0.5*(c(n)^2);
end

g2 = sum(g1)+ c0^2;
grms = sqrt(g2);

% total harmonic calculations
for k = 2:15
   g3(k) = c(k)^2;
end
g4 = sqrt(sum(g3));
thd = (g4/c(1))*100;
fprintf('The rms value is %8.3f\n', grms)
fprintf('The total harmonic distortion is %8.3f
percent\n',thd)
```

The MATLAB results are

```
The rms value is     3.536
The total harmonic distortion is   22.716 percent
```

Example 8.5 Average Power Dissipated in a Load

A voltage that is periodic was applied at the terminals of a 4-Ω resistor. If the voltage is periodic and is given as

$$v(t) = \frac{4}{\pi} \sum_{n=1}^{26} \frac{1}{(2n-1)} \sin\left[(2n-1)2\pi f_0 t\right] \tag{8.33}$$

where f_0 = 2000 Hz, find the average power dissipated in the resistor.

Solution

We calculate the rms value of the voltage. The average power P dissipated in the resistor R will be given by the expression

$$P = \frac{V_{rms}^2}{R}$$

where V_{rms} is the rms value of the voltage $v(t)$.

MATLAB script

```
% Power Calculation
% calculate coefficients of harmonics
c0 = 0.0;
for n = 1: 26
   c(n) = (4/pi)*(1/(2*n - 1));
end

% rms calculations
for n = 1: 26
   g1(n) = 0.5*(c(n)^2);
end

g2 = sum(g1)+ c0^2;
grms = sqrt(g2);

% Average Power Calculation
pwer = (grms^2)/4;
fprintf('The power dissipated in the resistor is %8.3f
Watts\n',pwer)
```

The result from MATLAB is

```
The power dissipated in the resistor is    0.248 Watts
```

8.3 Fourier Transforms

If $g(t)$ is a nonperiodic deterministic signal expressed as a function of time t, then the Fourier transform of $g(t)$ is given by the integral expression

$$G(f) = \int_{-\infty}^{\infty} g(t)\exp(-j2\pi ft)dt \tag{8.34}$$

where $j = \sqrt{-1}$ and f denotes frequency.

The value $g(t)$ can be obtained from the Fourier transform $G(f)$ by the inverse Fourier transform formula,

$$g(t) = \int_{-\infty}^{\infty} G(f)\exp(j2\pi ft)df \tag{8.35}$$

For a signal $g(t)$ to be Fourier transformable, it should satisfy the Dirichlet conditions that were discussed in Section 8.1. If $g(t)$ is continuous and nonperiodic, then $G(f)$ will be continuous and periodic. However, if $g(t)$ is continuous and periodic, then $G(f)$ will be discrete and nonperiodic, that is, if

$$g(t) = g(t \pm nT_p) \tag{8.36}$$

where T_p = period, then the Fourier transform of $g(t)$ is

$$G(f) = \frac{1}{T_p}\sum_{n=-\infty}^{\infty} c_n \delta\left(f - \frac{1}{T_p}\right) \tag{8.37}$$

where

$$c_n = \frac{1}{T_p}\int_{-t_p/2}^{T_p/2} g(t)\exp(-j2\pi nf_o t)dt \tag{8.38}$$

8.3.1 Properties of the Fourier Transform

If $g(t)$ and $G(f)$ are Fourier transform pairs, and they are expressed as

$$g(t) \Leftrightarrow G(f) \tag{8.39}$$

then the Fourier transform will have the following properties:

Linearity

$$ag_1(t) + bg_2(t) \Leftrightarrow aG_1(f) + bG_2(f) \tag{8.40}$$

where a and b are constants.

Time Scaling

$$g(at) \Leftrightarrow \frac{1}{|a|} G\left(\frac{f}{a}\right) \tag{8.41}$$

Duality

$$G(t) \Leftrightarrow g(-f) \tag{8.42}$$

Time Shifting

$$g(t - t_0) \Leftrightarrow G(f)\exp(-j2\pi f t_0) \tag{8.43}$$

Frequency Shifting

$$\exp(j2f_C t)g(t) \Leftrightarrow G(f - f_C) \tag{8.44}$$

Definition in the Time Domain

$$\frac{dg(t)}{dt} \Leftrightarrow j2\pi f G(f) \tag{8.45}$$

Integration in the Time Domain

$$\int_{-\infty}^{t} g(\tau)d\tau \Leftrightarrow \frac{1}{j2\pi f} G(f) + \frac{G(0)}{2}\delta(f) \tag{8.46}$$

Multiplication in the Time Domain

$$g_1(t)g_2(t) \Leftrightarrow \int_{-\infty}^{\infty} G_1(\lambda)G_2(f - \lambda)d\lambda \tag{8.47}$$

Convolution in the Time Domain

$$\int_{-\infty}^{\infty} g_1(\tau)g_2(t-\tau)d\tau \Leftrightarrow G_1(f)G_2(f) \tag{8.48}$$

8.4 Discrete and Fast Fourier Transforms

A Fourier series links a continuous time signal into the discrete-frequency domain. The periodicity of the time-domain signal forces the spectrum to be discrete. The discrete Fourier transform (DFT) of a discrete-time signal $g[n]$ is given as

$$G[k] = \sum_{n=0}^{N-1} g[n]\exp(-j2\pi nk / N) \qquad k = 0, 1, \ldots, N-1 \tag{8.49}$$

The inverse discrete Fourier transform, $g[n]$ is

$$g[n] = \sum_{k=0}^{N-1} G[k]\exp(j2\pi nk / N) \qquad n = 0, 1, \ldots, N-1 \tag{8.50}$$

where N is the number of time sequence values of $g[n]$. It is also the total number frequency sequence values in $G[k]$.

T is the time interval between two consecutive samples of the input sequence $g[n]$. F is the frequency interval between two consecutive samples of the output sequence $G[k]$.

N, T, and F are related by the expression

$$NT = \frac{1}{F} \tag{8.51}$$

NT is also equal to the record length. The time interval, T, between samples should be chosen such that Shannon's sampling theorem is satisfied. This means that T should be less than the reciprocal of $2f_H$, where f_H is the highest significant frequency component in the continuous time signal $g(t)$ from which the sequence $g[n]$ was obtained. Several fast DFT algorithms require N to be an integer power of 2.

A discrete-time function will have a periodic spectrum. In DFT, both the time function and frequency functions are periodic. Because of the periodicity of DFT, it is common to regard points from $n = 1$ through $n = N/2$ as positive and points from $n = N/2$ through $n = N - 1$ as negative frequencies. In addition, since both the time and frequency sequences are periodic, DFT values at points $n = N/2$ through $n = N - 1$ are equal to the DFT values at points $n = N/2$ through $n = 1$.

In general, if the time-sequence is real valued, then the DFTs will have real components that are even and imaginary components that are odd. Similarly, for an imaginary-valued time sequence, each DFT value will have an odd real component and an even imaginary component.

If we define the weighting function W_N as

$$W_N = e^{\frac{-j2\pi}{N}} = e^{-j2\pi FT} \tag{8.52}$$

Then Equation (8.49) and Equation (8.50) can be reexpressed as

$$G[k] = \sum_{n=0}^{N-1} g[n] W_N^{kn} \tag{8.53}$$

and

$$g[n] = \sum_{k=0}^{N-1} G[k] W_N^{-kn} \tag{8.54}$$

The fast Fourier transform, FFT, is an efficient method for computing the discrete Fourier transform. FFT reduces the number of computations needed for computing DFT. For example, if a sequence has N points, and N is an integral power of 2, then DFT requires N^2 operations, whereas FFT requires

$$\frac{N}{2}\log_2(N)$$

complex multiplications,

$$\frac{N}{2}\log_2(N)$$

complex additions, and

$$\frac{N}{2}\log_2(N)$$

subtractions. For $N = 1024$, the computational reduction from DFT to FFT is more than 200 to 1.

The FFT can be used to (a) obtain the power spectrum of a signal, (b) do digital filtering, and (c) obtain the correlation between two signals.

8.4.1 MATLAB Function fft

The MATLAB function for performing fast Fourier transforms is

$$fft(x)$$

where x is the vector to be transformed.

The MATLAB command

$$fft(x,N)$$

can be used to obtain N-point FFT. The vector x is truncated or zeros are added to N, if necessary.

The MATLAB function for performing inverse FFT is

$$ifft(x).$$

$$\left[z_m, z_p\right] = fftplot(x,ts)$$

is used to obtain fft and plot the magnitude z_m and z_p of DFT of x. The sampling interval is *ts*. Its default value is unity. The spectra are plotted vs. the digital frequency *F*. The following three examples illustrate usage of the MATLAB function fft.

Example 8.6 DFT and FFT of a Sequence

Given the sequence $x[n] = (1, 2, 1)$:

(a) Calculate the DFT of $x[n]$.

(b) Use the FFT algorithm to find DFT of $x[n]$.

(c) Compare the results of (a) and (b).

Solution

(a) From Equation (8.53),

$$G[k] = \sum_{n=0}^{N-1} g[n] W_N^{kn}$$

From Equation (8.52),

$$W_3^0 = 1$$

$$W_3^1 = e^{-\frac{j2\pi}{3}} = -0.5 - j0.866$$

$$W_3^2 = e^{-\frac{j4\pi}{3}} = -0.5 + j0.866$$

$$W_3^3 = W_3^0 = 1$$

$$W_3^4 = W_3^1$$

Using Equation (8.53), we have

$$G[0] = \sum_{n=0}^{2} g[n]W_3^0 = 1 + 2 + 1 = 4$$

$$G[1] = \sum_{n=0}^{2} g[n]W_3^n = g[0]W_3^0 + g[1]W_3^1 + g[2]W_3^2$$

$$= 1 + 2(-0.5 - j0.866) + (-0.5 + j0.866) = -0.5 - j0.866$$

$$G[2] = \sum_{n=0}^{2} g[n]W_3^{2n} = g[0]W_3^0 + g[1]W_3^2 + g[2]W_3^4$$

$$= 1 + 2(-0.5 + j0.866) + (-0.5 - j0.866) = -0.5 + j0.866$$

(b) The MATLAB program for performing the DFT of $x[n]$ is:

MATLAB script

```
% Program for performing DFT of x[n]
x = [1 2 1];
xfft = fft(x)
```

The results are

```
xfft =
        4.0000    -0.5000 - 0.8660i      -0.5000 + 0.8660i
```

(c) It can be seen that the answers obtained from parts (a) and (b) are identical.

Example 8.7 Fourier Transform and DFT of a Damped Exponential Sinusoid

Signal $g(t)$ is given as

$$g(t) = 4e^{-2t} \cos\left[2\pi(10)t\right] u(t) \tag{8.55}$$

(a) Find the Fourier transform of $g(t)$, i.e., $G(f)$.

(b) Find the DFT of $g(t)$ when the sampling interval is 0.05 sec with $N =$ 1000.

(c) Find the DFT of $g(t)$ when the sampling interval is 0.2 sec with $N =$ 250.

(d) Compare the results obtained from parts a, b, and c.

Solution

(a) $g(t)$ can be expressed as

$$g(t) = 4e^{-2t}\left[\frac{1}{2}e^{j20\pi t} + \frac{1}{2}e^{-j20\pi t}\right]u(t) \tag{8.56}$$

Using the frequency shifting property of the Fourier transform, we get

$$G(f) = \frac{2}{2 + j2\pi(f - 10)} + \frac{2}{2 + j2\pi(f + 10)} \tag{8.57}$$

(b, c) The MATLAB program for computing the DFT of $g(t)$ is:

MATLAB script

```
% DFT of g(t)
%  Sample 1, Sampling interval of 0.05 s
ts1 = 0.05;    % sampling interval
fs1 = 1/ts1;   %  Sampling frequency
n1 = 1000;      %  Total Samples
m1 = 1:n1;     %  Number of bins
sint1 = ts1*(m1 - 1);  %  Sampling instants
freq1 = (m1 - 1)*fs1/n1;  % frequencies
gb = (4*exp(-2*sint1)).*cos(2*pi*10*sint1);
gb_abs = abs(fft(gb));
subplot(121)
plot(freq1, gb_abs)
title('DFT of g(t), 0.05s Sampling interval')
xlabel('Frequency (Hz)')
```

```
% Sample 2, Sampling interval of 0.2 s
ts2 = 0.2;    % sampling interval
fs2 = 1/ts2;  %  Sampling frequency
n2 = 250;     %  Total Samples
m2 = 1:n2;    %  Number of bins
sint2 = ts2*(m2 - 1);  %  Sampling instants
freq2 = (m2 - 1)*fs2/n2;  % frequencies
gc = (4*exp(-2*sint2)).*cos(2*pi*10*sint2);
gc_abs = abs(fft(gc));
subplot(122)
plot(freq2, gc_abs)
title('DFT of g(t), 0.2s Sampling interval')
xlabel('Frequency (Hz)')
```

(d) The two plots are shown in Figure 8.7. From Figure 8.7, it can be seen that with the sample interval of 0.05 sec, there is no aliasing, and the spectrum of $G[k]$ in part (b) is almost the same as that of $G(f)$ of part (a). With a sampling interval of 0.2 sec (less than the Nyquist rate), there is aliasing and the spectrum of $G[k]$ is different from that of $G(f)$.

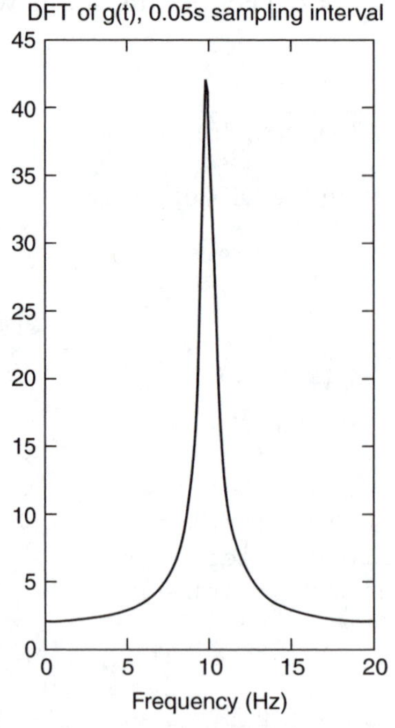

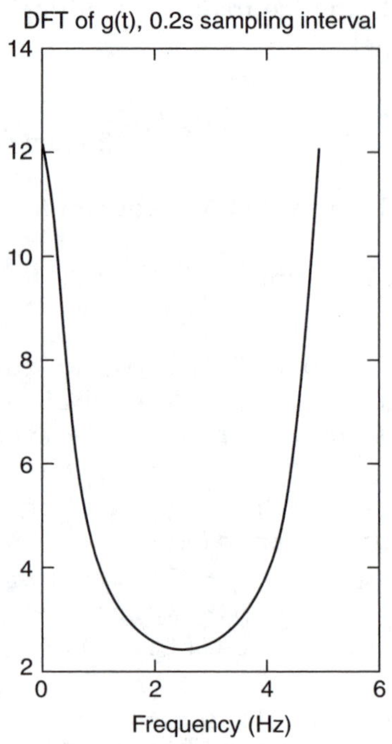

FIGURE 8.7
DFT of $g(t)$.

Example 8.8 Power Spectral Density of a Noisy Signal

Given a noisy signal

$$g(t) = \sin(2\pi f_1 t) + 0.5n(t) \tag{8.58}$$

where

$$f_1 = 100 \text{ Hz}$$

$n(t)$ is a normally distributed white noise. The duration of $g(t)$ is 0.5 sec. Use the MATLAB function randn to generate the noisy signal. Use MATLAB to obtain the power spectral density of $g(t)$.

Solution

A representative program that can be used to plot the noisy signal and obtain the power spectral density is:

MATLAB script

```
% power spectral estimation of noisy signal
t = 0.0:0.002:0.5;
f1 =100;

% generate the sine portion of signal
x = sin(2*pi*f1*t);

% generate a normally distributed white noise
n = 0.5*randn(size(t));

% generate the noisy signal
y = x+n;
subplot(211), plot(t(1:50),y(1:50)),
title('Noisy Time-domain Signal')

% power spectral estimation is done
yfft = fft(y,256);
len = length(yfft);
pyy = yfft.*conj(yfft)/len;
f = (500./256)*(0:127);

subplot(212), plot(f,pyy(1:128)),
title('Power Spectral Density'),
xlabel('Frequency in Hz')
```

The plot of the noisy signal and its spectrum is shown in Figure 8.8. The amplitude of the noise and the sinusoidal signal can be changed to observe their effects on the spectrum.

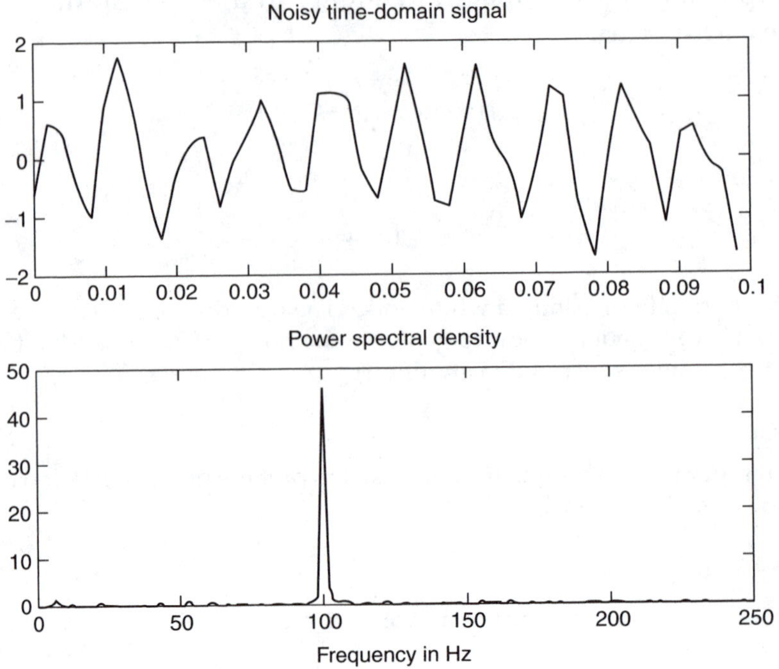

FIGURE 8.8
Noisy signal and its spectrum.

Bibliography

1. Alexander, C.K. and Sadiku, M.N.O., *Fundamentals of Electric Circuits*, 2nd ed., McGraw-Hill, New York, 2004.
2. Attia, J.O., *PSPICE and MATLAB for Electronics: An Integrated Approach*, CRC Press, Boca Raton, FL, 2002.
3. Biran, A. and Breiner, M., *MATLAB for Engineers*, Addison-Wesley, Reading, MA, 1995.
4. Chapman, S.J., *MATLAB Programming for Engineers*, Brook, Cole Thompson Learning, Pacific Grove, CA, 2000.
5. Dorf, R.C. and Svoboda, J.A., *Introduction to Electric Circuits*, 3rd ed., John Wiley & Sons, New York, 1996.
6. Etter, D.M., *Engineering Problem Solving with MATLAB*, 2nd ed., Prentice Hall, Upper Saddle River, NJ, 1997.
7. Etter, D.M., Kuncicky, D.C., and Hull, D., *Introduction to MATLAB 6*, Prentice Hall, Upper Saddle River, NJ, 2002.
8. Gottling, J.G., *Matrix Analysis of Circuits Using MATLAB*, Prentice Hall, Englewood Cliffs, NJ, 1995.
9. Johnson, D.E., Johnson, J.R., and Hilburn, J.L., *Electric Circuit Analysis*, 3rd ed., Prentice Hall, Upper Saddle River, NJ, 1997.

10. Meader, D.A., *Laplace Circuit Analysis and Active Filters*, Prentice Hall, Upper Saddle River, NJ, 1991.
11. Sigmor, K., *MATLAB Primer*, 4th ed., CRC Press, Boca Raton, FL, 1998.
12. *Using MATLAB, The Language of Technical Computing, Computation, Visualization, Programming*, Version 6, MathWorks, Inc., Natick, MA, 2000.
13. Vlach, J.O., Network Theory and CAD, *IEEE Trans. on Education*, 36, 23–27, 1993.

Problems

Problem 8.1

The triangular waveform shown in Figure P8.1 can be expressed as

$$g(t) = \frac{8A}{\pi^2} \sum_{n=1}^{\infty} \frac{(-1)^{n+1}}{4n^2 - 1} \cos\big((2n-1)w_0 t\big)$$

where

$$f_0 = \frac{1}{T_p} \text{ and } w_0 = 2\pi f_0$$

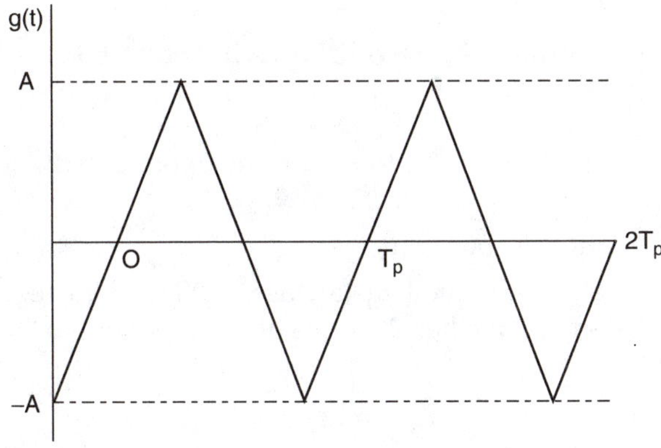

FIGURE P8.1
Triangular waveform.

If $A = 1$, $T_p = 8$ msec, and the sampling interval is 0.1 msec:

(a) Write a MATLAB program to resynthesize $g(t)$ if 20 terms are used.

(b) What is the root-mean-squared value of the function that is the difference between $g(t)$ and the approximation to $g(t)$ when 20 terms are used for the calculation of $g(t)$?

Problem 8.2

A periodic pulse train $g(t)$ is shown in Figure P8.2.

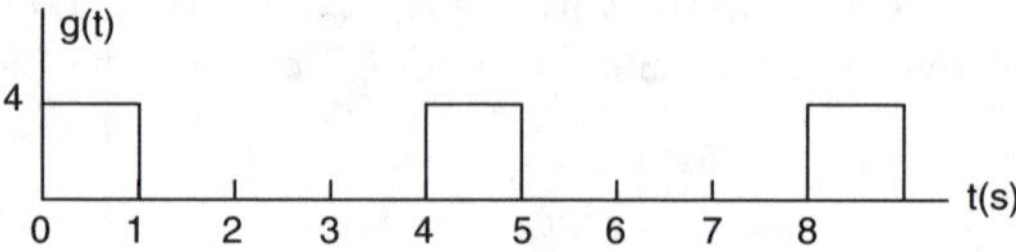

FIGURE P8.2
Periodic pulse train.

If $g(t)$ can be expressed by Equation (8.3):

(a) Derive expressions for determining the Fourier series coefficients a_n and b_n.
(b) Write a MATLAB program to obtain a_n and b_n for $n = 0, 1, \ldots, 10$ by using Equation (8.5) and Equation (8.6).
(c) Resynthesize $g(t)$ using 10 terms of the values a_n and b_n obtained from part (b).

Problem 8.3

For the half-wave rectifier waveform shown in Figure P8.3, with a period of 0.01 sec and a peak voltage of 17 V:

(a) Write a MATLAB program to obtain the exponential Fourier series coefficients c_n for $n = 0, 1, \ldots, 20$.
(b) Plot the amplitude spectrum.
(c) Using the values obtained in (a), use MATLAB to regenerate the approximation to $g(t)$ when 20 terms of the exponential Fourier series are used.

Problem 8.4

Figure P8.4(a) is a periodic triangular waveform.

(a) Derive the Fourier series coefficients a_n and b_n.
(b) With the signal $v(t)$ of the circuit shown in P8.4(b), derive the expression for the current $i(t)$.

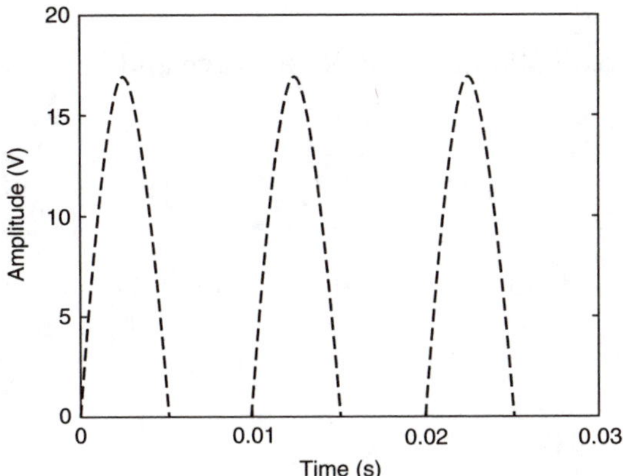

FIGURE P8.3
Half-wave rectifier waveform.

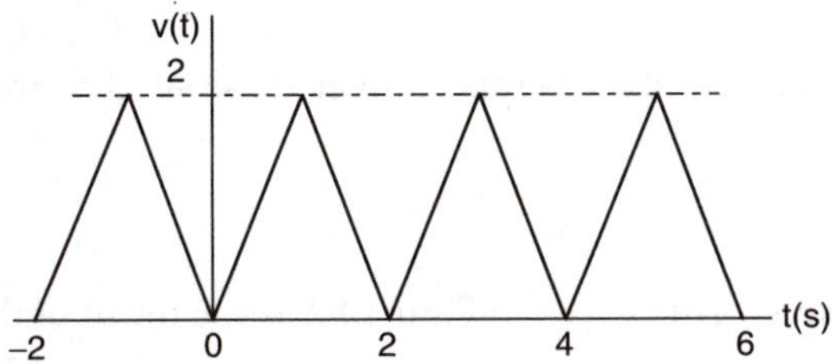

FIGURE P8.4(a)
Periodic triangular waveform.

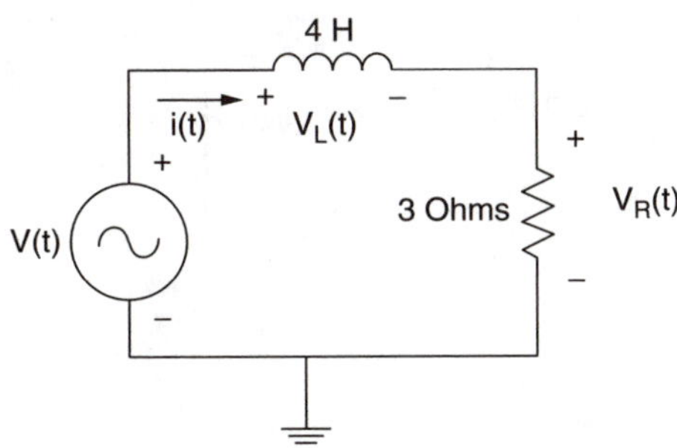

FIGURE P8.4(b)
Simple RL circuit.

Problem 8.5

$g_{app}(t)$ is an approximation of a sawtooth wave and

$$g_{app}(t) = 4 - \frac{8}{\pi}\sum_{n=1}^{20} \frac{1}{n}\sin\left[(2n\pi f_0 t)\right]$$

where $f_0 = 800$ Hz. Find the rms value and the total harmonic distortion of $g_{app}(t)$.

Problem 8.6

A voltage that is periodic was applied at the terminals of a 5-Ω resistor. If the voltage is given as

$$v(t) = 2.5 + \sum_{n=1}^{26} \frac{5}{n}\sin\left(\frac{n\pi}{4}\right)\cos\left[(2n\pi f_0 t)\right]$$

where $f_0 = 400$ Hz. Find the average power dissipated in the resistor.

Problem 8.7

If the periodic waveform shown in Figure 8.5 is the input of the circuit shown in Figure P8.5:

(a) Derive the mathematical expression for $v_C(t)$.
(b) Use MATLAB to plot the signals $g(t)$ and $v_C(t)$.

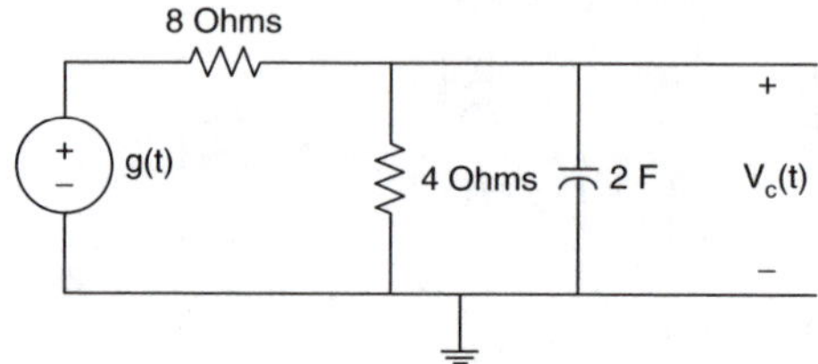

FIGURE P8.5
RC circuit.

Problem 8.8

The unit sample response of a filter is given as

$$h[n] = \begin{pmatrix} 0 & -1 & -1 & 0 & 1 & 1 & 0 \end{pmatrix}$$

(a) Find the discrete Fourier transform of $h[n]$; assume that the values of $h[n]$ not shown are zero.

(b) If the input to the filter is

$$x[n] = \sin\left[\frac{n}{8}\right]u[n]$$

find the output of the filter.

Problem 8.9

$$g(t) = \sin(200\pi t) + \sin(400\pi t)$$

(a) Generate 512 points of $g(t)$. Using the FFT algorithm, generate and plot the frequency content of $g(t)$. Assume a sampling rate of 1200 Hz. Find the power spectrum.

(b) Verify that the frequencies in $g(t)$ are observable in the FFT plot.

Problem 8.10

Find the DFT of

$$g(t) = e^{-5t}u(t)$$

(a) Find the Fourier transform of $g(t)$.

(b) Find the DFT of $g(t)$ using the sampling interval of 0.01 sec and time duration of 5 sec.

(c) Compare the results obtained from parts (a) and (b).

9

Diodes

In this chapter, the characteristics of diodes are presented. Diode circuit analysis techniques will be discussed. Problems involving diode circuits are solved using MATLAB.

9.1 Diode Characteristics

A diode is a two-terminal device. The electronic symbol of a diode is shown in Figure 9.1(a). Ideally, the diode conducts current in one direction. The current vs. voltage characteristics of an ideal diode are shown in Figure 9.1(b).

The current-voltage (I-V) characteristic of a semiconductor junction diode is shown in Figure 9.2. The characteristic is divided into three regions: forward-biased, reversed-biased, and the breakdown.

In the forward-biased and reversed-biased regions, the current, *i*, and the voltage, *v*, of a semiconductor diode are related by the diode equation

$$i = I_S\left[e^{(v/nV_T)} - 1\right] \tag{9.1}$$

where

I_S is reverse saturation current or leakage current
n is an empirical constant between 1 and 2
V_T is thermal voltage, given by

$$V_T = \frac{kT}{q} \tag{9.2}$$

and

k is Boltzmann's constant = 1.38×10^{-23} J/K
q is the electronic charge = 1.6×10^{-19} C
T is the absolute temperature in K

At room temperature (25°C), the thermal voltage is about 25.7 mV.

(a)

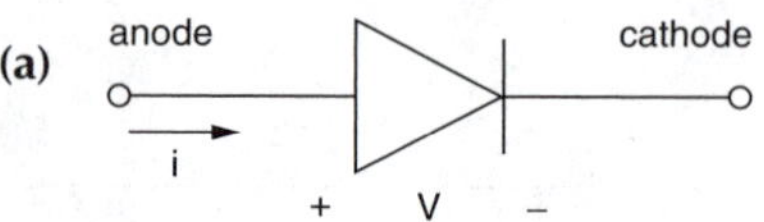

(b)

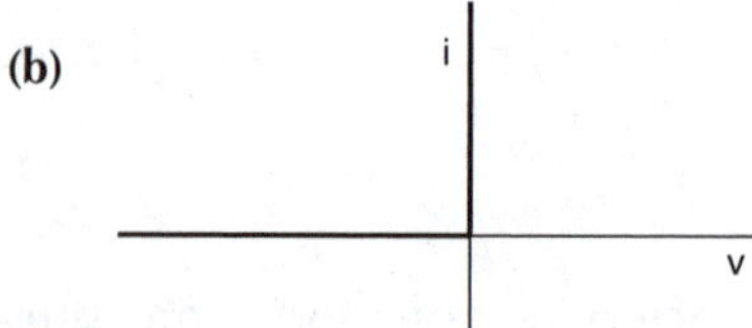

FIGURE 9.1
Ideal diode. (a) Electronic symbol; (b) I-V characteristics.

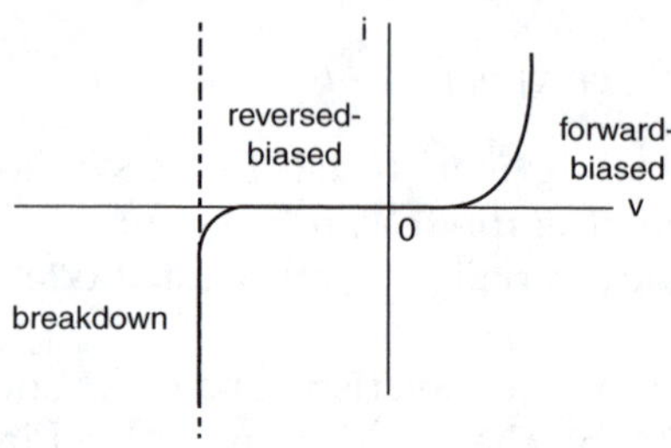

FIGURE 9.2
I-V characteristics of a semiconductor junction diode.

9.1.1 Forward-Biased Region

In the forward-biased region, the voltage across the diode is positive. If we assume that the voltage across the diode is greater than 0.1 V at room temperature, then Equation (9.1) simplifies to

$$i = I_S e^{(v/nV_T)} \tag{9.3}$$

For a particular operating point of the diode ($i = I_D$ and $v = V_D$), we have

$$i_D = I_S e^{(v_D/nV_T)} \tag{9.4}$$

To obtain the dynamic resistance of the diode at a specified operating point, we differentiate Equation (9.3) with respect to v, and we have

$$\frac{di}{dv} = \frac{I_s e^{(v/nV_T)}}{nV_T}$$

$$\left.\frac{di}{dv}\right|_{v=V_D} = \frac{I_s e^{(v_D/nV_T)}}{nV_T} = \frac{I_D}{nV_T}$$

and the dynamic resistance of the diode, r_d, is

$$r_d = \left.\frac{dv}{di}\right|_{v=V_D} = \frac{nV_T}{I_D} \tag{9.5}$$

From Equation (9.3), we have

$$\frac{i}{I_S} = e^{(v/nV_T)}$$

thus

$$\ln(i) = \frac{v}{nV_T} + \ln(I_S) \tag{9.6}$$

Equation (9.6) can be used to obtain the diode constants n and I_S, given the data that consists of the corresponding values of voltage and current. From Equation (9.6), a curve of v vs. $\ln(i)$ will have a slope given by

$$\frac{1}{nV_T}$$

and y-intercept of $\ln(I_S)$. The following example illustrates how to find n and I_S from experimental data. Since the example requires curve fitting, the MATLAB function **polyfit** will be covered before doing the example.

9.1.2 MATLAB Function polyfit

The **polyfit** function is used to compute the best fit of a set of data points to a polynomial with a specified degree. The general form of the function is

$$coeff\_xy = polyfit(x, y, n) \tag{9.7}$$

where

x and y are the data points.

n is the nth degree polynomial that will fit the vectors x and y.

coeff_xy is a polynomial that fits the data in vector y to x in the least square sense. *coeff_xy* returns n+1 coefficients in descending powers of x.

Thus, if the polynomial fit to data in vectors x and y is given as

$$coeff\_xy(x) = c_1 x^n + c_2 x^{n-1} + \ldots + c_m$$

The degree of the polynomial is n, the number of coefficients is $m = n + 1$, and the coefficients (c_1, c_2, ..., c_m) are returned by the MATLAB polyfit function.

Example 9.1 Determination of Diode Parameters from Data

A forward-biased diode has the following corresponding voltage and current. Use MATLAB to determine the reverse saturation current, I_S and diode parameter, n.

Forward Voltage, V	Forward Current, A
0.1	0.133E-12
0.2	1.79E-12
0.3	24.02E-12
0.4	0.321E-9
0.5	4.31E-9
0.6	57.69E-9
0.7	7.726E-7

Solution

MATLAB script

```
% Diode  parameters
vt = 25.67e-3;
v = [0.1 0.2 0.3 0.4 0.5 0.6 0.7];
i = [0.133e-12 1.79e-12 24.02e-12 321.66e-12 4.31e-9
57.69e-9 772.58e-9];

%
lni = log(i); % Natural log of current

% Coefficients of Best fit linear model is obtained
p_fit = polyfit(v,lni,1);

% linear equation is   y = m*x + b
b = p_fit(2);
m = p_fit(1);
ifit = m*v + b;

% Calculate Is and n
Is = exp(b)
n = 1/(m*vt)

% Plot v versus ln(i), and best fit linear model
plot(v,ifit,'b', v, lni,'ob')
```

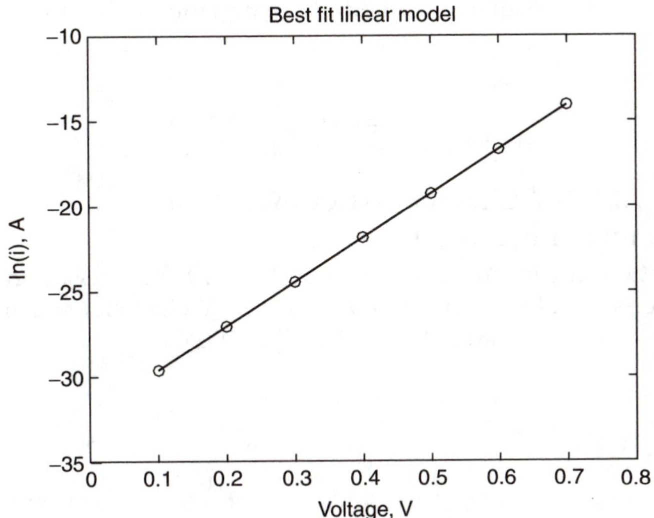

FIGURE 9.3
Best fit linear model of voltage vs. natural logarithm of current.

```
axis([0,0.8,-35,-10])
xlabel('Voltage, V')
ylabel('ln(i), A')
title('Best Fit Linear Model')
```

The results obtained from MATLAB are

```
Is  =   9.9525e-015

n   =   1.5009
```

Figure 9.3 shows the best fit linear model used to determine the reverse saturation current, I_S, and diode parameter, n.

9.1.3 Temperature Effects

From the diode equation [Equation (9.1)], the thermal voltage and the reverse saturation current are temperature dependent. The thermal voltage is directly proportional to temperature. This is expressed in Equation (9.2). The reverse saturation current I_S increases approximately 7.2%/°C for both silicon and germanium diodes. The expression for the reverse saturation current as a function of temperature is

$$I_S(T_2) = I_S(T_1)e^{[k_S(T_2-T_1)]} \tag{9.8}$$

where

$k_S = 0.072/°\text{C}$.

T_1 and T_2 are two different temperatures.

Since $e^{0.72}$ is approximately equal to two, Equation (9.8) can be simplified and rewritten as

$$I_S(T_2) = I_S(T_1)2^{(T_2-T_1)/10} \tag{9.9}$$

Example 9.2 I-V Characteristics of a Diode at Different Temperatures

The saturation current of a diode at 25°C is 10^{-12} A. Assuming that the emission constant of the diode is 1.9, plot the I-V characteristic of the diode at the following temperatures: T_1 = 0°C, T_2 = 100°C.

Solution

MATLAB script

```
% Temperature effects on diode characteristics
%
k = 1.38e-23; q = 1.6e-19;
t1 = 273 + 0;
t2 = 273 + 100;

ls1 = 1.0e-12;
ks = 0.072;
ls2 = ls1*exp(ks*(t2-t1))
v = 0.45:0.01:0.7;
l1 = ls1*exp(q*v/(k*t1));
l2 = ls2*exp(q*v/(k*t2));

plot(v,l1,'wo',v,l2,'w+')
axis([0.45,0.75,0,10])
title('Diode I-V Curve at two Temperatures')
xlabel('Voltage (V)')
ylabel('Current (A)')
text(0.5,8,'o is for 100 degrees C')
text(0.5,7, '+ is for 0 degree C')
```

Figure 9.4 shows the temperature effects of the diode forward characteristics.

9.2 Analysis of Diode Circuits

Figure 9.5 shows a diode circuit consisting of a dc source V_{DC}, resistance R, and a diode. We want to determine the diode current I_D and the diode voltage V_D.

Using the Kirchhoff voltage law, we can write the loadline equation

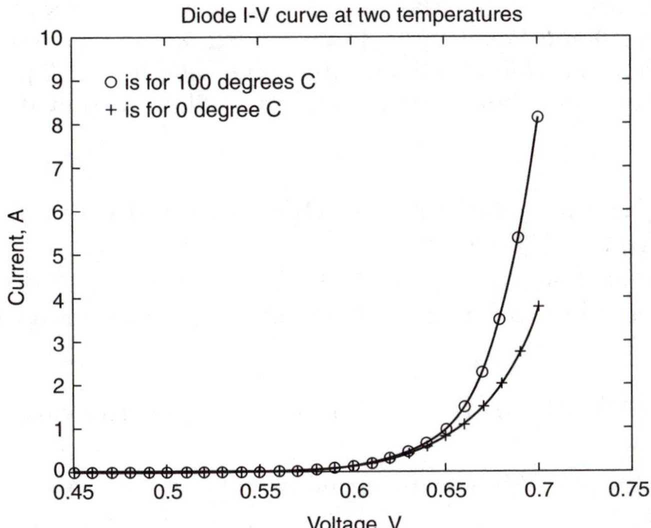

FIGURE 9.4
Temperature effects on the diode forward characteristics.

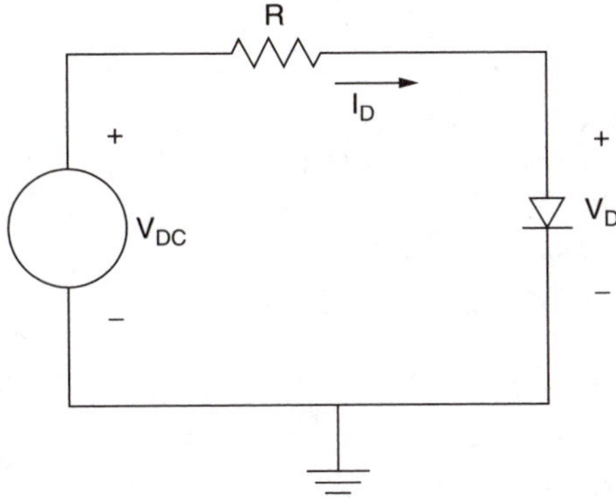

FIGURE 9.5
Basic diode circuit.

$$V_{DC} = RI_D + V_D \tag{9.10}$$

The diode current and voltage will be related by the diode equation

$$i_D = I_S e^{(v_D/nV_T)} \tag{9.11}$$

Equation (9.10) and Equation (9.11) can be used to solve for the current I_D and voltage V_D.

There are several approaches for solving I_D and V_D. In one approach, Equation (9.10) and Equation (9.11) are plotted and the intersection of the linear curve of Equation (9.10) and the nonlinear curve of Equation (9.11) will be the operating point of the diode. This is illustrated by the following example.

Example 9.3 Determining the Operating Point of a Diode by Using Graphical Techniques

For the circuit shown in Figure 9.5, if R = 10 kΩ, V_{DC} = 10 V, the reverse saturation current of the diode is 10^{-12}A, and n = 2.0 (assume a temperature of 25°C):

(a) Use MATLAB to plot the diode forward characteristic curve and the loadline.

(b) From the plot, estimate the operating point of the diode.

Solution

MATLAB script

```
% Determination of operating point using
% graphical technique
%
% diode equation
k = 1.38e-23;q = 1.6e-19;
t1 = 273 + 25; vt = k*t1/q;
v1 = 0.25:0.05:1.1;
i1 = 1.0e-12*exp(v1/(2.0*vt));

% load line 10=(1.0e4)i2 + v2
vdc = 10;
r = 1.0e4;
v2 = 0:2:10;
i2 = (vdc - v2)/r;

% plot
plot(v1,i1,'b', v2,i2,'b')
axis([0,2, 0, 0.0015])
title('Graphical Method - Operating Point')
xlabel('Voltage, V')
ylabel('Current, A')
text(0.4,1.05e-3,'Loadline')
text(1.08,0.3e-3,'Diode curve')
```

Figure 9.6 shows the intersection of the diode forward characteristics and the loadline.

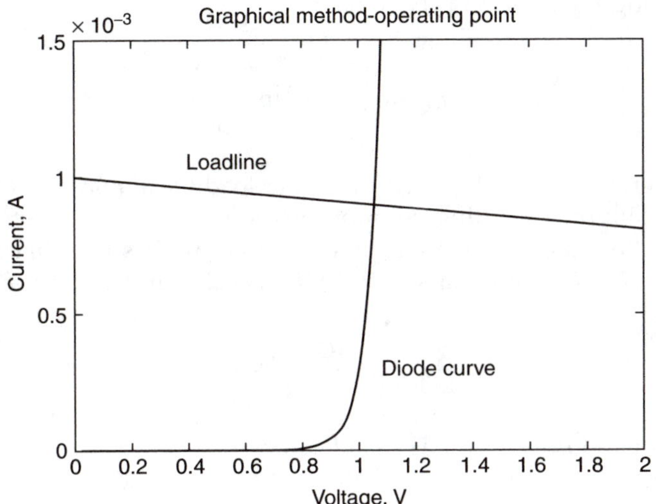

FIGURE 9.6
Loadline and diode forward characteristics.

From Figure 9.6, the operating point of the diode is the intersection of the loadline and the diode forward characteristic curve. The operating point is approximately

$$I_D = 0.9 \text{ mA}$$

$$V_D = 0.7 \text{ V}$$

The second approach for obtaining the diode current I_D and diode voltage V_D of Figure 9.5 is to use iteration. Assume that (I_{D1}, V_{D1}) and (I_{D2}, V_{D2}) are two corresponding points on the diode forward characteristics. Then, from Equation (9.3), we have

$$i_{D1} = I_S e^{(v_{D1}/nV_T)} \tag{9.12}$$

$$i_{D2} = I_S e^{(v_{D2}/nV_T)} \tag{9.13}$$

Dividing Equation (9.13) by Equation (9.12), we have

$$\frac{I_{D2}}{I_{D1}} = e^{(V_{D2} - V_{D1}/nV_T)} \tag{9.14}$$

Simplifying Equation (9.14), we have

$$v_{D2} = v_{D1} + nV_T \ln\left(\frac{I_{D2}}{I_{D1}}\right) \tag{9.15}$$

Using iteration, Equation (9.15) and the loadline Equation (9.10) can be used to obtain the operating point of the diode.

To show how the iterative technique is used, we assume that $I_{D1} = 1$ mA and $V_{D1} = 0.7$ V. Using Equation (9.10), I_{D2} is calculated by

$$I_{D2} = \frac{V_{DC} - V_{D1}}{R} \tag{9.16}$$

Using Equation (9.15), V_{D2} is calculated by

$$V_{D2} = V_{D1} + nV_T \ln\left(\frac{I_{D2}}{I_{D1}}\right) \tag{9.17}$$

Using Equation (9.10), I_{D3} is calculated by

$$I_{D3} = \frac{V_{DC} - V_{D2}}{R} \tag{9.18}$$

Using Equation (9.15), V_{D3} is calculated by

$$V_{D3} = V_{D1} + nV_T \ln\left(\frac{I_{D3}}{I_{D1}}\right) \tag{9.19}$$

Similarly, I_{D4} and V_{D4} are calculated by

$$I_{D4} = \frac{V_{DC} - V_{D3}}{R} \tag{9.20}$$

$$V_{D4} = V_{D1} + nV_T \ln\left(\frac{I_{D4}}{I_{D1}}\right) \tag{9.21}$$

The iteration is stopped when V_{Dn} is approximately equal to V_{Dn-1} or I_{Dn} is approximately equal to I_{Dn-1} to the desired decimal points. The iteration technique is particularly facilitated by using computers. Example 9.4 illustrates the use of MATLAB for doing the iteration technique.

Example 9.4 Determining the Operating Point of a Diode by Using the Iterative Technique

Redo Example 9.3 using the iterative technique. The iteration can be stopped when the current and previous value of the diode voltage are different by 10^{-7} volts.

Solution

MATLAB script

```
% Determination of diode operating point using
% iterative method
k = 1.38e-23;q = 1.6e-19;
t1 = 273 + 25; vt = k*t1/q;
vdc = 10;
r = 1.0e4;
n = 2;
id(1) = 1.0e-3; vd(1) = 0.7;
reltol = 1.0e-7;
i = 1;
vdiff = 1;
while vdiff > reltol
  id(i+1) = (vdc - vd(i))/r;
  vd(i+1) = vd(i) + n*vt*log(id(i+1)/id(i));
  vdiff = abs(vd(i+1) - vd(i));
  i = i+1;
end
k = 0:i-1;
% operating point of diode is (vdiode, idiode)
idiode = id(i)
vdiode = vd(i)
% Plot the voltages during iteration process
plot(k, vd, k,vd,'bo')
axis([-1,5,0.6958,0.701])
title('Diode Voltage during Iteration')
xlabel('Iteration Number')
ylabel('Voltage, V')
```

From the MATLAB program, we have

```
idiode =    9.3037e-004

vdiode =    0.6963
```

Thus $I_D = 0.9304$ mA and $V_D = 0.6963$ V. Figure 9.7 shows the diode voltage during the iteration process.

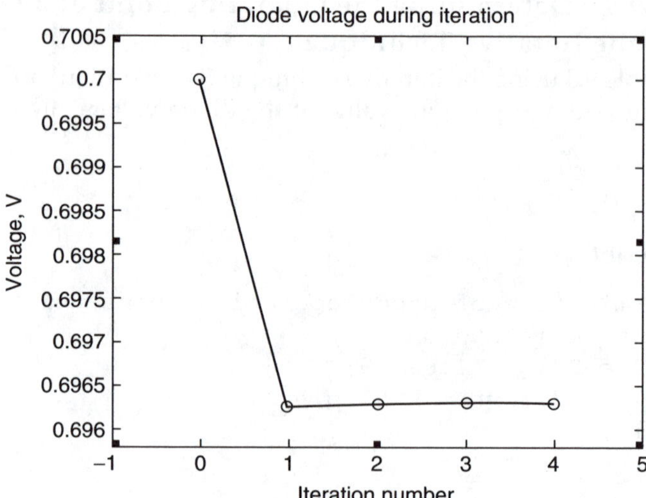

FIGURE 9.7
Diode voltage during iteration process.

9.3 Half-Wave Rectifier

A half-wave rectifier circuit is shown in Figure 9.8. It consists of an alternating current (ac) source, a diode, and a resistor. Assuming that the diode is ideal, the diode conducts when source voltage is positive, making

$$v_0 = v_S \text{ when } v_S \geq 0 \tag{9.22}$$

When the source voltage is negative, the diode is cut off, and the output voltage is

$$v_0 = 0 \text{ when } v_S < 0 \tag{9.23}$$

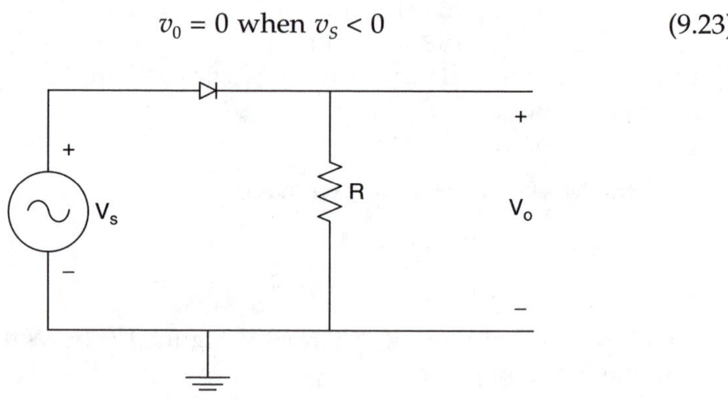

FIGURE 9.8
Half-wave rectifier circuit.

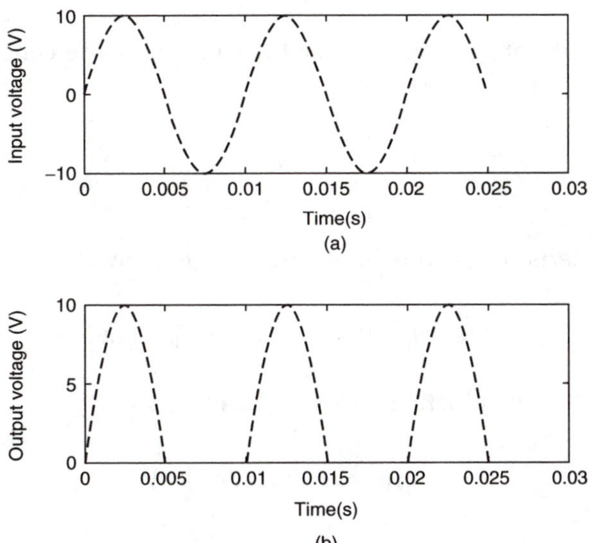

FIGURE 9.9
(a) Input and (b) output waveforms of a half-wave rectifier circuit.

Figure 9.9 shows the input and output waveforms when the input signal is a sinusoidal signal.

The battery charging circuit, explored in the following example, consists of a source connected to a battery through a resistor and a diode.

Example 9.5 Battery Charging Circuit — Current, Conduction Angle, and Peak Current

A battery charging circuit is shown in Figure 9.10. The battery voltage is $V_B = 11.8$ V. The source voltage is $v_S(t) = 18\sin(120\pi t)$ V and the resistance is $R = 100\ \Omega$. Use MATLAB (a) to sketch the input voltage, (b) to plot the current flowing through the diode, (c) to calculate the conduction angle of the diode, and (d) to calculate the peak current. (Assume that the diode is ideal.)

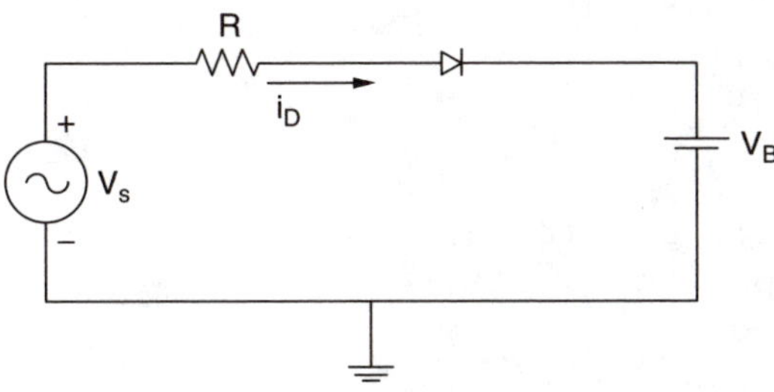

FIGURE 9.10
A battery charging circuit.

Solution

When the input voltage v_S is greater than V_B, the diode conducts and the diode current, i_d, is given as

$$i_d = \frac{V_S - V_B}{R} \tag{9.24}$$

The diode starts conducting at an angle θ, given by $v_S \geq V_B$, i.e.,

$$18\sin\theta_1 = 18\sin(120\pi t_1) = V_B = 11.8$$

The diode stops conducting current when $v_s \leq V_B$

$$18\sin\theta_2 = 18\sin(120\pi t_2) = V_B$$

due to the symmetry

$$\theta_2 = \pi - \theta_1$$

MATLAB script

```
% Battery charging circuit
period = 1/60;
 period2 = period*2;
inc =period/100;
npts = period2/inc;
vb = 11.8;

t = [];
for i = 1:npts
  t(i) = (i-1)*inc;
  vs(i) = 18*sin(120*pi*t(i));
    if vs(i) > vb
     idiode(i) = (vs(i) -vb)/r;
    else
     idiode(i) = 0;
    end
end

subplot(211), plot(t,vs)
%title('Input Voltage')
xlabel('Time, s')
ylabel('Voltage, V')
text(0.027,10, 'Input Voltage')
subplot(212), plot(t,idiode)
```

```
%title('Diode Current')
xlabel('Time, s')
ylabel('Current, A')
text(0.027, 0.7e-3, 'Diode Current')

% conduction angle
theta1 = asin(vb/18); theta2 = pi - theta1;
acond = (theta2 -theta1)/(2*pi)
% peak current
pcurrent = (18*sin(pi/2) - vb)/r
% pcurrent = max(idiode)
```

The conduction angle, acond, and the peak current, pcurrent, are

```
acond =    0.2724

pcurrent =   0.0620
```

Figure 9.11 shows the input voltage and diode current.

The output of the half-wave rectifier circuit of Figure 9.8 can be smoothed by connecting a capacitor across the load resistor. The smoothing circuit is shown in Figure 9.12.

When the amplitude of the source voltage V_S is greater than the output voltage, the diode conducts and the capacitor is charged. When the source voltage becomes less than the output voltage, the diode is cut off and the capacitor discharges with the time constant *CR*. The output voltage and the diode current waveforms are shown in Figure 9.13.

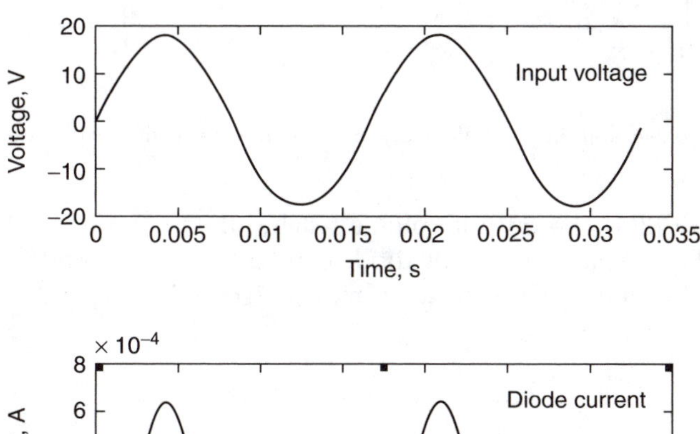

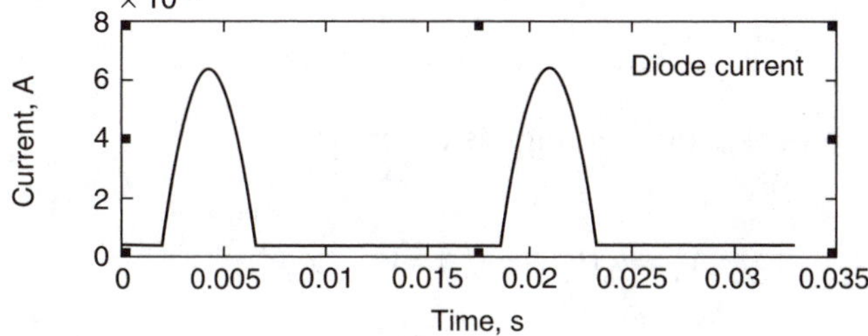

FIGURE 9.11
Input voltage and diode current.

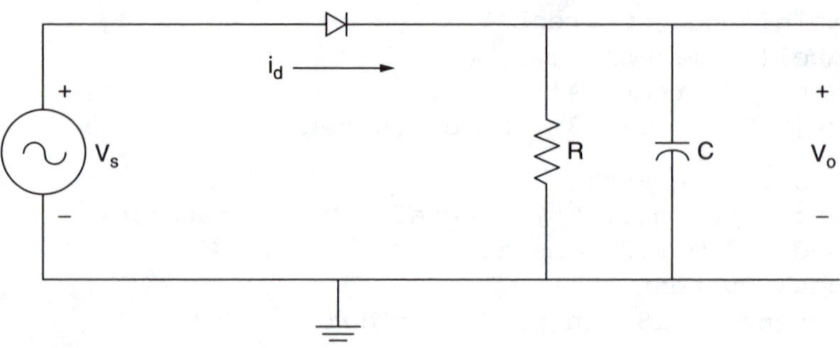

FIGURE 9.12
Capacitor smoothing circuit.

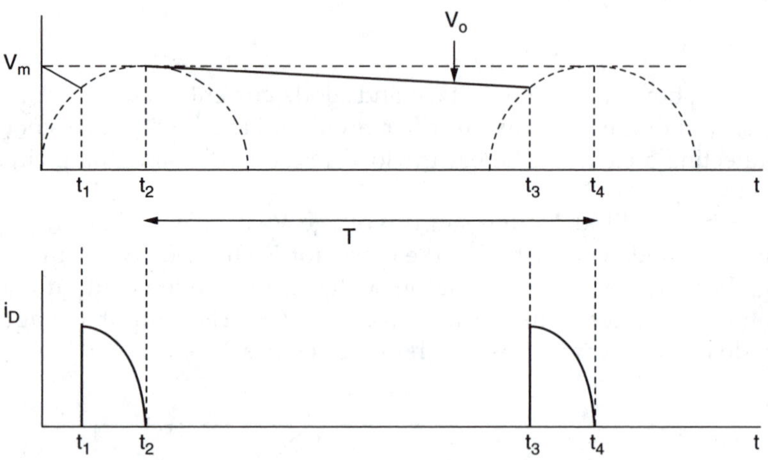

FIGURE 9.13
(Top) Output voltage and (bottom) diode current for half-wave rectifier with smoothing capacitor filter.

In Figure 9.12(a), the output voltage reaches the maximum voltage V_m, at time $t = t_2$. From $t = t_2$ to $t = t_3$, the diode conduction ceases, and the capacitor discharges through R. The output voltage between times t_2 and t_3 is given as

$$v_0(t) = V_m e^{-\left(\frac{t-t_2}{RC}\right)} \qquad t_2 < t < t_3 \tag{9.25}$$

The peak-to-peak ripple voltage is defined as

$$\begin{aligned} V_r &= v_0(t_2) - v_0(t_3) = V_m - V_m e^{-\left(\frac{t_3-t_2}{RC}\right)} \\ &= V_m\left[1 - e^{-\left(\frac{t_3-t_2}{RC}\right)}\right] \end{aligned} \tag{9.26}$$

For large values of C such that $CR >> (t_3 - t_2)$, we can use the well-known exponential series approximation

$$e^{-x} \cong 1 - x \quad \text{for } |x| << 1$$

Thus, Equation (9.26) approximates to

$$V_r = \frac{V_m(t_3 - t_2)}{RC} \tag{9.27}$$

The discharging time for the capacitor, $(t_3 - t_2)$, is approximately equal to the period of the input ac signal, provided the time constant is large. That is,

$$t_3 - t_2 \cong T = \frac{1}{f_0} \tag{9.28}$$

where f_0 is the frequency of the input ac source voltage.

Using Equation (9.28), Equation (9.27) becomes

$$V_{r(peak\text{-}to\text{-}peak)} = \frac{V_m}{f_0 CR} \tag{9.29}$$

For rectifier circuits, because $RC >> T$, the output voltage decays for a small fraction of its fully charged voltage, and the output voltage may be regarded as linear. Therefore, the output waveform of Figure 9.12 is approximately triangular. The *rms* value of the triangular wave is given by

$$V_{rms} = \frac{V_{peak\text{-}to\text{-}peak}}{2\sqrt{3}} = \frac{V_m}{2\sqrt{3} f_o CR} \tag{9.30}$$

The approximate dc voltage of the output waveform is

$$V_{dc} = V_m - \frac{V_r}{2} = V_m - \frac{V_m}{2 f_o CR} \tag{9.31}$$

9.3.1 MATLAB Function fzero

The MATLAB function **fzero** is used to obtain the zero of a function of one variable. The general form of the **fzero** function is

fzero('function', x1)

fzero('function', x1, tol)

where $fzero('funct', x1)$ finds the zero of the function $funct(x)$ that is near the point $x1$ and $fzero('funct', x1, tol)$ returns a zero of the function $funct(x)$ accurate to within a relative error of tol.

The MATLAB function fzero is used in the following example.

Example 9.6 Capacitor Smoothing Circuit — Calculation of Critical Times

For a capacitor smoothing circuit of Figure 9.12, if R = 10 kΩ, C = 100 μF, and $v_S(t) = 120\sqrt{2}\sin(120\pi t)$:

(a) Use MATLAB to calculate the times t_2, t_3, of Figure 9.13.
(b) Compare the capacitor discharge time with period of the input signal.

Solution

The maximum value of $v_S(t)$ is $120\sqrt{2}$, and it occurs at $120\pi t_2 = \pi/2$, thus

$$t_2 = \frac{1}{240} = 0.00417 \text{ sec}$$

The capacitor discharge waveform is given by

$$v_C(t) = 120\sqrt{2}\exp\left(-\frac{(t-t_2)}{RC}\right) \qquad t_2 < t < t_3$$

at $t = t_3$ $v_C(t) = v_S(t)$.

If t_p is the period of the sinusoidal voltage of the source $v_S(t)$ and defining $v(t)$ as

$$v(t) = 120\sqrt{2}\sin\left(120\pi(t-t_p)\right) - 120\sqrt{2}\exp\left(-\frac{(t-t_2)}{RC}\right)$$

Then,

$$v(t_3) = 0 = 120\sqrt{2}\sin\left(120\pi(t_3-t_p)\right) - 120\sqrt{2}\exp\left(-\frac{(t_3-t_2)}{RC}\right)$$

Thus,

$$v(t_3) = 0 = \sin\left(120\pi(t_3-t_p)\right) - \exp\left(-\frac{(t_3-t_2)}{RC}\right) \tag{9.32}$$

MATLAB is used to solve Equation (9.32).

MATLAB script

```
% Capacitance discharge time for smoothing capacitor
% filter circuit
vm = 120*sqrt(2);
f0 = 60; r =10e3; c = 100e-6;
t2 = 1/(4*f0);
tp = 1/f0;

% use MATLAB function fzero to find the zero of a
% function of one variable
rc = r*c;
t3 = fzero('sinexpf1',4.5*t2);
tdis_cap = t3- t2;
fprintf('The value of t2 is %9.5f  s\n', t2)
fprintf('The value of t3 is %9.5f s\n', t3)
fprintf('The capacitor discharge time is %9.5f  s\n',
tdis_cap)
fprintf('The period of input signal is %9.5f  s\n', tp)

%
function y = sinexpf1(t)
t2 = 1/240; tp = 1/60;
rc = 10e3*100e-6;
y = sin(120*pi*(t-tp)) - exp(-(t-t2)/rc);
```

The results are

```
The value of t2 is   0.00417 s
The value of t3 is   0.02036 s
The capacitor discharge time is   0.01619 s
The period of input signal is   0.01667 s
```

9.4 Full-Wave Rectification

A full-wave rectifier that uses a center-tapped transformer is shown in Figure 9.14.

When $v_S(t)$ is positive, the diode D1 conducts but diode D2 is off, and the output voltage $v_0(t)$ is given as

$$v_0(t) = v_S(t) - V_D \tag{9.33}$$

where V_D is a voltage drop across a diode.

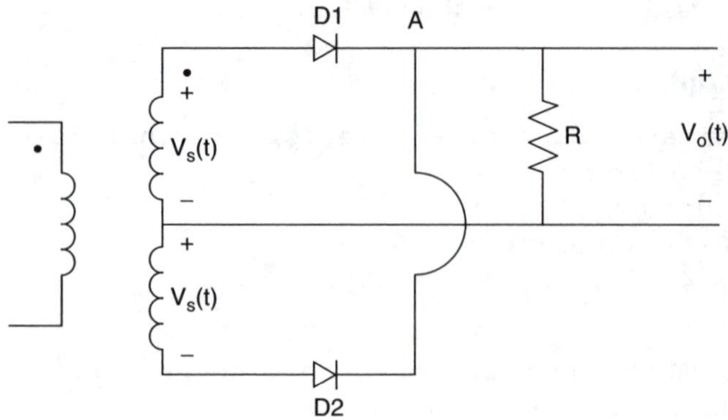

FIGURE 9.14
Full-wave rectifier circuit with center-tapped transformer.

When $v_S(t)$ is negative, diode D1 is cut off but diode D2 conducts. The current flowing through the load R enters it through node A. The output voltage is

$$v(t) = |v_S(t)| - V_D \tag{9.34}$$

One full-wave rectifier that does not require a center-tapped transformer is the bridge rectifier of Figure 9.15.

When $v_S(t)$ is positive, diodes D1 and D3 conduct, but diodes D2 and D4 do not. The current enters the load resistance R at node A. In addition, when $v_S(t)$ is negative, the diodes D2 and D4 conduct, but diodes D1 and D3 do not. The current entering the load resistance R enters it through node A. The output voltage is

$$v(t) = |v_S(t)| - 2V_D \tag{9.35}$$

Figure 9.16 shows the input and output waveforms of a full-wave rectifier circuit assuming ideal diodes.

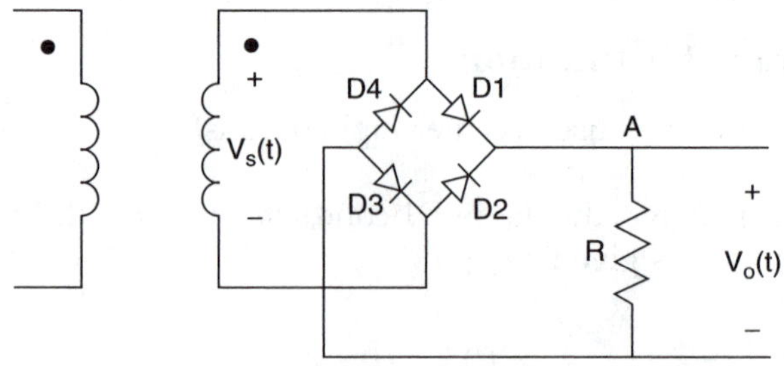

FIGURE 9.15
Bridge rectifier.

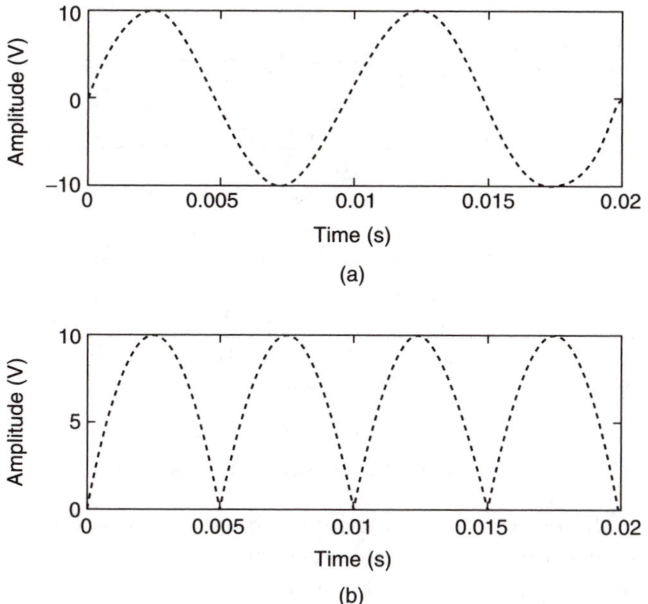

FIGURE 9.16
(a) Input and (b) output voltage waveforms for full-wave rectifier circuit.

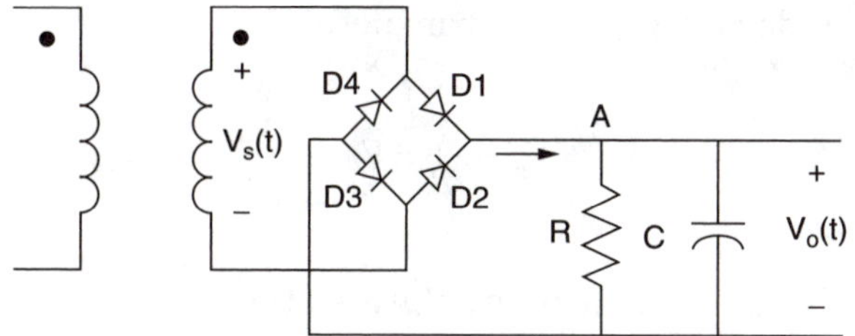

FIGURE 9.17
Full-wave rectifier with capacitor smoothing filter.

Connecting a capacitor across the load can smooth the output voltage of a full-wave rectifier circuit. The resulting circuit is shown in Figure 9.17.

The output voltage and the current waveforms for the full-wave rectifier with RC filter are shown in Figure 9.18.

From Figures 9.13 and 9.18, it can be seen that the frequency of the ripple voltage is twice that of the input voltage. The capacitor in Figure 9.17 has only half the time to discharge. Therefore, for a given time constant, *CR*, the ripple voltage will be reduced, and it is given by

$$V_{r(peak\text{-}to\text{-}peak)} = \frac{V_m}{2f_oCR} \tag{9.36}$$

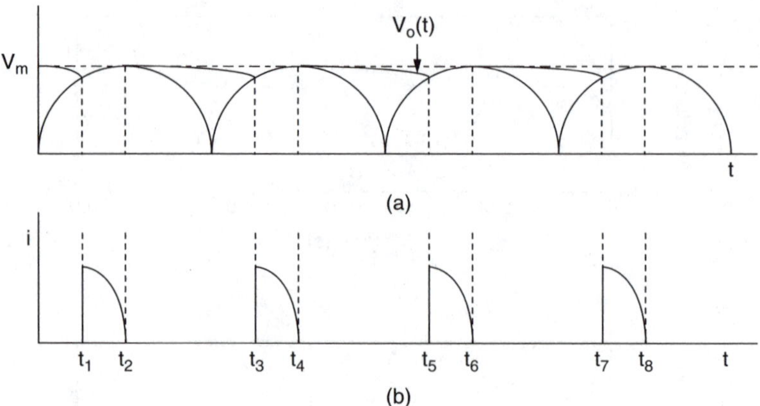

FIGURE 9.18
(a) Voltage and (b) current waveform of a full-wave rectifier with RC filter.

where V_m is the peak value of the input sinusoidal waveform and f_0 is the frequency of the input sinusoidal waveform.

The rms value of the ripple voltage is

$$V_{rms} = \frac{V_m}{4\sqrt{3} f_o CR} \tag{9.37}$$

and the output dc voltage is approximately

$$V_{dc} = V_m - \frac{V_r}{2} = V_m - \frac{V_m}{4 f_o CR} \tag{9.38}$$

Example 9.7 Calculations of Parameters of a Full-Wave Rectifier

For the full-wave rectifier with RC filter shown in Figure 9.17, if $v_S(t) = 20\sin(120\pi t)$ and $R = 10$ kΩ, $C = 100$ μF, use MATLAB to find the:

(a) Peak-to-peak value of ripple voltage
(b) dc output voltage
(c) Discharge time of the capacitor
(d) Period of the ripple voltage

Solution

Peak-to-peak ripple voltage and dc output voltage can be calculated using Equation (9.36) and Equation (9.37), respectively. The discharge time of the capacitor is the time $(t_3 - t_1)$ of Figure 9.19.

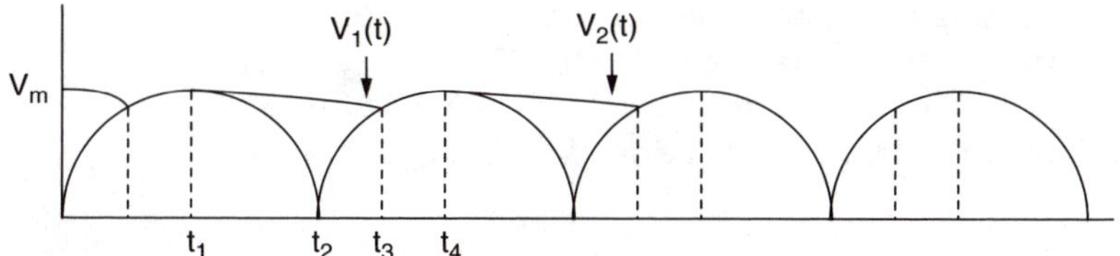

FIGURE 9.19
Diagram for calculating capacitor discharge time.

$$v_1(t) = V_m \exp\left[-\frac{(t-t_1)}{RC}\right] \tag{9.39}$$

$$v_2(t) = \left|V_m \sin\left[2\pi(t-t_2)\right]\right| \tag{9.40}$$

$v_1(t)$ and $v_2(t)$ intersect at time t_3.

The period of input waveform, $v_S(t)$, is $T = \frac{1}{240}$ sec.

Thus,

$$t_1 = \frac{T}{4} = \frac{1}{240} \text{ sec,} \quad \text{and} \quad t_2 = \frac{T}{2} = \frac{1}{120} \text{ sec} \tag{9.41}$$

MATLAB script

```
% Full-wave rectifier
period = 1/60;
t1 = period/4;
vripple = 20/(2*60*10e3*100e-6);
vdc = 20 - vripple/2;
t3 = fzero('sinexpf2',0.7*period);
tdis_cap = t3 - t1;
fprintf('Ripple value (peak-peak) is %9.5f V\n',
vripple)
fprintf('DC output voltage is %9.5f V\n', vdc)
fprintf('Capacitor discharge time is %9.5f s\n',
tdis_cap)
fprintf('Period of ripple voltage is %9.5f s\n',
0.5*period)
```

```
function y = sinexpf2(t)
t1 = 1/240; t2 = 2*t1; rc = 10e3*100e-6;
y = 20*(sin(120*pi*(t - t2))) - exp(-(t-t1)/rc);
```

The results are

```
Ripple value (peak-peak) is   0.16667 V
DC output voltage is  19.91667 V
Capacitor discharge time is   0.00430 s
Period of ripple voltage is   0.00833 s
```

9.5 Zener Diode Voltage Regulator Circuits

The zener diode is a pn junction diode with controlled reverse-biased breakdown voltage. Figure 9.20 shows the electronic symbol and the current-voltage characteristics of the zener diode.

I_{ZK} is the minimum current needed for the zener to break down. I_{ZM} is the maximum current that can flow through the zener without being destroyed. It is obtained by

$$I_{ZM} = \frac{P_Z}{V_Z} \tag{9.42}$$

where P_Z is the zener power dissipation.

The incremental resistance of the zener diode at the operating point is specified by

$$r_Z = \frac{\Delta V_Z}{\Delta I_Z} \tag{9.43}$$

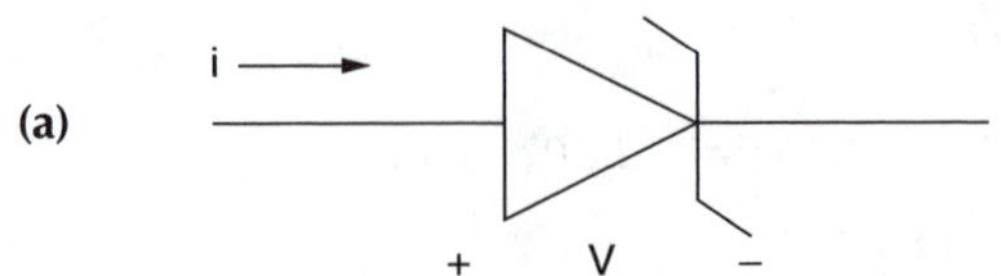

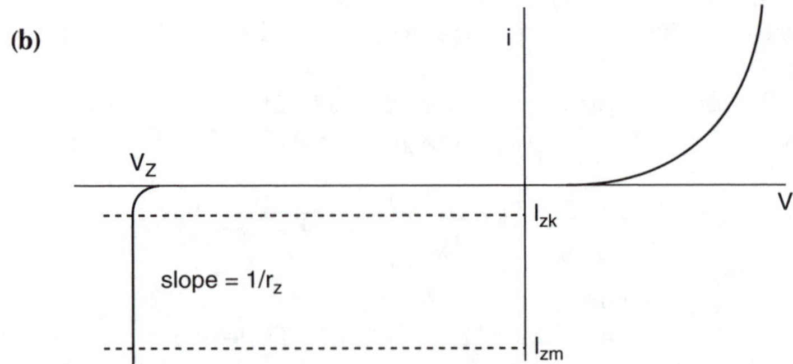

FIGURE 9.20
Zener diode: (a) electronic symbol; (b) I-V characteristics.

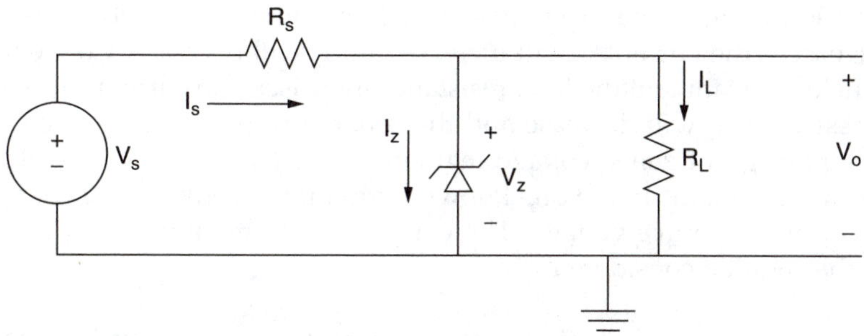

FIGURE 9.21
Zener diode shunt voltage regulator circuit.

One of the applications of a zener diode is its use in the design of voltage reference circuits. A zener diode shunt voltage regulator circuit is shown in Figure 9.21.

The circuit is used to provide an output voltage, V_0, which is nearly constant. When the source voltage is greater than the zener breakdown voltage, the zener will break down and the output voltage will be equal to the zener breakdown voltage. Thus,

$$V_0 = V_Z \tag{9.44}$$

From Kirchhoff current law, we have

$$I_S = I_Z + I_L \tag{9.45}$$

and from Ohm's law, we have

$$I_S = \frac{V_S - V_Z}{R_S} \tag{9.46}$$

and

$$I_L = \frac{V_O}{R_L} \tag{9.47}$$

Assuming the load resistance R_L is held constant and V_S (which was originally greater than V_Z) is increased, the source current I_S will increase; and since I_L is constant, the current flowing through the zener will increase. Conversely, if R is constant and V_S decreases, the current flowing through the zener will decrease since the breakdown voltage is nearly constant; the output voltage will remain almost constant with changes in the source voltage V_S.

Now assuming the source voltage is held constant and the load resistance is decreased, the current I_L will increase and I_Z will decrease. Conversely, if V_S is held constant and the load resistance increases, the current through the load resistance I_L will decrease and the zener current I_Z will increase.

In the design of zener voltage regulator circuits, it is important that the zener diode remains in the breakdown region irrespective of the changes in the load or the source voltage. There are two extreme input/output conditions that will be considered:

1. The diode current I_Z is minimum when the load current I_L is maximum and the source voltage V_S is minimum.
2. The diode current I_Z is maximum when the load current I_L is minimum and the source voltage V_S is maximum.

From condition 1 and Equation (9.46), we have

$$R_S = \frac{V_{S,\min} - V_Z}{I_{L,\max} + I_{Z,\min}} \tag{9.48}$$

Similarly, from condition 2, we get

$$R_S = \frac{V_{S,\max} - V_Z}{I_{L,\min} + I_{Z,\max}} \tag{9.49}$$

Equating Equation (9.48) and Equation (9.49), we get

$$(V_{S,\min} - V_Z)(I_{L,\min} + I_{Z,\max}) = (V_{S,\max} - V_Z)(I_{L,\max} + I_{Z,\min}) \tag{9.50}$$

We use the rule of thumb that the maximum zener current is about ten times the minimum value, that is,

$$I_{Z,\min} = 0.1 I_{Z,\max} \tag{9.51}$$

Substituting Equation (9.49) into Equation (9.51) and solving for $I_{Z,\max}$, we obtain

$$I_{Z,\max} = \frac{I_{L,\min}(V_Z - V_{S,\min}) + I_{L,\max}(V_{S,\max} - V_Z)}{V_{S,\min} - 0.9V_Z - 0.1V_{S,\max}} \tag{9.52}$$

Knowing $I_{Z,\max}$, we can use Equation (9.49) to calculate R_S. The following example uses MATLAB to solve a zener voltage regulator problem.

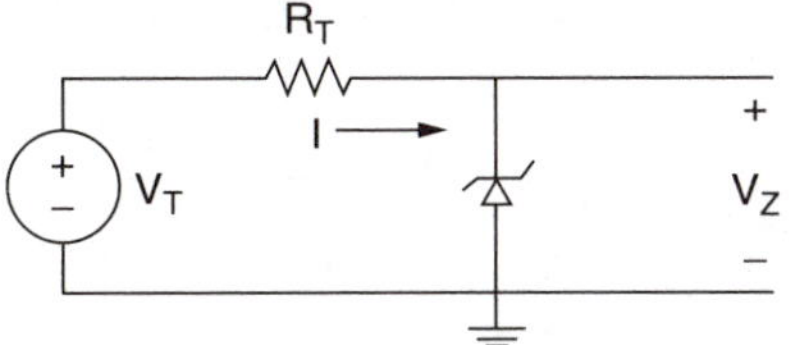

FIGURE 9.22
Equivalent circuit of voltage regulator circuit.

Example 9.8 A Zener Diode Voltage Regulator

A zener diode voltage regulator circuit of Figure 9.21 has the following data:

$$30 \le V_S \le 35 \text{ V}; R_L = 10 \text{ k}\Omega, R_S = 2 \text{ k}\Omega$$

$$V_Z = -20 + 0.05\, I \text{ for } -100 \text{ mA} \le I < 0 \tag{9.53}$$

Use MATLAB to (a) plot the zener breakdown characteristics, (b) plot the loadline for $V_S = 30$ V and $V_S = 35$ V, and (c) determine the output voltage when $V_S = 30$ V and $V_S = 35$ V.

Solution

Using the Thevenin theorem, Figure 9.21 can be simplified into the form shown in Figure 9.22.

$$V_T = \frac{V_S R_L}{R_L + R_S} \tag{9.54}$$

and

$$R_T = R_L \| R_S \tag{9.55}$$

Since $R_L = 10$ K, $R_S = 2$ K, and $R_T = (10)(2 \text{ K})/12 \text{ K} = 1.67 \text{ k}\Omega$:

when $V_S = 30$ V, $V_T = (30)(10)/12 = 25$ V
when $V_S = 35$ V, $V_T = (35)(10)/12 = 29.17$ V

The loadline equation is

$$V_T = R_T I + V_Z \tag{9.56}$$

Equation (9.53) and Equation (9.56) are two linear equations solving for I, so we get

$$V_Z = V_T - R_T I = -20 + 0.05I$$

$$\Rightarrow \quad I = \frac{(V_T + 20)}{R_T + 0.05} \tag{9.57}$$

From Equation (9.56) and Equation (9.57), the output voltage (which is also the zener voltage) is

$$V_Z = V_T - R_T I = V_T - \frac{R_T(V_T + 20)}{R_T + 0.05} \tag{9.58}$$

MATLAB script

```
% Zener diode voltage regulator
vs1 = -30; vs2 = -35; rl =10e3; rs = 2e3;
i = -50e-3: 5e-3 :0;
vz = -20 + 0.05*i;
m = length(i);
i(m+1) = 0; vz(m+1) = -10;
i(m+2) = 0; vz(m+2) = 0;
% loadlines
vt1 = vs1*rl/(rl+rs);
vt2 = vs2*rl/(rl+rs);
rt = rl*rs/(rl+rs);
l1 = vt1/20;
l2 = vt2/20;
v1 = vt1:abs(l1):0;
i1 = (vt1 - v1)/rt;
v2 = vt2:abs(l2):0;
i2 = (vt2 - v2)/rt;
% plots of Zener characteristics, loadlines
plot(vz,i,'b',v1,i1,'b',v2,i2,'b')
axis([-30,0,-0.03,0.005])
title('Zener Voltage Regulator Circuit')
xlabel('Voltage, V')
ylabel('Current, A')
text(-19.5,-0.025,'Zener Diode Curve')
text(-15.6,-0.016, 'Loadline for 35 V Source')
text(-14.7,-0.005,'Loadline for 30 V Source')
% output voltage when vs = -30v
ip1 = (vt1 + 20)/(rt + 0.05)
vp1 = vt1 - rt*(vt1+20)/(rt + 0.05)
% output voltage when vs = -35v
ip2 = (vt2 + 20)/(rt + 0.05)
vp2 = vt2 - rt*(vt2+20)/(rt + 0.05)
```

The results obtained are

```
ip1 =   -0.0030

vp1 =   -20.0001

ip2 =   -0.0055

vp2 =   -20.000
```

When the source voltage is 30 V, the output voltage is 20.0001 V. In addition, when the source voltage is 35 V, the output voltage is 20.0003 V.

The zener breakdown characteristics and the loadlines are shown in Figure 9.23.

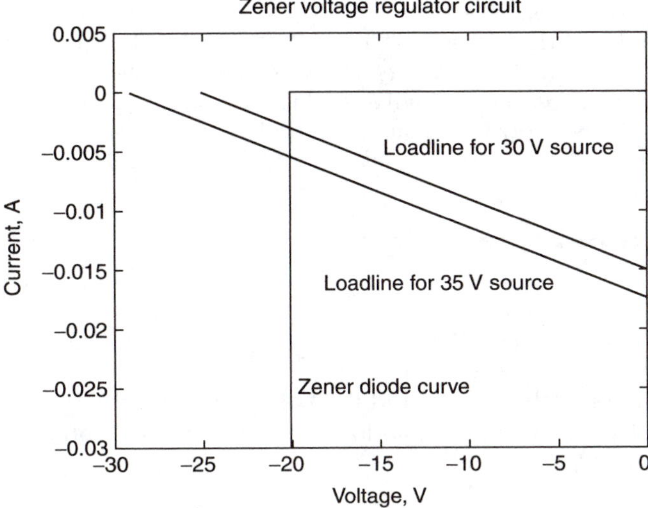

FIGURE 9.23
Zener characteristics and loadlines.

Bibliography

1. Alexander, C.K. and Sadiku, M.N.O., *Fundamentals of Electric Circuits*, 2nd ed., McGraw-Hill, New York, 2004.
2. Attia, J.O., *PSPICE and MATLAB for Electronics: An Integrated Approach*, CRC Press, Boca Raton, FL, 2002.
3. Biran, A. and Breiner, M., *MATLAB for Engineers*, Addison-Wesley, Reading, MA, 1995.
4. Chapman, S.J., *MATLAB Programming for Engineers*, Brook, Cole Thompson Learning, Pacific Grove, CA, 2000.
5. Dorf, R.C. and Svoboda, J.A., *Introduction to Electric Circuits*, 3rd ed., John Wiley & Sons, New York, 1996.

6. Etter, D.M., *Engineering Problem Solving with MATLAB*, 2nd ed., Prentice Hall, Upper Saddle River, NJ, 1997.
7. Etter, D.M., Kuncicky, D.C., and Hull, D., *Introduction to MATLAB 6*, Prentice Hall, Upper Saddle River, NJ, 2002.
8. Ferris, C.D., *Elements of Electronic Design*, West Publishing Co., New York, 1995.
9. Gottling, J.G., *Matrix Analysis of Circuits Using MATLAB*, Prentice Hall, Englewood Cliffs, NJ, 1995.
10. Johnson, D.E., Johnson, J.R., and Hilburn, J.L., *Electric Circuit Analysis*, 3rd ed., Prentice Hall, Upper Saddle River, NJ, 1997.
11. Meader, D.A., *Laplace Circuit Analysis and Active Filters*, Prentice Hall, Upper Saddle River, NJ, 1991.
12. Rashid, M.H., *Microelectronic Circuits, Analysis and Design*, PWS Publishing Company, Englewood Cliffs, NJ, 1999.
13. Sedra, A.S. and Smith, K.C., *Microelectronic Circuits*, Oxford University Press, 4th ed., New York, 1998.
14. Sigmor, K., *MATLAB Primer*, 4th ed., CRC Press, Boca Raton, FL, 1998.
15. Using *MATLAB, The Language of Technical Computing, Computation, Visualization, Programming*, Version 6, MathWorks, Inc., Natick, MA, 2000.
16. Vlach, J.O., Network theory and CAD, *IEEE Trans. on Education*, 36, 23–27, 1993.

Problems

Problem 9.1

Use the iteration technique to find the voltage V_D and the I_D of Figure P9.1. Assume that $T = 25°C$, $n = 1.5$, and $I_S = 10^{-16}$ A. The iteration can stop when $|V_n - V_{n-1}| < 10^{-9}$ V.

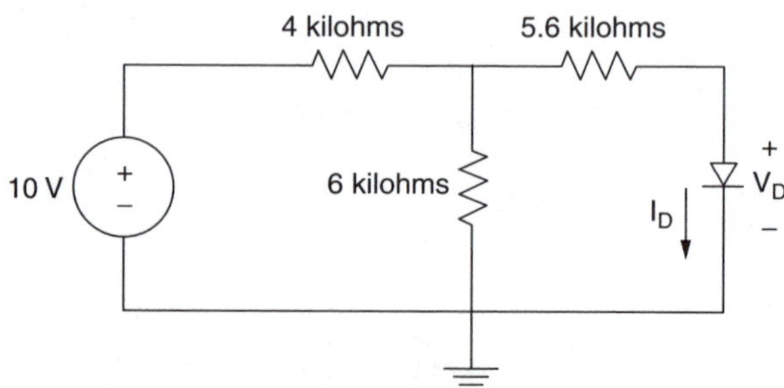

FIGURE P9.1
A diode circuit.

Problem 9.2

For the circuit shown in Figure 9.5, $R = 1000\ \Omega$, $V_{DC} = 5$ V, $n = 2.0$, and the reverse saturation current of the diode $I_S = 10^{-12}$ A. Use the iteration technique to find the voltage V_D. The iteration can stop when $|V_n - V_{n-1}| < 10^{-9}$ V. Assume that $T = 25°C$.

Problem 9.3

A zener diode has the following I-V characteristics:

Reverse Voltage (V)	Reverse Current (A)
–2	–1.0E–10
–4	–1.0E–10
–6	–1.0E–8
–8	–1.0E–5
–8.5	–2.0E–5
–8.7	–15.0E–3
–8.9	–43.5E–3

(a) Plot the reverse characteristics of the diode. (b) Determine the dynamic resistance of the diode in its breakdown region.

Problem 9.4

A forward-biased diode has the following corresponding voltage and current.

(a) Plot the static I-V characteristics.
(b) Determine the diode parameters I_S and n.
(c) Calculate the dynamic resistance of the diode at $V = 0.5$ V.

Forward Voltage, V	Forward Current, A
0.2	7.54E–7
0.3	6.55E–6
0.4	5.69E–5
0.5	4.94E–4
0.6	4.29E–3
0.7	3.73E–2

Problem 9.5

For Figure P9.5,

(a) Use iteration to find the current through the diode. The iteration can be stopped when $|I_{dn} - I_{dn-1}| < 10^{-12}$ A.
(b) How many iterations were performed before the required result was obtained? Assume a temperature of 25°C, emission coefficient, n, of 1.5, and a reverse saturation current, I_S, of 10^{-16} A.

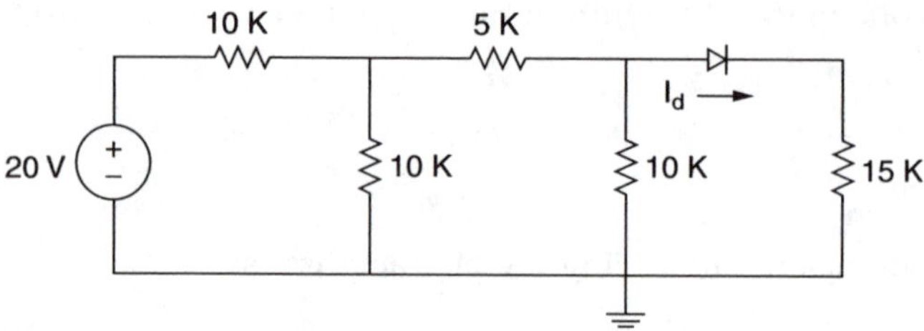

FIGURE P9.5
Diode circuit.

Problem 9.6

For the battery charging circuit shown in Figure 9.10, the battery voltage is $V_B = 12.25$ V. The source voltage is $v_S(t) = 25\sin(240\pi t)$ V and $R = 200\ \Omega$. Use MATLAB to calculate the conduction angle and the peak current of the diode. Assume that the diode is ideal.

Problem 9.7

For the full-wave rectifier ircuit with smoothing capacitor shown in Figure 9.17, if $v_S(t) = 100\sin(120\pi t)$ V, $R = 50$ kΩ, and $C = 250$ μF:

(a) Plot the input and output voltages when the capacitor is disconnected from the load resistance R.

(b) When the capacitor is connected across load resistance R, determine the conduction time of the diode.

(c) What is the diode conduction time?

Problem 9.8

For the capacitor smoothing circuit shown in Figure 9.12, $R = 2$ kΩ and $C =$ 5 μF. If $v_S(t) = 20\sin(2000\pi t)$ V, determine the discharge time.

Problem 9.9

For the voltage regulator circuit shown in Figure 9.21, assume that $50 < V_S <$ 60 V, $R_L = 50$ K, $R_S = 5$ K, and $V_z = -40 + 0.01$ I. Use MATLAB to

(a) Plot the zener diode breakdown characteristics.

(b) Plot the loadline for $V_S = 50$ V and $V_S = 60$ V.

(c) Determine the output voltage and the current flowing through the source resistance R_S when $V_S = 50$ V and $V_S = 60$ V.

Problem 9.10

For the zener voltage regulator shown in Figure 9.21, If $V_S = 35$ V, $R_S = 1$ kΩ, $V_Z = -25 + 0.02I$, and $5 \text{ K} < R_L < 50 \text{ K}$, use MATLAB to

(a) Plot the zener breakdown characteristics.
(b) Plot the loadline when $R_L = 5$ K and $R_L = 50$ K.
(c) Determine the output voltage when $R_L = 5$ kΩ and $R_L = 50$ kΩ.
(d) What is the power dissipation of the diode when $R_L = 50$ kΩ?

10

Semiconductor Physics

In this chapter, a brief description of the basic concepts governing the flow of current in a pn junction described in Section 10.3 are discussed. Both intrinsic and extrinsic semiconductors are discussed. The characteristics of depletion and diffusion capacitance are explored through the use of example problems solved with MATLAB. The effect of doping concentration on the breakdown voltage of pn junctions is examined.

10.1 Intrinsic Semiconductors

10.1.1 Energy Bands

According to the planetary model of an isolated atom, the nucleus, which contains protons and neutrons, constitutes most of the mass of the atom. Electrons surround the nucleus in specific orbits. The electrons are negatively charged and the nucleus is positively charged. If an electron absorbs energy (in the form of a photon), it moves to orbits further from the nucleus. An electron transition from a higher energy orbit to a lower energy orbit emits a photon for a direct band gap semiconductor.

The energy levels of the outer electrons form energy bands. In insulators, the lower energy band (valence band) is completely filled and the next energy band (conduction band) is completely empty. The valence and conduction bands are separated by a forbidden energy gap.

In conductors, the valence band partially overlaps the conduction band with no forbidden energy gap between the valence and conduction bands. In semiconductors, the forbidden gap is less than 1.5 eV. Some semiconductor materials are silicon (Si), germanium (Ge), and gallium arsenide (GaAs). Figure 10.1 shows the energy level diagram of silicon, germanium, and insulator (carbon).

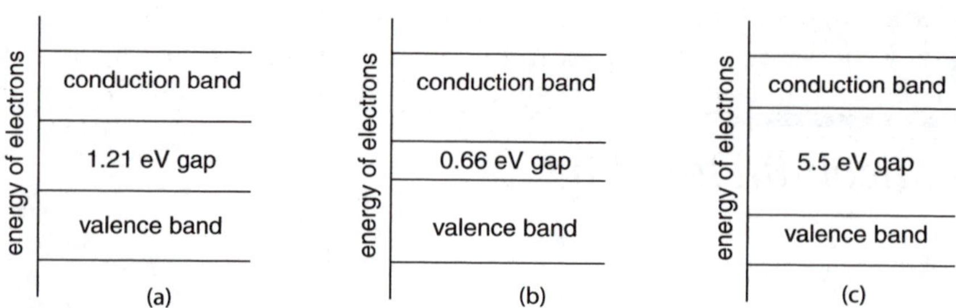

FIGURE 10.1
Energy level diagram of (a) silicon, (b) germanium, and (c) insulator (carbon).

10.1.2 Mobile Carriers

Silicon is the most commonly used semiconductor material in the integrated circuit industry. Silicon has four valence electrons, and its atoms are bound together by covalent bonds. At absolute zero temperature, the valence band is completely filled with electrons, and no current flow can take place. As the temperature of a silicon crystal is raised, there is an increased probability of breaking covalent bonds and freeing electrons. The vacancies left by the freed electrons are called holes. The process of creating free electron-hole pairs is called ionization. The free electrons move in the conduction band. The average number of carriers (mobile electrons or holes) that exist in an intrinsic semiconductor material may be found from the mass-action law:

$$n_i = AT^{1.5}e^{[-E_g/(kT)]} \tag{10.1}$$

where T is the absolute temperature in K, k is Boltzmann's constant (k = 1.38×10^{-23} J/K or 8.62×10^{-5} eV/K), and E_g is the width of the forbidden gap in eV. E_g is 1.21 and 1.1 eV for Si at 0 and 300 K, respectively. It is given as

$$E_g = E_c - E_v \tag{10.2}$$

A is a constant dependent on a given material; it is given as

$$A = \frac{2}{h^3}(2\pi m_0 k)^{3/2}\left(\frac{m_n^*}{m_0}\frac{m_p^*}{m_o}\right)^{3/4} \tag{10.3}$$

where h is Planck's constant ($h = 6.62 \times 10^{-34}$ J sec or 4.14×10^{-15} eV sec), m_o is the rest mass of an electron, $m_n{}^*$ is the effective mass of an electron in a material, and $m_p{}^*$ is the effective mass of a hole in a material.

The mobile carrier concentrations are dependent on the width of the energy gap, E_g, measured with respect to the thermal energy kT. For small values of T ($kT << E_g$), n_i is small, implying that there are less mobile carriers. For silicon, the equilibrium intrinsic concentration at room temperature is

$$n_i = 1.52 \times 10^{10} \text{ electrons/cm}^3 \tag{10.4}$$

Of the two carriers that we find in semiconductors, electrons have a higher mobility than holes do. For example, intrinsic silicon at 300 K has electron mobility of 1350 cm^2/V-sec and hole mobility of 480 cm^2/V-sec. The conductivity of an intrinsic semiconductor is given by

$$\sigma_i = q(n_i\mu_n + p_i\mu_p) \tag{10.5}$$

where q is the electronic charge (1.6×10^{-19} C), n_i is the electron concentration, p_i is the hole concentration, $p_i = n_i$ for the intrinsic semiconductor, μ_n is electron mobility in the semiconductor material, and μ_p is hole mobility in the semiconductor material.

Since electron mobility is about three times hole mobility in silicon, the electron current is considerably more than the hole current. The following example illustrates the dependence of electron concentration on temperature.

Example 10.1 Electron Concentration vs. Temperature

Given that at T = 300 K, the electron concentration in silicon is 1.52×10^{10} electrons/cm^3 and E_g = 1.1 eV at 300 K,

(a) Find the constant A of Equation (10.1).

(b) Use MATLAB to plot the electron concentration vs. temperature.

Solution

From Equation (10.1), we have

$$1.52 \times 10^{10} = A(300)^{1.5} e^{[-1.1/300*8.62*10^{-5})]}$$

We use MATLAB to solve for A. The width of energy gap with temperature is given as[1]

$$E_g(T) = 1.17 - 4.37 \times 10^{-4}\left(\frac{T^2}{T+636}\right) \tag{10.6}$$

Using Equation (10.1) and Equation (10.6), we can calculate the electron concentration at various temperatures.

MATLAB script

```
% Electron Concentration vs. Temperature
% Calculation of the constant A
k = 8.62e-5;
na = 1.52e10;    ta = 300;
ega = 1.1;
ka  = -ega/(k*ta);
t32a = ta.^1.5;
A = na/(t32a*exp(ka));
fprintf('constant A is %10.5e \n', A)
% Electron Concentration vs. temperature

for i = 1:10
    t(i)  = 273 + 10*(i-1);
    eg(i) = 1.17 - 4.37e-4*(t(i)*t(i))/(t(i) + 636);
    t32(i) = t(i).^1.5;
    ni(i) = A*t32(i)*exp(-eg(i)/(k*t(i)));
end
semilogy(t,ni)
title('Electron Concentration vs. Temperature')
xlabel('Temperature, K')
ylabel('Electron Concentration, cm-3')
```

Result for part (a)

```
Constant A is 8.70225e+024
```

Figure 10.2 shows the plot of the electron concentration vs. temperature.

10.2 Extrinsic Semiconductor

10.2.1 Electron and Hole Concentrations

Extrinsic semiconductors are formed by adding specific amounts of impurity atoms to the silicon crystal. An n-type semiconductor is formed by doping the silicon crystal with elements of group V of the periodic table (antimony, arsenic, and phosphorus). The impurity atom is called a donor. The majority carriers are electrons and the minority carriers are holes. A p-type semiconductor is formed by doping the silicon crystal with elements of group III of the periodic table (aluminum, boron, gallium, and indium). The impurity atoms are called acceptor atoms. The majority carriers are holes and minority carriers are electrons.

In a semiconductor material (intrinsic or extrinsic), the law of mass action states that

$$pn = \text{constant} \tag{10.7}$$

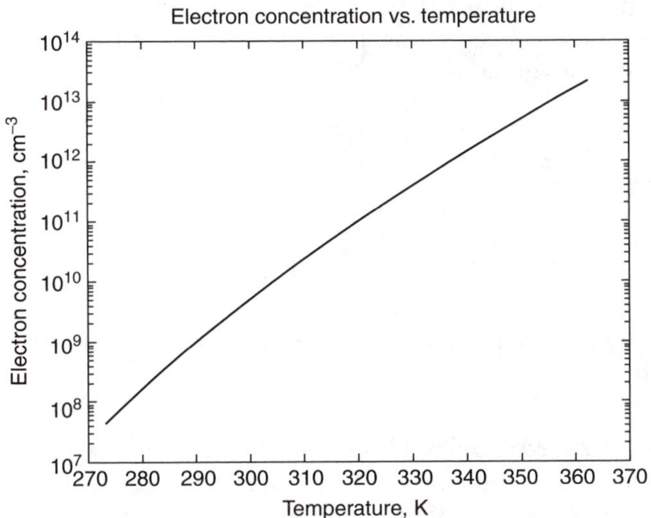

FIGURE 10.2
Electron concentration vs. temperature.

where p is the hole concentration and n is the electron concentration.

For intrinsic semiconductors,

$$p = n = n_i \tag{10.8}$$

and Equation (10.5) becomes

$$pn = n_i^2 \tag{10.9}$$

and n_i is given by Equation (10.1).

The law of mass action enables us to calculate the majority and minority carrier density in an extrinsic semiconductor material. The charge neutrality condition of a semiconductor implies that

$$p + N_D = n + N_A \tag{10.10}$$

where N_D is the donor concentration, N_A is the acceptor concentration, p is the hole concentration, and n is the electron concentration.

In an n-type semiconductor, the donor concentration is greater than the intrinsic electron concentration, i.e., N_D is typically 10^{17} cm^{-3} and $n_i = 1.5 \times 10^{10}$ cm^{-3} in Si at room temperature. Thus, the majority and minority concentrations are given by

$$n_n \cong N_D \tag{10.11}$$

$$p \cong \frac{n_i^2}{N_D} \tag{10.12}$$

In a p-type semiconductor, the acceptor concentration N_A is greater than the intrinsic hole concentration $p_i = n_i$. Thus, the majority and minority concentrations are given by

$$p_p \cong N_A \tag{10.13}$$

$$n \cong \frac{n_i^2}{N_A} \tag{10.14}$$

The following example gives the minority carrier as a function of doping concentration.

Example 10.2 Minority Carriers in a Doped Semiconductor

For an n-type semiconductor at 300 K, if the doping concentration is varied from 10^{13} to 10^{18} atoms/cm^3, determine the minority carriers in the doped semiconductors.

Solution

From Equation (10.11) and Equation (10.12),

$$\text{Electron concentration} = N_D$$

and

$$\text{Hole concentration} = \frac{n_i^2}{N_D}$$

where

$$n_i = 1.52 \times 10^{10} \text{ electrons/cm}^3$$

The MATLAB program is as follows:

```
% Minority Carriers in Doped Semiconductor
% hole concentration in a n-type semiconductor
nd = logspace(13,18);
n = nd;
ni = 1.52e10;
ni_sq = ni*ni;
p = ni_sq./nd;
semilogx(nd,p,'b',nd, p,'ob')
title('Hole Concentration')
xlabel('Doping Concentration, cm-3')
ylabel('Hole Concentration, cm-3')
```

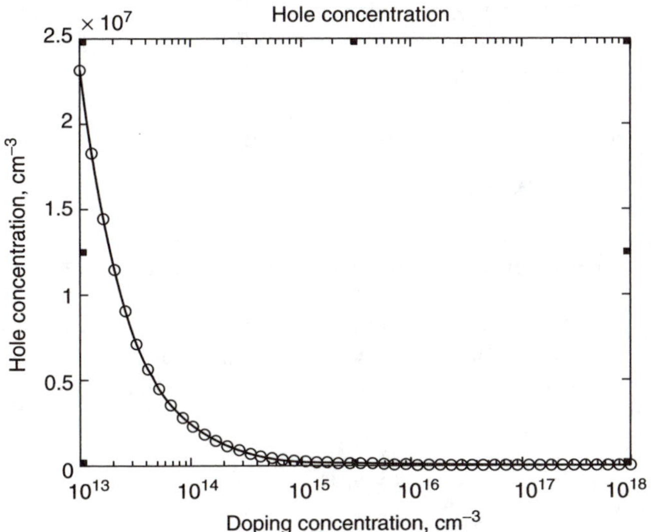

FIGURE 10.3
Hole concentration in n-type semiconductor (Si).

Figure 10.3 shows the hole concentration vs. doping.

10.2.2 Fermi Level

The Fermi level, E_F, is a chemical energy of a material. It is used to describe the energy level of the electronic state at which an electron has the probability of 0.5 of occupying that state. It is given as

$$E_F = \frac{1}{2}(E_C + E_V) - \frac{4}{3}KT\ln\left(\frac{m_n^*}{m_p^*}\right) \tag{10.15}$$

where E_C = energy in the conduction band, E_V = energy in the valence band, and k, T, $m_n{}^*$, and $m_p{}^*$ were defined in Section 10.1.

In an intrinsic semiconductor (Si and Ge), m_n^* and m_p^* are of the same order of magnitude and typically, $E_F >> kT$. Equation (10.15) simplifies to

$$E_F = E_i \cong \frac{1}{2}(E_C + E_V) \tag{10.16}$$

Equation (10.16) shows that the Fermi energy occurs near the center of the energy gap in an intrinsic semiconductor. In addition, the Fermi energy can be thought of as the average energy of mobile carriers in a semiconductor material.

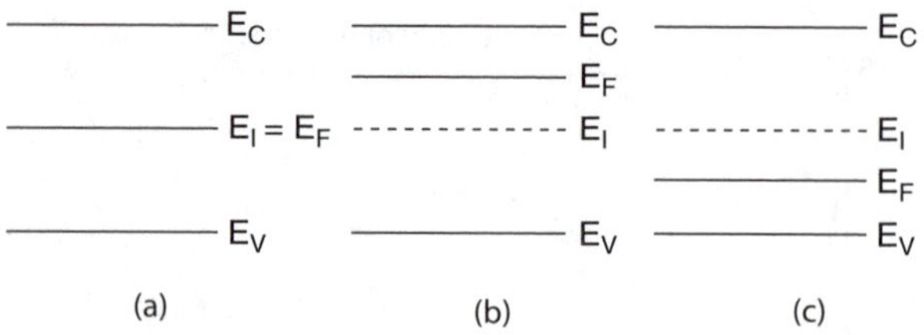

FIGURE 10.4
Energy-band diagram of (a) intrinsic, (b) n-type, and (c) p-type semiconductors.

In an n-type semiconductor, there is a shift of the Fermi level toward the edge of the conduction band. The upward shift is dependent on how much the doped electron density has exceeded the intrinsic value. The relevant equation is

$$n = n_i e^{\left[(E_F - E_i)/kT\right]} \tag{10.17}$$

where n is the total electron carrier density, n_i is the intrinsic electron carrier density, E_F is the doped Fermi level, and E_i is the intrinsic Fermi level.

In the case of a p-type semiconductor, there is a downward shift in the Fermi level. The total hole density will be given by

$$p = n_i e^{\left[(E_i - E_F)/kT\right]} \tag{10.18}$$

Figure 10.4 shows the energy band diagram of intrinsic and extrinsic semiconductors.

10.2.3 Current Density and Mobility

Two mechanisms account for the movement of carriers in a semiconductor material: drift and diffusion. Drift current is caused by the application of an electric field, whereas diffusion current is obtained when there is a net flow of carriers from a region of high concentration to a region of low concentration. The total drift current density in an extrinsic semiconductor material is

$$J = q(n\mu_n + p\mu_p)\mathrm{E} \tag{10.19}$$

where J is current density, n is mobile electron density, p is hole density, μ_n is mobility of an electron, μ_p is mobility of a hole, q is the electron charge, and E is the electric field.

The total conductivity is

$$\sigma = \frac{J}{q} = q(n\mu_n + p\mu_p)\mathrm{E} \tag{10.20}$$

Assuming that there is a diffusion of holes from an area of high concentration to that of low concentration, then the current density of holes in the x-direction is

$$J_p = -qD_p \frac{dp}{dx} \quad \text{A/cm}^2 \tag{10.21}$$

where q is the electronic charge, D_p is the hole diffusion constant, and p is the hole concentration.

Equation (10.21) also assumes that, although the hole concentration varies along the x-direction, it is constant in the y- and z-directions. Similarly, the electron current density, J_n, for diffusion of electrons is

$$J_n = qD_n \frac{dn}{dx} \quad \text{A/cm}^2 \tag{10.22}$$

where D_n is the electron diffusion constant and n is the electron concentration.

For silicon, D_p = 13 cm²/sec and D_n = 200 cm²/sec. Under steady-state conditions, the diffusion and mobility constants are related by the Einstein relation

$$\frac{D_n}{\mu_n} = \frac{D_p}{\mu_p} = \frac{kT}{q} \tag{10.23}$$

The diffusion process depends on temperature. In the case of diffusion of doping atoms, the diffusion coefficient D is related to temperature by the expression

$$D = D_O e^{\left(\frac{-E_A}{KT}\right)} \tag{10.24}$$

where T is the absolute temperature in K, K is the Boltzmann's constant, E_A is the activation energy, and D_O is the frequency factor.

The parameters E_A and D_O have to be determined for a semiconductor material and for specific doping species. The dependency of the diffusion process with temperature is explored in the following example.

Example 10.3 Diffusion Coefficient vs. Temperature

For boron at 1000°C, $D_O = 0.76$ cm²/sec, $E_A = 3.46$ eV, and $K = 8.62 \times 10^{-5}$ eV/K. Calculate the diffusion constants for temperatures ranging from 1000 to 1250°C. Plot the diffusion coefficient vs. temperature.

Solution

The MATLAB program is as follows:

```
% Diffusion Coefficient vs. Temperature
% Constants
K = 8.62e-5;
EA = 3.46;
Do = 0.76;
% Diffusion coefficient with temperature

for i = 1:11
   t(i) = 273 + 1000 + 25*(i-1);
   Dcoef(i) = Do*exp(-EA/(k*t(i)));
end
semilogy(t,Dcoef)
title('Diffusion Coefficient  vs. Temperature')
xlabel('Temperature, K')
ylabel('Diffusion Coefficient , cm2/s ')
```

The plot of the diffusion coefficients with respect to temperature is shown in Figure 10.5.

The following two examples show the effects of doping concentration on mobility and resistivity.

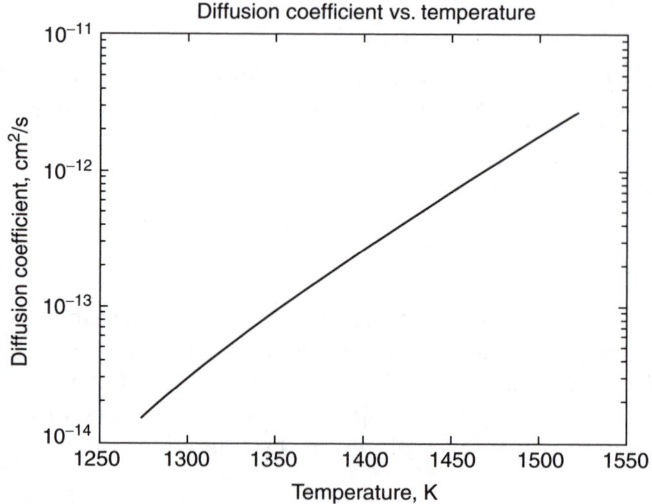

FIGURE 10.5
Diffusion coefficient vs. temperature.

Example 10.4 Electron and Hole Mobilities vs. Doping Concentration

From measured data, an empirical relationship between electron (μ_n) and hole (μ_p) mobilities vs. doping concentration at 300 K is given as[2]

$$\mu_n(N_D) = \frac{5.1 \times 10^{18} + 92 N_D^{0.91}}{3.75 \times 10^{15} + N_D^{0.91}} \tag{10.25}$$

$$\mu_p(N_A) = \frac{2.9 \times 10^{15} + 47.7 N_A^{0.76}}{5.86 \times 10^{12} + N_A^{0.76}} \tag{10.26}$$

where N_D and N_A are donor and acceptor concentration per cm^3, respectively.

Plot the $\mu_n\,(N_D)$ and $\mu_p\,(N_A)$ for doping concentrations from 10^{14} to 10^{20} cm^{-3}.

Solution

MATLAB script

```
% Electron and Hole Mobilities with Doping Concentration
% nc - is doping concentration
%
nc = logspace(14,20);
un = (5.1e18 + 92*nc.^0.91)./(3.75e15 + nc.^0.91);
up = (2.90e15 + 47.7*nc.^0.76)./(5.86e12 + nc.^0.76);
semilogx(nc,un,'b',nc,up,'b')
text(8.0e16,1000,'Electron Mobility')
text(5.0e14,560,'Hole Mobility')
title('Mobility versus Doping')
xlabel('Doping Concentration in cm-3')
ylabel('Bulk Mobility (cm2/v.s)')
```

Figure 10.6 shows the plot of mobility vs. doping concentration.

Example 10.5 Resistivity vs. Doping

At the temperature of 300 K, the resistivity of silicon doped by phosphorus is given as[3]

$$\rho_n = \frac{3.75 \times 10^{15} + N_D^{0.91}}{1.47 \times 10^{-17} N_D^{1.91} + 8.15 \times 10^{-1} N_D} \tag{10.27}$$

A similar relation for silicon doped with boron is given as[4]

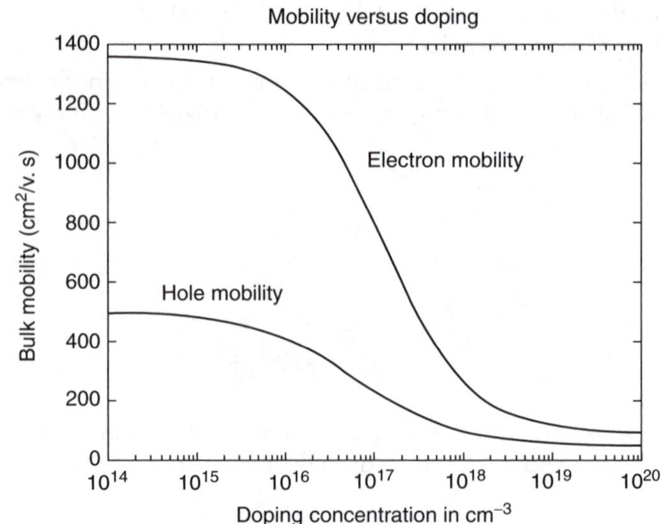

FIGURE 10.6
Mobility vs. doping concentration.

$$\rho_p = \frac{5.86 \times 10^{12} + N_A^{0.76}}{7.63 \times 10^{-18} N_A^{1.76} + 4.64 \times 10^{-4} N_A} \tag{10.28}$$

where N_D and N_A are donor and acceptor concentration, respectively.

Use MATLAB to plot the resistivity vs. doping concentration (cm^{-3}).

Solution

MATLAB script

```
% Resistivity vs. Doping Concentration
% nc is doping concentration
% rn - resistivity of n-type
% rp - resistivity of p-type

nc = logspace(14,20);
rn = (3.75e15 + nc.^0.91)./(1.47e-17*nc.^1.91 + 8.15e-
1*nc);
rp = (5.86e12 + nc.^0.76)./(7.63e-18*nc.^1.76 + 4.64e-
4*nc);

semilogx(nc,rn,'b',nc,rp,'b')
axis([1.0e14, 1.0e17,0,140])
title('Resistivity versus Doping')
ylabel('Resistivity, Ohm-cm')
xlabel('Doping Concentration, cm-3')
text(1.1e14,12,'N-type')
text(3.0e14,50,'P-type')
```

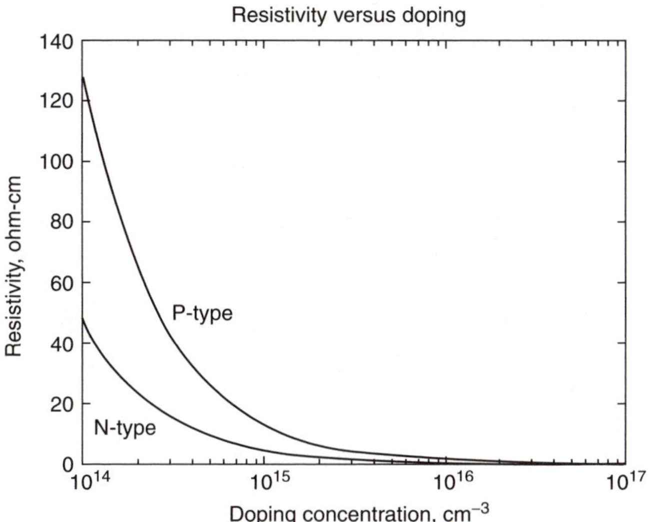

FIGURE 10.7
Resistivity vs. doping concentration.

Figure 10.7 shows the resistivity of n- and p-type silicon.

10.3 pn Junction: Contact Potential, Junction Current

10.3.1 Contact Potential

An ideal pn junction is obtained when a uniformly doped p-type material abruptly changes to n-type material. This is shown in Figure 10.8.

Practical pn junctions are formed by diffusing into an n-type semiconductor a p-type impurity atom, or vice versa. Because the p-type semiconductor has many free holes and the n-type semiconductor has many free electrons, there is a strong tendency for the holes to diffuse from p-type to n-type semiconductors. Similarly, electrons diffuse from n-type to p-type material. When holes cross the junction into n-type material, they recombine with the free electrons in the n-type. Similarly, when electrons cross the junction into the p-type region, they recombine with free holes. In the junction, a transition region or depletion region is created.

In the depletion region, the free holes and electrons are many magnitudes lower than holes in p-type material and electrons in the n-type material. As electrons and holes recombine in the transition region, the region near the junction within the n-type semiconductor is left with a net positive charge. The region near the junction within the p-type material will be left with a net negative charge. This is illustrated in Figure 10.9.

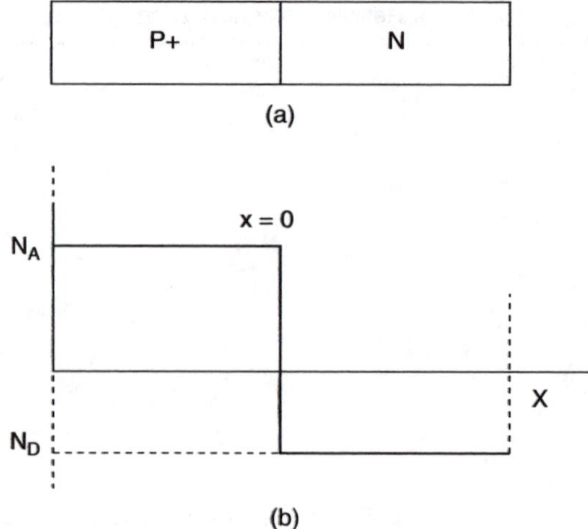

FIGURE 10.8
Ideal pn junction: (a) structure; (b) concentration of donors (N_D) and acceptor (N_A) impurities.

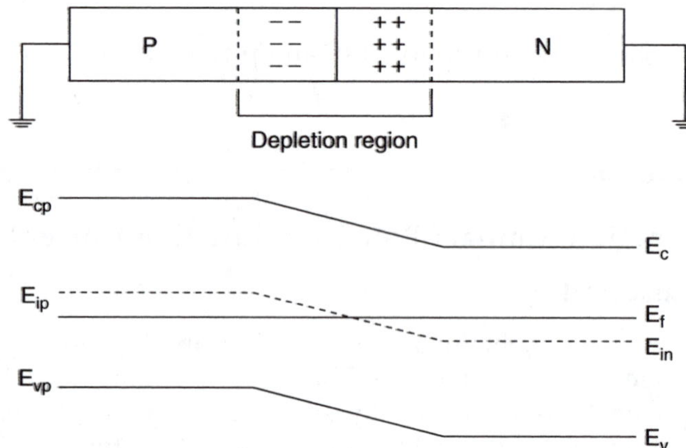

FIGURE 10.9
Pn junction: (top) depletion region with positive and negative ions; (bottom) energy band diagram near a pn junction.

Because of the positive and negative fixed ions in the transition region, an electric field is established across the junction. The electric field creates a potential difference across the junction, the potential barrier. The latter is also called diffusion potential or contact potential, V_C. The potential barrier prevents the flow of majority carriers across the junction under equilibrium conditions.

The contact potential, V_C, may be obtained from the relations

$$\frac{n_n}{n_p} = e^{\left(\frac{qV_C}{kT}\right)} = \frac{p_p}{p_n} \tag{10.29}$$

or

$$V_C = \frac{kT}{q}\ln\left(\frac{n_n}{n_p}\right) = \frac{kT}{q}\ln\left(\frac{p_p}{p_n}\right) \tag{10.30}$$

But, noting that $p_p \cong N_A$, $n_p \cong \frac{n_i^2}{N_A}$, $n_n \cong N_D$, $p_n \cong \frac{n_i^2}{N_D}$, Equation (10.30) becomes

$$V_C = \frac{kT}{q}\ln\left(\frac{N_A N_D}{n_i^2}\right) \tag{10.31}$$

The contact potential can also be obtained from the band-bending diagram of the pn junction shown in Figure 10.9. That is, from Figure 10.9

$$V_C = \frac{E_{in} - E_{ip}}{q} \tag{10.32}$$

or

$$V_C = -\left(\phi_{fn} + \left|\phi_{fp}\right|\right) \tag{10.33}$$

where ϕ_{FN} and ϕ_{FP} are the electron and hole Fermi potentials, respectively. They are given as

$$\phi_{FN} = \frac{E_F - E_{IN}}{q} = \frac{kT}{q}\ln\left(\frac{N_D}{n_i}\right) \tag{10.34}$$

and

$$\phi_{FP} = \frac{E_F - E_{IP}}{q} = \frac{kT}{q}\ln\left(\frac{N_A}{n_i}\right) \tag{10.35}$$

Using Equation (10.32) to Equation (10.35), we have

$$V_C = \frac{kT}{q} \ln\left(\frac{N_A N_D}{n_i^2}\right) \tag{10.36}$$

It should be noted that Equation (10.31) and Equation (10.36) are identical. Typically, V_C is from 0.5 to 0.8 V for the silicon pn junction. For germanium, V_C is approximately 0.1 to 0.2 V, and for gallium arsenide it is 1.5 V.

When a positive voltage V_S is applied to the p-side of the junction and the n-side is grounded, holes are pushed from the p-type material into the transition region. In addition, electrons are attracted to the transition region. The depletion region decreases, and the effective contact potential is reduced. This allows majority carriers to flow through the depletion region. Equation (10.28) modifies to

$$\frac{n_n}{n_p} = e^{\left[\frac{q(V_C - V_S)}{kT}\right]} = \frac{p_p}{p_n} \tag{10.37}$$

When a negative voltage V_S is applied to the p-side of a junction and the n-side is grounded, the applied voltage adds directly to the contact potential. The depletion region increases, and it becomes more difficult for the majority carriers to flow across the junction. The current flow is mainly due to the flow of minority carriers. Equation (10.29) modifies to

$$\frac{n_n}{n_p} = e^{\left[\frac{q(V_C + V_S)}{kT}\right]} = \frac{p_p}{p_n} \tag{10.38}$$

Figure 10.10 shows the potential across the diode when a pn junction is forward-biased and reverse-biased.

The following example illustrates the effect of source voltage on the junction potential.

Example 10.6 Junction Potential vs. Voltage

For a silicon pn junction with $N_D = 10^{14}$ cm^{-3} and $N_A = 10^{17}$ cm^{-3} and with $n_i^2 = 1.04 \times 10^{26}$ cm^{-6} at $T = 300$ K:

(a) Calculate the contact potential.

(b) Plot the junction potential when the source voltage V_S of Figure 10.10 increases from –1.0 to 0.7 V.

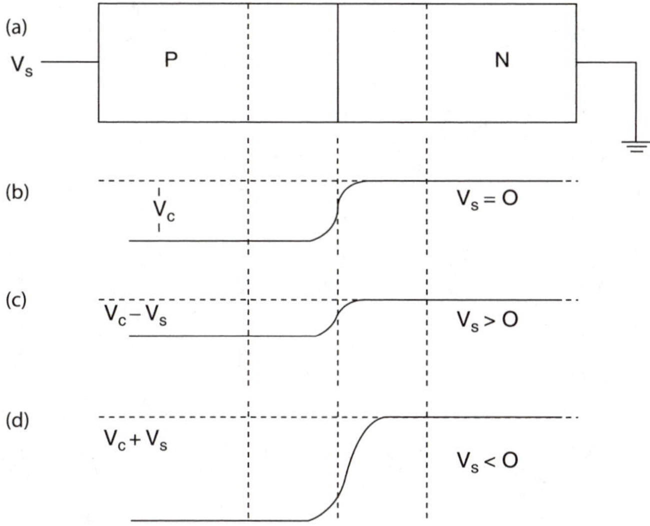

FIGURE 10.10
Pn junction with (a) depletion layer and source connection; (b) contact potential with no source voltage ($V_S = 0$); (c) junction potential for forward-biased pn junction ($V_S > 0$); and (d) junction potential for reverse-biased pn junction ($V_S < 0$).

Solution

MATLAB script

```
% Junction potential versus source voltage
% using equation(10.37) contact potential is

t = 300;
na = 1.0e17;
nd = 1.0e14;
nisq = 1.04e20;
q = 1.602e-19;
k = 1.38e-23;

% calculate contact potential
vc = (k*t/q)*(log(na*nd/nisq))
vs = -1.0:0.1:0.7;
jct_pot = vc - vs;

% plot curve
plot(vs, jct_pot,'b', vs,jct_pot,'ob')
title('Junction Potential vs. Source Voltage')
xlabel('Source Voltage, V')
ylabel('Junction Potential, V')
```

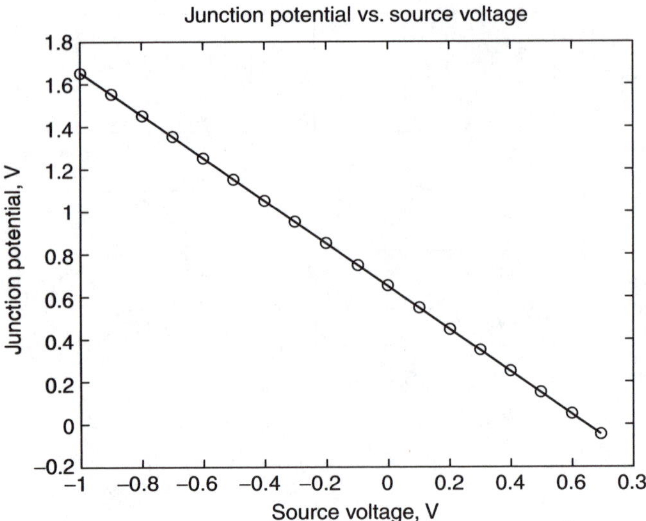

FIGURE 10.11
Junction potential vs. source voltage.

(a) The contact potential is

```
vc =
         0.6535
```

(b) Figure 10.11 shows the graph of the junction potential vs. the source voltage.

10.3.2 Junction Current

The pn junction current is given as

$$I = I_s \left[e^{\left(\frac{qV_S}{kT}\right)} - 1 \right] \tag{10.39}$$

where V_S is the voltage across the pn junction, q is the electronic charge, T is the absolute temperature, k is Boltzmann's constant, and I_S is the reverse saturation current. It is given as

$$I_S = qA\left(\frac{D_p p_n}{L_p} + \frac{D_n n_p}{L_n} \right) \tag{10.40}$$

where A is the diode cross-sectional area, L_p, L_n are the hole and electron diffusion lengths, p_n, n_p are the equilibrium minority carrier concentrations, and D_p, D_n are the hole and electron diffusion coefficients, respectively.

Since

$$p_n \cong \frac{n_i^2}{N_D}$$

and

$$n_p \cong \frac{n_i^2}{N_A}$$

Equation (10.40) becomes

$$I_S = qA\left(\frac{D_p}{L_p N_D} + \frac{D_n}{L_n N_A}\right)n_i^2 \tag{10.41}$$

The diffusion coefficient and diffusion length are related by the expressions

$$L_p = \sqrt{D_p \tau_p} \tag{10.42}$$

and

$$L_n = \sqrt{D_n \tau_n} \tag{10.43}$$

where τ_p, τ_n are the hole minority and electron minority carrier lifetime, respectively.

Equation (10.39) is the diode equation. It is applicable for forward-biased ($V_S > 0$) and reverse-biased ($V_S < 0$) pn junctions.

Using Equation (10.1) and Equation (10.40), the reverse saturation current can be rewritten as

$$I_S = k_1 T^3 e^{\left[-E_g/(kT)\right]} \tag{10.44}$$

where k_1 is a proportionality constant

$$\frac{dI_S}{dT} = 3k_1 T^2 e^{-\frac{E_g}{kT}} + k_1 T^3 \frac{-E_g}{kT^2} e^{-\frac{E_g}{kT}}$$

Thus

$$\frac{1}{I_S}\frac{dI_S}{dT} = \frac{3}{T} + \frac{1}{T}\frac{E_g}{kT} = \frac{3}{T} + \frac{1}{T}\frac{V_g}{V_T}$$

where

$$V_T = \frac{kT}{q}$$

and

$$V_g = \frac{E_g}{q}$$

For silicon at room temperature,

$$\frac{V_g}{V_T} = 44.4$$

Thus,

$$\frac{dI_S}{dT} = (3 + \frac{V_g}{V_T})\frac{dT}{T} = 47.4\frac{dT}{T} \tag{10.45}$$

At room temperature (300 K), the saturation current approximately doubles every 5°C [5]. The following example shows how I_S is affected by temperature.

Example 10.7 Effects of Temperature on Reverse Saturation Current

A silicon diode has $I_S = 10^{-15}$ A at 25°C. Assuming I_S increases by 15% per degree Celsius rise in temperature, find and plot the value of I_S from 25 to 125°C.

Solution

From the information given above, the reverse saturation current can be expressed as

$$I_S = 10^{-15}(1.15)^{(T-25)}$$

MATLAB is used to find I_S at various temperatures.

MATLAB script

```
% Reverse Saturation Current vs. temperature
%
t = 25:5:125;
is = 1.0e-15*(1.15).^(t-25);
plot(t,is,'b', t,is,'ob')
title('Reverse Saturation Current vs. Temperature')
xlabel('Temperature, C')
ylabel('Current, A')
```

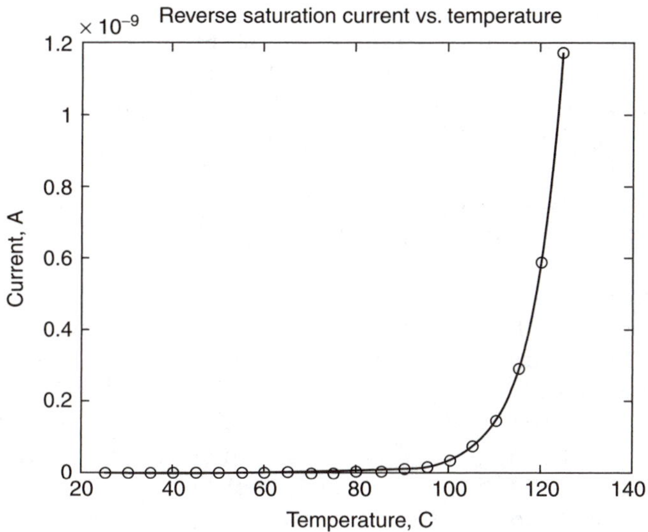

FIGURE 10.12
Reverse saturation current vs. temperature.

Figure 10.12 shows the effect of temperature on the reverse saturation current.

Example 10.8 Equation of Best Fit for Diode Data

A forward-biased diode has the corresponding voltage and current shown in Table 10.1. (a) Determine the equation of best fit. (b) For the voltage of 0.58 V, what is the diode current?

TABLE 10.1

Voltage vs. Current of a Diode

Forward-Biased Voltage, V	Forward Current, A
0.1	1.33E–13
0.2	1.79E–12
0.3	24.02E–12
0.4	0.321E–9
0.5	4.31E–9
0.6	57.69E–9
0.7	7.72E–7

Solution

MATLAB script

```
% Equation of Best Fit for Diode Data
% Diode parameters
```

```
vt = 25.67e-3;
vd =  [0.1 0.2  0.3  0.4  0.5  0.6  0.7];
id = [1.33e-13  1.79e-12  24.02e-12  ...
      0.321e-9  4.31e-9  57.69e-9  7.72e-7]; %
lnid = log(id);% Natural log of current
% Determine coefficients
pfit = polyfit (vd, lnid, 1);% curve fitting
% Linear equation is y= mx + b
b = pfit(2);
m = pfit(1);
ifit = m*vd + b;
% Calculate current when diod voltage is 0.64 V
Ix = m*0.58 + b;
I_58v = exp(Ix);
% Plot v versys ln(i) and best fit linear model
plot(vd, ifit, 'b', vd, lnid, 'ob')
xlabel ('Voltage, V')
ylabel ('ln(i), A')
title ('Best Fit Linear Model')
fprintf('Diode current for voltage of 0.58V is
%9.3e\n', I_58v)
```

The plot is shown in Figure 10.13.
The MATLAB result is

```
Diode current for voltage of 0.58V is 3.433e-008 A
```

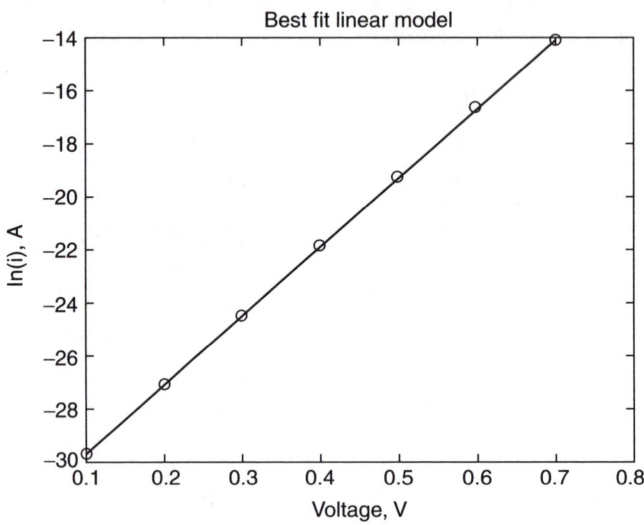

FIGURE 10.13
Best fit linear model for diode data.

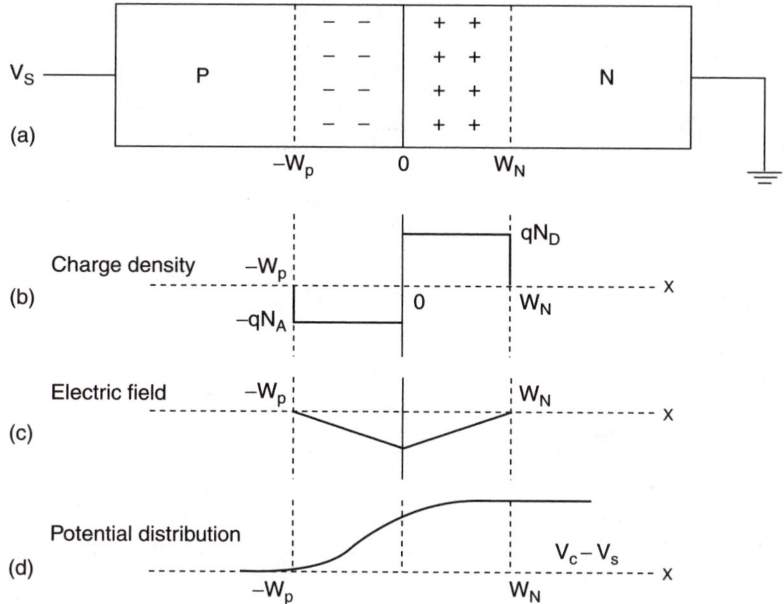

FIGURE 10.14
PN junction with abrupt junction: (a) depletion region; (b) charge density; (c) electric field; and (d) potential distribution.

10.4 Depletion and Diffusion Capacitances

10.4.1 Depletion Capacitance

As mentioned previously, a pn junction is formed when a p-type material is joined to an n-type region. During device fabrication, a pn junction can be formed using processes such as ion-implantation diffusion or epitaxy. The doping profile at the junction can take several shapes. Two popular doping profiles are abrupt (step) junction and linearly graded junction.

In the abrupt junction, the doping of the depletion region on either side of the metallurgical junction is a constant. This gives rise to constant charge densities on either side of the junction. This is shown in Figure 10.14.

For charge equality

$$qN_A W_P = qN_D W_N \qquad (10.46)$$

it can be shown [6] that the depletion width in the p-type (W_P) and that of the n-type material (W_N) can be given as

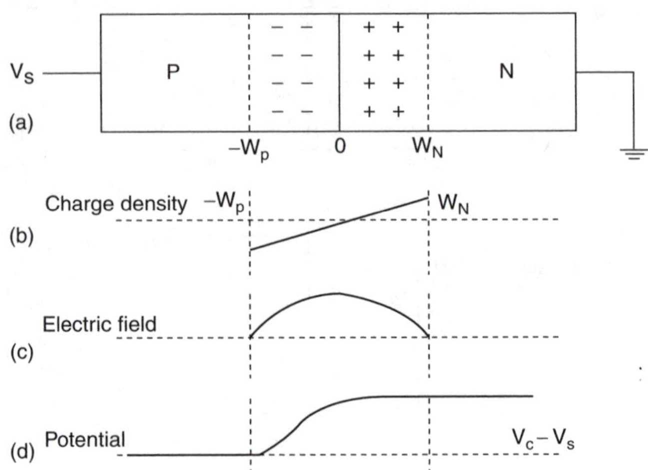

FIGURE 10.15
PN junction with linearly graded junction: (a) depletion region; (b) charge density; (c) electric field; and (d) potential distribution.

$$W_P = \sqrt{\frac{2\varepsilon N_D(V_C - V_s)}{qN_A(N_D + N_A)}} \tag{10.47}$$

$$W_N = \sqrt{\frac{2\varepsilon N_A(V_C - V_s)}{qN_D(N_D + N_A)}} \tag{10.48}$$

where ε is the relative dielectric constant ($\varepsilon = 12\varepsilon_0$ for Si, and $\varepsilon_0 = 8.85 \times 10^{-12}$ F/m), N_D is donor concentration, N_A is acceptor concentration, q is electronic charge, V_C is contact potential obtained from Equation (10.31), and V_S is source voltage.

If the doping density on one side of the metallurgical junction is greater than that on the other side (i.e., $N_A >> N_D$ or $N_D >> N_A$), then the junction properties are controlled entirely by the lightly doped side. This condition is termed the one-sided step junction approximation. This is the practical model for shallow junctions formed by a heavily doped diffusion into a lightly doped region of opposite polarity.[7]

In a linearly graded junction, the ionized doping charge density varies linearly across the depletion region. The charge density passes through zero at the metallurgical junction. Figure 10.15 shows the profile of the linearly graded junction.

For a linearly graded junction, the depletion width in the p-type and n-type material, on either side of the metallurgical junction, can be shown to be

$$W_N = |W_P| = \left[\frac{12\varepsilon(V_C - V_S)}{qa}\right]^{\frac{1}{3}} \tag{10.49}$$

where a is the slope of the graded junction impurity profile.

The contact potential is given as[6]

$$V_C = \frac{kT}{q}\ln\left(\frac{aW_N}{2n_i}\right) \tag{10.50}$$

The depletion capacitance, C_j, is due to the charge stored in the depletion region. It is generally given as

$$C_j = \frac{\varepsilon A}{W_T} \tag{10.51}$$

where

$$W_T = W_N + |W_P| \tag{10.52}$$

A is the cross-sectional area of the pn junction.

For an abrupt junction, the depletion capacitance is given as

$$C_j = A\sqrt{\frac{\varepsilon q N_A N_D}{2(N_D + N_A)(V_C - V_S)}} \tag{10.53}$$

For a linearly graded junction, the depletion capacitance is given as

$$C_j = 0.436(aq)^{1/3}\varepsilon^{2/3}A(V_C - V_S)^{-1/3}$$

$$C_j = 0.436A\left[\frac{aq\varepsilon^2}{(V_C - V_S)}\right]^{1/3} \tag{10.54}$$

In general, we may express the depletion capacitance of a pn junction by

$$C_j = \frac{C_{j0}}{\left[1 - \frac{V_S}{V_C}\right]^m} \qquad \frac{1}{3} \le m \le \frac{1}{2} \tag{10.55}$$

where

$$m = \frac{1}{3}$$

for a linearly graded junction and

$$m = \frac{1}{2}$$

for a step junction.

C_{j0} = zero-biased junction capacitance. It can be obtained from Equation (10.53) and Equation (10.54) by setting V_S equal to zero.

Equation (10.53) to Equation (10.55) are, strictly speaking, valid under the conditions of reverse-biased $V_S < 0$. The equations can, however, be used when $V_S < 0.2$ V. The positive voltage, V_C, is the contact potential of the pn junction. As the pn junction becomes more reversed biased ($V_S < 0$), the depletion capacitance decreases. However, when the pn junction becomes slightly forward biased, the capacitance increases rapidly. This is illustrated by the following example.

Example 10.9 Depletion Capacitance of a pn Junction

For a certain pn junction, with contact potential 0.065 V, the junction capacitance is 4.5 pF for $V_S = -10$ and C_j is 6.5 pF for $V_S = -2$ V.

(a) Find m and C_{j0} of Equation (10.55).

(b) Use MATLAB to plot the depletion capacitance from –30 to 0.4 V.

Solution

From Equation (10.55),

$$C_{j1} = \frac{C_{j0}}{\left[1 - \frac{V_{S1}}{V_C}\right]^m}$$

$$C_{j2} = \frac{C_{j0}}{\left[1 - \frac{V_{S2}}{V_C}\right]^m}$$

Therefore,

$$\frac{C_{j1}}{C_{j2}} = \left[\frac{V_C - V_{S2}}{V_C - V_{S1}}\right]^m$$

$$m = \frac{\log_{10}\left[\frac{C_{j1}}{C_{j2}}\right]}{\log_{10}\left[\frac{V_C - V_{S2}}{V_C - V_{S1}}\right]} \tag{10.56}$$

and

$$C_{j0} = C_{j1}\left[1 - \frac{V_{S1}}{V_C}\right]^m \tag{10.57}$$

MATLAB is used to find m and C_{j0}. It is also used to plot the depletion capacitance.

MATLAB script

```
% depletion capacitance
%
cj1 = 4.5e-12; vs1 = -10;
cj2 = 6.5e-12; vs2 = -2;
vc = 0.65;

num = cj1/cj2;
den = (vc-vs2)/(vc-vs1);
m = log10(num)/log10(den);
cj0 = cj1*(1 - (vs1/vc))^m;
vs = -30:0.2:0.4;
k = length(vs);
for i = 1:k
 cj(i) = cj0/(1-(vs(i)/vc))^m;
end
plot(vs,cj,'b')
xlabel('Voltage, V')
ylabel('Capacitance, F')
title('Depletion Capacitance vs. Voltage')
axis([-30,2,1e-12,14e-12])
m
cj0
```

(a) The values of m, C_{j0} are

```
m   =   0.02644

cj0 =   9.4246e-012
```

(b) Figure 10.16 shows the depletion capacitance vs. the voltage across the junction.

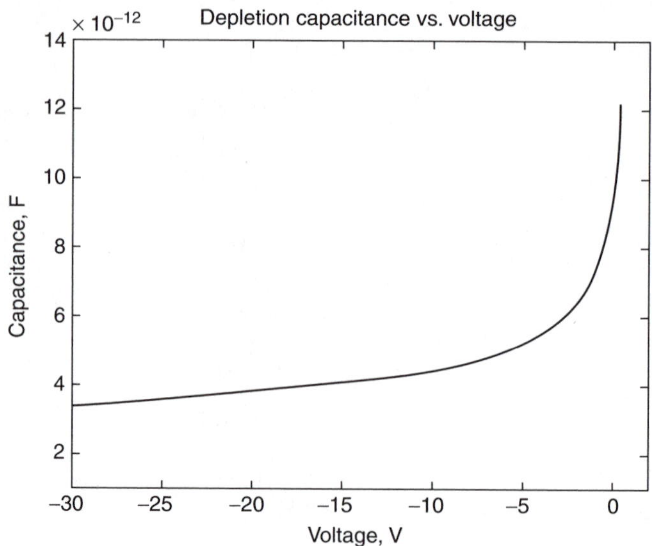

FIGURE 10.16
Depletion capacitance of a pn junction.

10.4.2 Diffusion Capacitance

When a pn junction is forward biased, holes are injected from the p-side of the metallurgical junction into the n-type material. The holes are momentarily stored in the n-type material before they recombine with the majority carriers (electrons) in the n-type material. Similarly, electrons are injected into and temporarily stored in the p-type material. The electrons then recombine with the majority carriers (holes) in the p-type material. The diffusion capacitance, C_d, is due to the buildup of minority carriers' charge around the metallurgical junction as the result of forward biasing the pn junction. Changing the forward current or forward voltage, ΔV, will result in the change in the value of the stored charge ΔQ. The diffusion capacitance, C_d, can be found from the general expression

$$C_d = \frac{\Delta Q}{\Delta V} \tag{10.58}$$

It turns out that the diffusion capacitance is proportional to the forward-biased current. That is,

$$C_d = K_d I_{DF} \tag{10.59}$$

where K_d is constant at a given temperature and I_{DF} is forward-biased diode current.

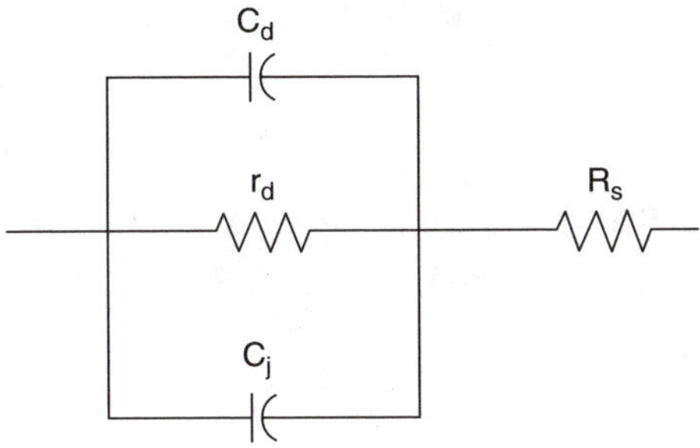

FIGURE 10.17
Small-signal model of a forward-biased pn junction.

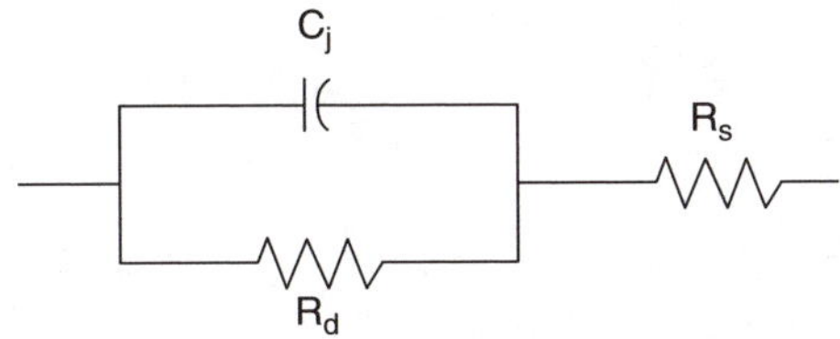

FIGURE 10.18
Model of a reverse-biased pn junction.

The diffusion capacitance is usually larger than the depletion capacitance [1,6]. Typical values of C_d range from 10 to 500 pF. A small signal model of the diode is shown in Figure 10.17.

In Figure 10.17, C_d and C_j are the diffusion and depletion capacitance, respectively. R_S is the semiconductor bulk and contact resistance. The dynamic resistance, r_d, of the diode is given as

$$r_d = \frac{nkT}{qI_{DF}} \tag{10.60}$$

where n is constant, k is Boltzmann's constant, T is temperature in Kelvin, and q is electronic charge.

When a pn junction is reversed biased, $C_d = 0$. The model of the diode is shown in Figure 10.18. In Figure 10.18, C_j is the depletion capacitance. The diffusion capacitance is zero. The resistance R_d is reverse resistance of the pn junction (normally in the mega-ohms range).

Example 10.10 Diffusion and Depletion Capacitances as a Function of Voltage

A certain diode has contact potential; $V_C = 0.55$ V, C_{j0} = diffusion capacitance at zero biased is 8 pF; the diffusion capacitance at 1 mA is 100 pF. Use MATLAB to plot the diffusion and depletion capacitance for forward-biased voltages from 0.0 to 0.7 V. Assume that $I_S = 10^{-14}$ A, $n = 2.0$ and step-junction profile.

Solution

Using Equation (10.39) and Equation (10.59), we write the MATLAB program to obtain the diffusion and depletion capacitances.

MATLAB script

```
%
% Diffusion and depletion Capacitance
%
cd1 = 100e-12; id1 = 1.0e-3; cj0 = 8e-12; vc =0.55;
m = 0.5;
is = 1.0e-14; nd = 2.0;
k = 1.38e-23; q = 1.6e-19; T = 300;
kd = cd1/id1;
vt = k*T/q;
v = 0.0:0.05:0.55;
nv = length(v);

for i = 1:nv
 id(i) = is*exp(v(i)/(nd*vt));
 cd(i) = kd*id(i);
 ra(i) = v(i)/vc;
 cj(i) = cj0/((1 - ra(i)).^m);
end

subplot(121)
plot(v,cd)
title('Diffusion Cap.')
xlabel('Voltage, V'), ylabel('Capacitance, F')
subplot(122)
plot(v,cj)
title('Depletion Cap.')
xlabel('Voltage, V'), ylabel('Capacitance, F')
```

Figure 10.19 shows the depletion and diffusion capacitances of a forward-biased pn junction.

10.5 Breakdown Voltages of pn Junctions

The electric field E is related to the charge density through the Poisson's equation

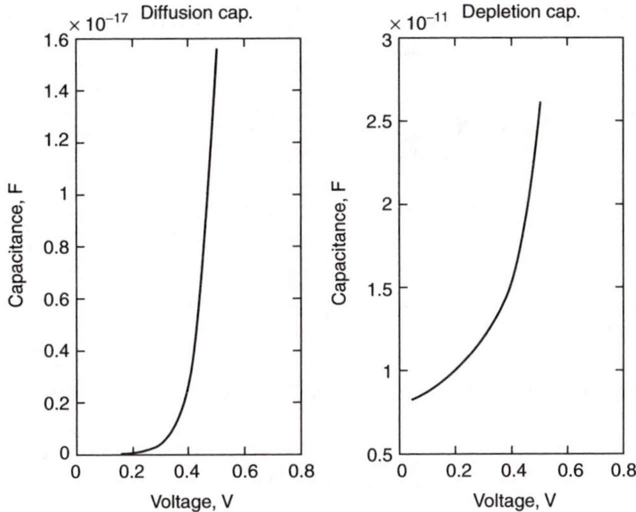

FIGURE 10.19
(Left) Depletion and (right) diffusion capacitance.

$$\frac{dE(x)}{dx} = \frac{\rho(x)}{\varepsilon_S \varepsilon_0} \tag{10.61}$$

where ε_S is the semiconductor dielectric constant, ε_0 is the permittivity of free space, 8.86×10^{-14} F/cm, and $\rho(x)$ is the charge density.

For the abrupt junction with charge density shown in Figure 10.12, the charge density

$$\begin{aligned} \rho(x) &= -qN_A \qquad & -W_P < x < 0 \\ &= qN_D \qquad & 0 < x < W_N \end{aligned} \tag{10.62}$$

The maximum electric field

$$|E_{\max}| = \frac{qN_A W_P}{\varepsilon_s \varepsilon_0} = \frac{qN_D W_N}{\varepsilon_s \varepsilon_0} \tag{10.63}$$

Using Equation (10.47) or Equation (10.48), Equation (10.63) becomes

$$|E_{\max}| = \sqrt{\frac{2qN_D N_A (V_C - V_S)}{\varepsilon_S \varepsilon_0 (N_A + N_D)}} \tag{10.64}$$

For a linearly graded junction, the charge density, $\rho(x)$, is given as (see Figure 10.13)

$$\rho(x) = ax \qquad -\frac{W}{2} < x < \frac{W}{2} \tag{10.65}$$

and the maximum electric field can be shown to be

$$\left|E_{\max}\right| = \frac{aq}{8\varepsilon_S\varepsilon_0}W^2 \tag{10.66}$$

where a is slope of charge density, W is the width of the depletion layer, and

$$\frac{W}{2} = W_N = W_P$$

The width of the depletion region, W, can be obtained from Equation (10.49).

Equation (10.64) indicates that as the reverse voltage increases, the magnitude of the electric field increases. The large electric field accelerates the carriers crossing the junction. At a critical field, E_{crit}, the accelerated carriers in the depletion region have sufficient energy to create new electron-hole pairs as they collide with other atoms. The secondary electrons can in turn create more carriers in the depletion region. This is termed the avalanche breakdown process. For silicon with an impurity concentration of 10^{16} cm^{-3}, the critical electric field is about 2.0×10^5 V/cm.

In a highly doped pn junction, where the impurity concentration is about 10^{18} cm^{-3}, the critical electric field is about 10^6 V/cm. This high electric field is able to strip electrons away from the outer orbit of the silicon atoms, thus creating hole–electron pairs in the depletion region. This mechanism of breakdown is called zener breakdown. This breakdown mechanism does not involve any multiplication effect. Normally, when the breakdown voltage is less than 6 V, the mechanism is a zener breakdown process. For breakdown voltages beyond 6 V, the mechanism is generally an avalanche breakdown process.

For an abrupt junction, where one side is heavily doped, the electrical properties of the junction are determined by the lightly doped side. Experimentally, the breakdown voltage of a semiconductor step junction (n^+p or p^+n) as the function of doping concentration in the lightly doped side is given as [7]

$$V_{BR} = k\left[\frac{N_B}{10^{16}}\right]^{-0.75} \tag{10.67}$$

where $k = 25$ V for Ge and 60 V for Si, and N_B is the doping concentration of the lightly doped side.

The following example shows the effect of doping concentration on breakdown voltage.

Example 10.11 Effect of Doping Concentration on the Breakdown Voltage of a pn Junction

Use MATLAB to plot the breakdown voltage vs. doping concentration for a one-sided step junction for silicon and germanium, and using doping concentrations from 10^{14} to 10^{19} cm^{-3}.

Solution

Using Equation (10.67), we calculate the breakdown voltage for various doping concentrations.

MATLAB script

```
%
% Breakdown voltage
%
k1 = 25;
 k2 = 60;
nb = logspace(14,19);
n = length(nb);

for i = 1:n
 vbr1(i) = k1*(nb(i)/1.0e16)^(-0.75); % Ge breakdown
voltage
 vbr2(i) = k2*(nb(i)/1.0e16)^(-0.75); % Si breakdown
voltage
end

semilogx(nb,vbr1,'w', nb,vbr2,'w')
xlabel('Impurity Concentration, cm-3')
ylabel('Breakdown Voltage,V')
title('Breakdown Voltage vs. Impurity Concentration')
axis([1.0e14,1.0e17,0,2000])
text(2.0e14,270,'Ge')
text(3.0e14,1000,'Si')
```

Figure 10.18 shows the plot of the breakdown voltage of a one-sided abrupt junction.

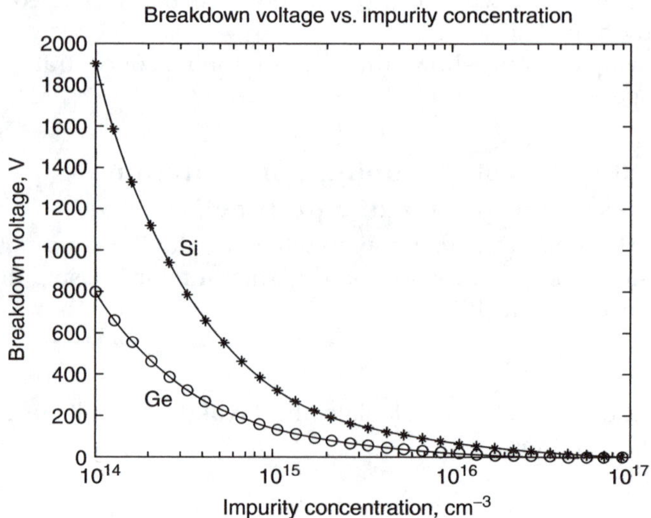

FIGURE 10.20
Breakdown voltage vs. impurity concentration.

References

1. Singh, J., *Semiconductor Devices*, McGraw-Hill, New York, 1994.
2. Jacoboni, C., Canli, C., Ottaviani, G., and Quaranta, A.A., Review of some charge transport properties of silicon, *Solid State Electronics*, 20, 77–89, 1977.
3. Mousty, F., Ostoga, P., and Passari, L., Relationship between resistivity and phosphorus concentration in silicon, *Journal of Applied Physics*, 45, 4576–4580, 1974.
4. Caughey, D.M. and Thomas, R.F., Carrier mobilities in silicon empirically related to doping and field, *Proc. IEEE*, 55, 2192–2193, 1967.
5. Hodges, D.A. and Jackson, H.G., *Analysis and Design of Digital Integrated Circuits*, McGraw-Hill, New York, 1988.
6. Neudeck, G.W., *The PN Junction Diode, Modular Series on Solid State Devices*, Vol. II, Addison-Wesley, Reading, MA, 1983.
7. Beadle, W.E., Tsai, J.C., and Plummer, R.D. (Editors), *Quick Reference Manual for Silicon Integrated Circuits Technology*, John Wiley & Sons, New York, 1985.
8. McFarlane, G.G., McLean, J.P., Quarrington, J.E., and Roberts, V., Fine structure in the absorption edge spectrum of silicon, *Physics Review*, 111, 1245–1254, 1958.
9. Sze, S.M. and Gibbons, G., Avalanche breakdown voltages of abrupt and linearly graded p-n junctions in Ge, Si, GaAs and GaP, *Applied Physics Letters*, 8, 111–113, 1966.
10. Pierret, R.F., *Semiconductor Device Fundamentals*, Addison-Wesley Publishing Company, Reading, MA, 1995.

Problems

Problem 10.1

In the case of silicon for temperatures below 700 K, the density of intrinsic created carriers, n_i, can be approximated as[8]

$$n_i = 3.87 \times 10^{16} T^{3/2} e^{-\left(\frac{7.02\times 10^3}{T}\right)} \tag{10.68}$$

(a) Use MATLAB to plot the intrinsic carrier concentration vs. (1000/T) where T is temperature in Kelvin.

(b) Compare the above relation for intrinsic concentration with that of Example 10.1. Plot the difference between n_i obtained from Equation (10.1) and Equation (10.68). For Equation (10.1), assume that the constant A = 8.702e + 024.

Problem 10.2

Between the temperatures of 275 and 375 K, the intrinsic carrier concentration with respect to temperature can be approximated as[10]:

$$n_i = (9.15\times 10^{19})\left(\frac{T}{300}\right)^2 e^{-0.5928/KT} \tag{10.69}$$

Use MATLAB to plot the intrinsic carrier concentration for the above-mentioned temperature range.

Problem 10.3

Assuming that at 300 K the mobile carrier concentrations of intrinsic germanium and silicon semiconductor materials are 2.390 × 10^{13} and 1.52 × 10^{10}, respectively, use MATLAB to plot the $E_F - E_i$ vs. donor concentration for Ge and Si. Assume donor concentrations from 10^{11} to 10^{16}.

Problem 10.4

Another empirical relationship between electron (μ_n) and hole (μ_p) mobilities vs. doping concentration at 300 K is given as[10]:

$$\mu_n = 92 + \frac{1268}{1 + \left(\frac{N_D}{1.3 \times 10^{17}} \right)^{0.91}} \tag{10.70}$$

$$\mu_p = 54.3 + \frac{406.9}{1 + \left(\frac{N_A}{2.35 \times 10^{17}} \right)^{0.88}} \tag{10.71}$$

Calculate μ_n and μ_p for doping concentrations from 10^{14} to 10^{20} cm^{-3}. Compare your results with those obtained from Equation (10.25) and Equation (10.26).

Problem 10.5

For power devices with breakdown voltages above 100 V and resistivities greater than 1 ohm-cm (n-type silicon) and 3 ohm-cm (p-type silicon), the resistivity vs. doping concentrations can be simplified to

$$\rho_n = 4.596 \times 10^{15} N_D^{-1}$$

$$\rho_{pn} = 1.263 \times 10^{16} N_A^{-1}$$

(a) Use MATLAB to plot resistivity vs. doping concentration (from 10^{12} to 10^{18} cm^{-3}).

(b) Compare your results with those obtained in Example 10.5.

Problem 10.6

For a Ge pn junction with $N_A = 10^{18}$ cm^{-3}, $N_D = 10^{15}$ cm^{-3}, and n_i at 300 K = 2.39×10^{13}:

(a) Calculate the contact potential.

(b) Plot the junction potential for source voltages of –1.0 V to 0.3 V.

Problem 10.7

A forward-biased diode has the corresponding voltage and current shown in Table P10.7. (a) Determine the equation of best fit. (b) For the voltage of 0.32 V, what is the diode current?

TABLE P10.7

Voltage vs. Current of a Diode

Forward-Biased Voltage, V	Forward Current, A
0.1	4.22E–8
0.15	2.84E–7
0.20	1.76E–6
0.25	1.31E–5
0.30	8.52E–5
0.35	4.23E–4

Problem 10.8

For the small signal model of the forward-biased pn junction, shown in Figure 10.17, $R_S = 5\ \Omega$, $r_d = 10\ \Omega$, and $C_d = 110$ pF at I_{DF} of 1 mA. Use MATLAB to plot the equivalent input impedance (magnitude and phase) for frequencies from 10^4 to 10^{10} Hz.

Problem 10.9

Empirically, the breakdown voltage of a linearly graded junction can be approximated as [9]

$$V_{BR} = k\left[\frac{a^{-4}}{10^{21}}\right]^{-0.75}$$

where k = 18 V for Ge or 40 V for Si.

Use MATLAB to plot the breakdown voltage vs. impurity gradient of Ge and Si. Use impurity gradient values from 10^{19} to 10^{24}.

11

Operational Amplifiers

The operational amplifier (op amp) is one of the versatile electronic circuits. It can be used to perform the basic mathematical operations of addition, subtraction, multiplication, and division. Op amps can also be used to do integration and differentiation. Several electronic circuits use an op amp as an integral element, including amplifiers, filters, and oscillators. In this chapter, the basic properties of op amps will be discussed. The nonideal characteristics of the op amp will be illustrated, whenever possible, with example problems solved using MATLAB.

11.1 Properties of the Op Amp

The op amp, from a signal point of view, is a three-terminal device: two inputs and one output. Its symbol is shown in Figure 11.1. The inverting input is designated by the "–" sign and noninverting input by the "+" sign.

An ideal op amp has an equivalent circuit shown in Figure 11.2. It is a difference amplifier, with output equal to the amplified difference of the two inputs.

An ideal op amp has the following properties:

- Infinite input resistance
- Zero output resistance
- Zero offset voltage
- Infinite frequency response
- Infinite common-mode rejection ratio
- Infinite open-loop gain, A

A practical op amp will have large but finite open-loop gain in the range from 10^5 to 10^9 Ω. It also has a very large input resistance 10^6 to 10^{10} Ω. The output resistance might be in the range of 50 to 125 Ω. The offset voltage is small but finite and the frequency response will deviate considerably from

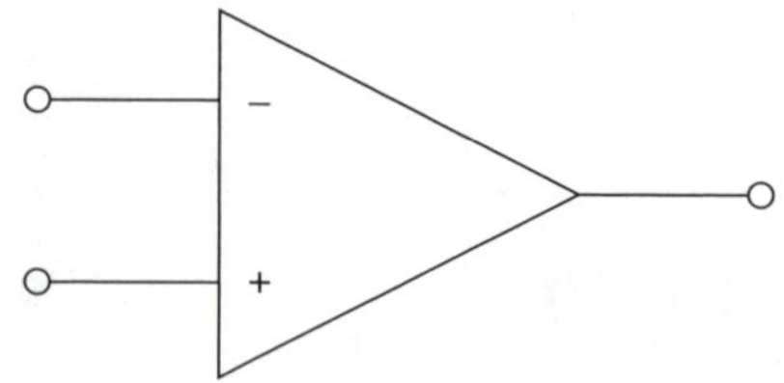

FIGURE 11.1
Op amp circuit symbol.

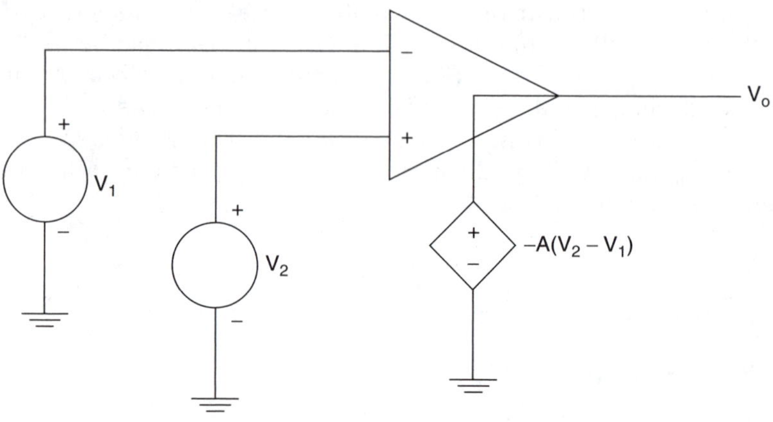

FIGURE 11.2
Equivalent circuit of an ideal op amp.

TABLE 11.1
Properties of 741 Op Amp

Property	Value (Typical)
Open loop gain	2×10^5
Input resistance	2.0 MΩ
Output resistance	75 Ω
Offset voltage	1 mV
Input bias current	30 nA
Unity-gain bandwidth	1 MHz
Common-mode rejection ratio	95 dB
Slew rate	0.7 V/μV

the infinite frequency response. The common-mode rejection ratio is not infinite but finite. Table 11.1 shows the properties of the general-purpose 741 op amp.

Whenever there is a connection from the output of the op amp to the inverting input as shown in Figure 11.3, we have a negative feedback connection.

With negative feedback and finite output voltage, Figure 11.2 shows that

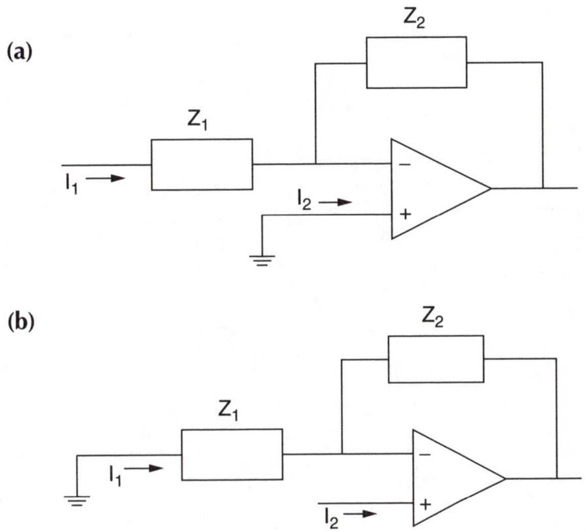

FIGURE 11.3
Negative feedback connections for op amp (a) inverting and (b) noninverting configurations.

$$V_O = A(V_2 - V_1) \tag{11.1}$$

Since the open-loop gain is very large,

$$(V_2 - V_1) = \frac{V_O}{A} \cong 0 \tag{11.2}$$

Equation (11.2) implies that the two input voltages are also equal. This condition is termed the concept of the virtual short circuit. In addition, because of the large input resistance of the op amp, the latter is assumed to take no current for most calculations.

11.2 Inverting Configuration

An op amp circuit connected in an inverted closed-loop configuration is shown in Figure 11.4.

Using nodal analysis at node A, we have

$$\frac{V_a - V_{in}}{Z_1} + \frac{V_a - V_O}{Z_2} + I_1 = 0 \tag{11.3}$$

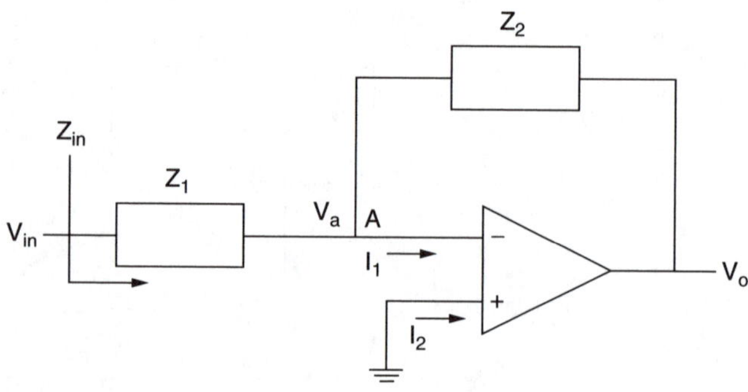

FIGURE 11.4
Inverting configuration of an op amp.

From the concept of a virtual short circuit,

$$V_a = V_b = 0 \tag{11.4}$$

and because of the large input resistance, $I_1 = 0$. Thus, Equation (11.3) simplifies to

$$\frac{V_O}{V_{IN}} = -\frac{Z_2}{Z_1} \tag{11.5}$$

The minus sign implies that V_{IN} and V_0 are out of phase by 180°. The input impedance, Z_{IN}, is given as

$$Z_{IN} = \frac{V_{IN}}{I_1} = Z_1 \tag{11.6}$$

If $Z_1 = R_1$ and $Z_2 = R_2$, we have an inverting amplifier, as shown in Figure 11.5.

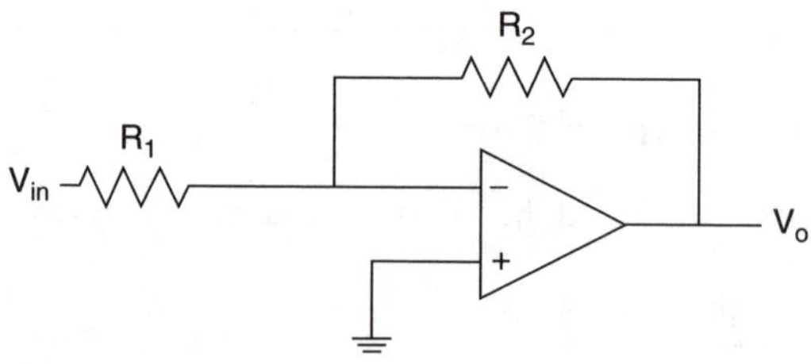

FIGURE 11.5
Inverting amplifier.

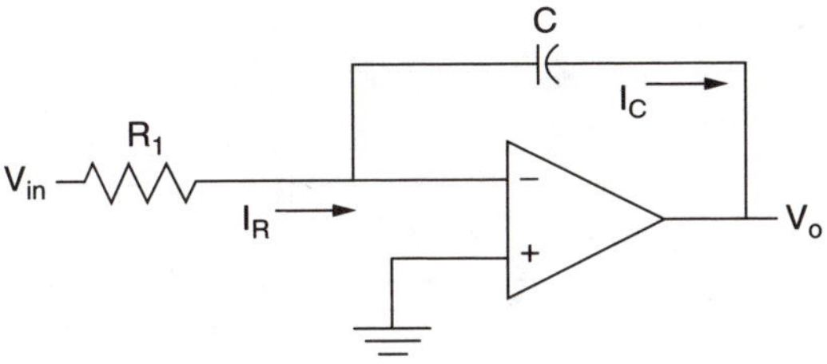

FIGURE 11.6
Op amp inverting integrator.

The closed-loop gain of the amplifier is

$$\frac{V_O}{V_{IN}} = -\frac{R_2}{R_1} \tag{11.7}$$

and the input resistance is R_1. Normally, $R_2 > R_1$ such that $|V_0| > |V_{IN}|$. With the assumptions of very large open-loop gain and high input resistance, the closed-loop gain of the inverting amplifier depends on the external components R_1 and R_2 and is independent of the open-loop gain.

For Figure 11.4, if $Z_1 = R_1$ and $Z_2 = 1/jwC$, we obtain an integrator circuit, as shown in Figure 11.6. The closed-loop gain of the integrator is

$$\frac{V_O}{V_{IN}} = -\frac{1}{jwCR_1} \tag{11.8}$$

In the time domain,

$$\frac{V_{IN}}{R_1} = I_R \text{ and } I_C = -C\frac{dV_O}{dt} \tag{11.9}$$

Since $I_R = I_C$,

$$V_O(t) = -\frac{1}{R_1 C}\int_0^t V_{IN}(t)d\tau + V_O(0) \tag{11.10}$$

The above circuit is termed the Miller integrator. The integrating time constant is CR_1. It behaves as a lowpass filter, passing low frequencies and attenuating high frequencies. However, at zero frequency (dc) the capacitor becomes open circuited and there is no longer a negative feedback from the output to the input. The output voltage then saturates. To provide finite closed-loop gain at dc, a resistance R_2 is connected in parallel with the

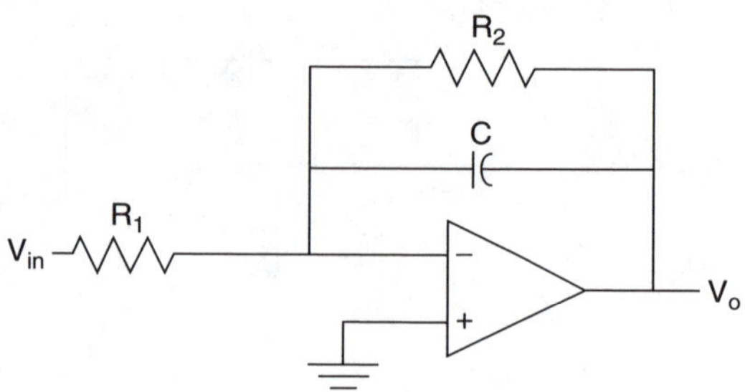

FIGURE 11.7
Miller integrator with finite closed-loop gain at DC.

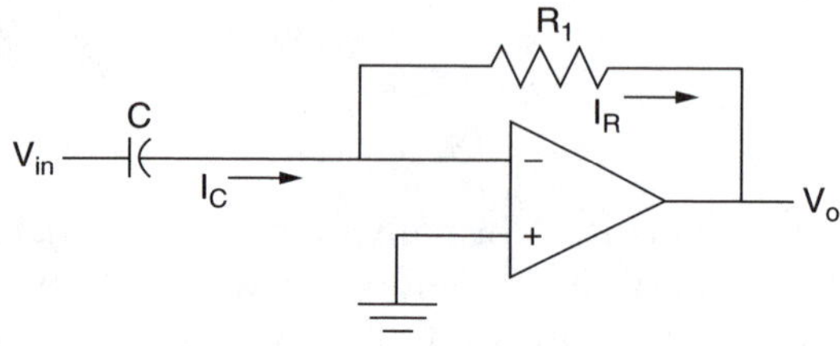

FIGURE 11.8
Op amp differentiator circuit.

capacitor. The circuit is shown in Figure 11.7. The resistance R_2 is chosen such that R_2 is far greater than R_1.

For Figure 11.4, if

$$Z_1 = \frac{1}{jwC}$$

and $Z_2 = R$, we obtain a differentiator circuit, as shown in Figure 11.8. From Equation (11.5), the closed-loop gain of the differentiator is

$$\frac{V_O}{V_{IN}} = -jwCR \tag{11.11}$$

In the time domain,

$$I_C = C\frac{dV_{IN}}{dt}, \text{ and } V_O(t) = -I_R R_1 \tag{11.12}$$

Since

$$I_C(t) = I_R(t)$$

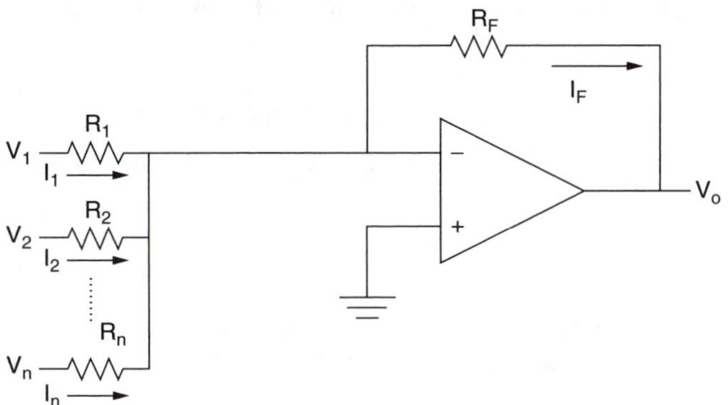

FIGURE 11.9
Weighted summer circuit.

we have

$$V_O(t) = -CR_1 \frac{dV_{IN}(t)}{dt} \tag{11.13}$$

Differentiator circuits will differentiate input signals. This implies that if an input signal is rapidly changing, the output of the differentiator circuit will appear "spike-like."

The inverting configuration can be modified to produce a weighted summer. This circuit is shown in Figure 11.9.

From Figure 11.9,

$$I_1 = \frac{V_1}{R_1},\ I_2 = \frac{V_2}{R_2},\ \ldots,\ I_n = \frac{V_n}{R_n} \tag{11.14}$$

Also,

$$I_F = I_1 + I_2 + \ldots I_N \tag{11.15}$$

$$V_O = -I_F R_F \tag{11.16}$$

Substituting Equation (11.14) and Equation (11.15) into Equation (11.16), we have

$$V_O = -\left(\frac{R_F}{R_1} V_1 + \frac{R_F}{R_2} V_2 + \ldots \frac{R_F}{R_N} V_N \right) \tag{11.17}$$

The frequency response of a Miller integrator, with finite closed-loop gain at dc, is obtained in the following example.

Example 11.1 Frequency Response of a Miller Integrator
For Figure 11.7:

(a) Derive the expression for the transfer function

$$\frac{V_o}{V_{in}}(jw)$$

(b) If $C = 1$ nF and $R_1 = 2$ kΩ, plot the magnitude response for R_2 equal to (i) 100 kΩ, (ii) 300 kΩ, and (iii) 500 kΩ.

Solution

$$Z_2 = R_2 \| \frac{1}{sC_2} = \frac{R_2}{1 + sC_2R_2} \tag{11.18}$$

$$Z_1 = R_1 \tag{11.19}$$

$$\frac{V_o}{V_{in}}(s) = \frac{-\,{R_2}/{R_1}}{1 + sC_2R_2} \tag{11.20}$$

$$\frac{V_o}{V_{in}}(s) = \frac{-\,{1}/{C_2R_1}}{s + {1}/{C_2R_2}} \tag{11.21}$$

MATLAB script

```
% Frequency response of lowpass circuit
c = 1e-9; r1 = 2e3;
r2 = [100e3, 300e3, 500e3];
n1 = -1/(c*r1); d1 = 1/(c*r2(1));
num1 = [n1]; den1 = [1 d1];
w = logspace(-2,6);
h1 = freqs(num1,den1,w);
f = w/(2*pi);
d2 = 1/(c*r2(2)); den2 = [1 d2];
h2 = freqs(num1, den2, w);
d3 = 1/(c*r2(3)); den3 = [1 d3];
h3 = freqs(num1,den3,w);
semilogx(f,abs(h1),'b',f,abs(h2),'b',f,abs(h3),'b')
xlabel('Frequency, Hz')
```

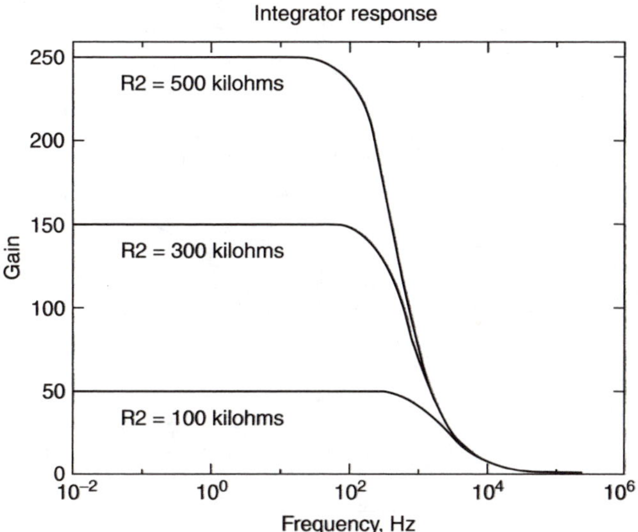

FIGURE 11.10
Frequency response of Miller integrator with finite closed-loop gain at DC.

```
ylabel('Gain')
axis([1.0e-2,1.0e6,0,260])
text(5.0e-2,35,'R2 = 100 Kilohms')
text(5.0e-2,135,'R2 = 300 Kilohms')
text(5.0e-2,235,'R2 = 500 Kilohms')
title('Integrator Response')
```

Figure 11.10 shows the frequency response of Figure 11.7.

11.3 Noninverting Configuration

An op amp connected in a noninverting configuration is shown in Figure 11.11.

Using nodal analysis at node A,

$$\frac{V_a}{Z_1} + \frac{V_a - V_O}{Z_2} + I_1 = 0 \tag{11.22}$$

From the concept of a virtual short circuit,

$$V_{IN} = V_a \tag{11.23}$$

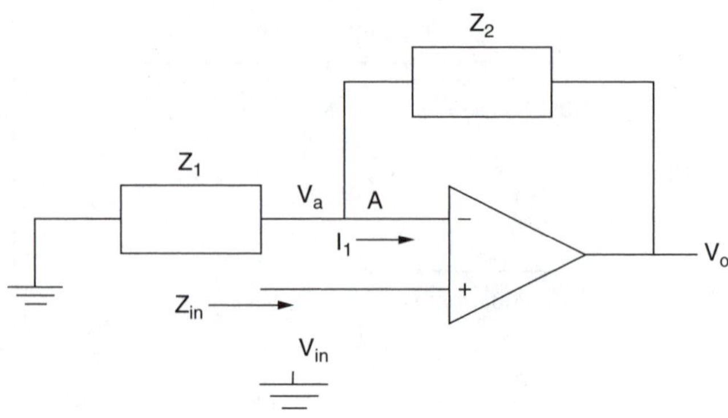

FIGURE 11.11
Noninverting configuration.

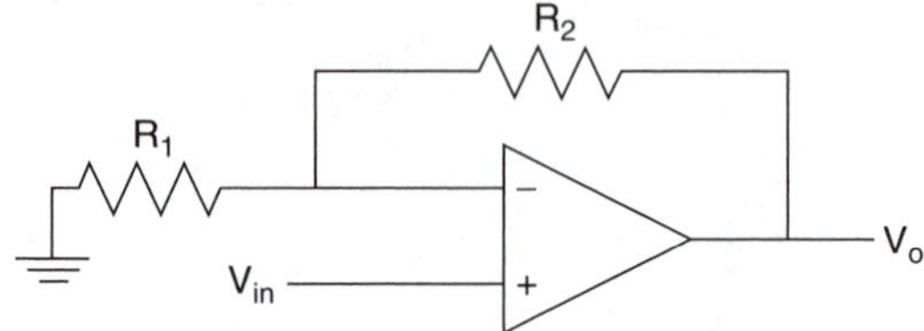

FIGURE 11.12
Voltage follower with gain.

and because of the large input resistance ($i_1 = 0$), Equation (11.22) simplifies to

$$\frac{V_O}{V_{IN}} = 1 + \frac{Z_2}{Z_1} \tag{11.24}$$

The gain of the inverting amplifier is positive. The input impedance of the amplifier Z_{IN} approaches infinity, since the current that flows into the positive input of the op amp is almost zero.

If $Z_1 = R_1$ and $Z_2 = R_2$, Figure 11.10 becomes a voltage follower with gain. This is shown in Figure 11.11.

The voltage gain is

$$\frac{V_O}{V_{IN}} = \left(1 + \frac{R_2}{R_1}\right) \tag{11.25}$$

The **zero,** pole, and frequency response of a noninverting configuration are obtained in Example 11.2.

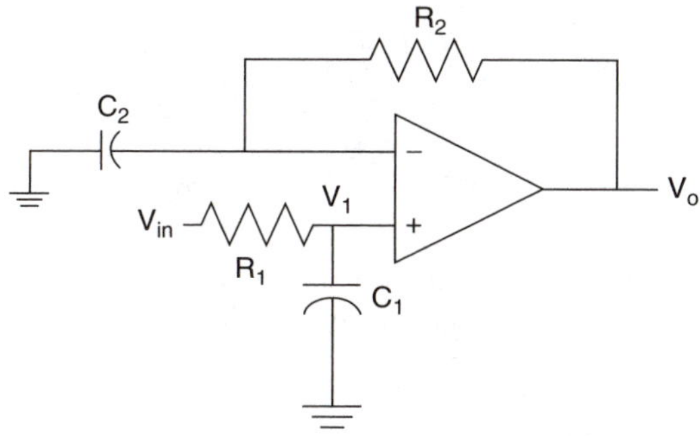

FIGURE 11.13
Noninverting configuration.

Example 11.2 Transfer Function, Pole, and Zero of a Noninverting Op Amp Circuit

For the Figure 11.13 (a) Derive the transfer function. (b) Use MATLAB to find the pole and zero. (c) Plot the magnitude and phase response; assume that $C_1 = 0.1\ \mu F$, $C_2 = 1000$ uF, $R_1 = 10\ k\Omega$, and $R_2 = 10\ \Omega$.

Solution

Using voltage division,

$$\frac{V_1}{V_{IN}}(s) = \frac{1/sC_1}{R_1 + 1/sC_1} \tag{11.26}$$

From Equation (11.24),

$$\frac{V_O}{V_1}(s) = 1 + \frac{R_2}{1/sC_2} \tag{11.27}$$

Using Equation (11.26) and Equation (11.27), we have

$$\frac{V_O}{V_{IN}}(s) = \left(\frac{1 + sC_2R_2}{1 + sC_1R_1}\right) \tag{11.28}$$

The above equation can be rewritten as

$$\frac{V_O}{V_{IN}}(s) = \frac{C_2R_2\left(s + \frac{1}{C_2R_2}\right)}{C_1R_1\left(s + \frac{1}{C_1R_1}\right)} \tag{11.29}$$

The MATLAB program that can be used to find the poles and zero **and** plot the frequency response is as follows:

```
% Pole and zero, frequency response of  a
%        Non-inverting op amp circuit
%
c1 = 1e-7; c2 = 1e-3; r1 = 10e3; r2 = 10;
% poles and zeros
b1 = c2*r2;
a1 = c1*r1;
num = [b1 1];
den = [a1 1];
disp('The zero is')
z = roots(num)
disp('The pole is')
p = roots(den)
% the frequency response
w = logspace(-2,6);
h = freqs(num,den,w);
gain = 20*log10(abs(h));
f = w/(2*pi);
phase = angle(h)*180/pi;
subplot(211),semilogx(f,gain,'b');
xlabel('Frequency, Hz')
ylabel('Gain, dB')
axis([1.0e-2,1.0e6,0,22])
text(2.0e-2,15,'Magnitude Response')
subplot(212),semilogx(f,phase,'b')
xlabel('Frequency, Hz')
ylabel('Phase in Degrees')
axis([1.0e-2,1.0e6,0,75])
text(2.0e-2,60,'Phase Response')
```

The results are:

```
The zero is       z =    -100
The pole is    p =     -1000
```

The magnitude and phase plots are shown in Figure 11.14.

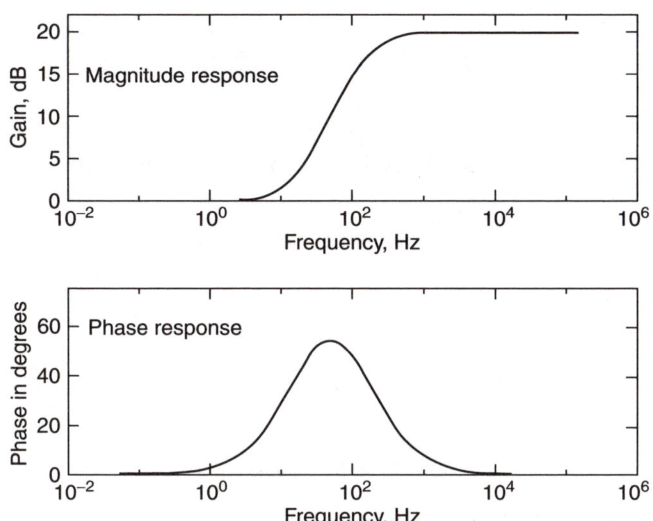

FIGURE 11.14
Frequency response of Figure 11.13.

11.4 Effect of Finite Open-Loop Gain

For the inverting amplifier shown in Figure 11.15, if we assume a finite open-loop gain A, the output voltage V_0 can be expressed as

$$V_O = A(V_2 - V_1)$$

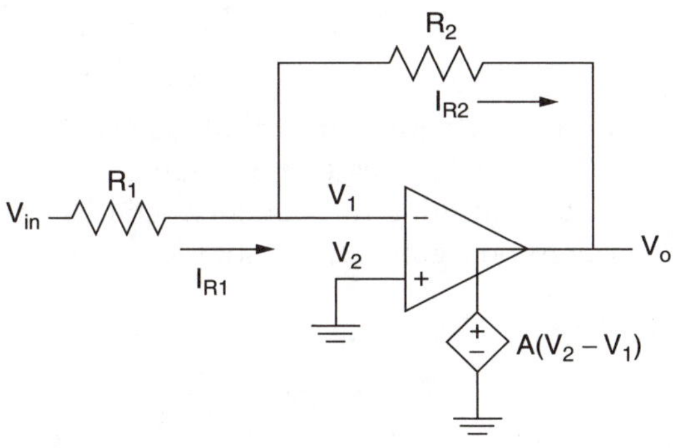

FIGURE 11.15
Inverter with finite open-loop gain.

Since $V_2 = 0$,

$$V_1 = -\frac{V_O}{A} \tag{11.30}$$

Because the op amp has a very high input resistance, we have

$$I_{R1} = I_{R2} \tag{11.31}$$

But

$$I_{R1} = \frac{V_{IN} - V_0/A}{R_1} \tag{11.32}$$

Also

$$V_O = V_1 - I_{R2}R_2 \tag{11.33}$$

Using Equation (11.30), Equation (11.31), and Equation (11.32), Equation (11.33) becomes

$$V_O = -\frac{V_O}{A} - \frac{R_2}{R_1}\left(V_{IN} + V_O/A\right) \tag{11.34}$$

Simplifying Equation (11.34), we get

$$\frac{V_O}{V_{IN}} = -\frac{R_2/R_1}{1+\left(1+R_2/R_1\right)/A} \tag{11.35}$$

It should be noted that as the open-loop gain approaches infinity, the closed-loop gain becomes

$$\frac{V_O}{V_{IN}} \cong -\frac{R_2}{R_1}$$

The above expression is identical to Equation (11.7). In addition, from Equation (11.30), the voltage V_1 goes to zero as the open-loop gain goes to infinity. Furthermore, to minimize the dependence of the closed-loop gain on the value of the open-loop gain, A, we should make

$$\left(1+\frac{R_2}{R_1}\right) << A \tag{11.36}$$

This is illustrated by the following example.

Example 11.3 Effect of Finite Open-Loop Gain

In Figure 11.15, $R_1 = 500\ \Omega$ and $R_2 = 50\ \text{k}\Omega$. Plot the closed-loop gain as the open-loop gain increases from 10^2 to 10^8.

Solution

```
% Effect of finite open-loop gain
a = logspace(2,8);
r1 = 500; r2 = 50e3; r21 = r2/r1;
g = [];
n = length(a);
for i = 1:n
 g(i) = r21/(1+(1+r21)/a(i));
end
semilogx(a,g,'b')
xlabel('Open Loop Gain')
ylabel('Closed Loop Gain')
title('Effect of Finite Open Loop Gain')
axis([1.0e2,1.0e8,40,110])
```

Figure 11.16 shows the characteristics of the closed-loop gain as a function of the open-loop gain.

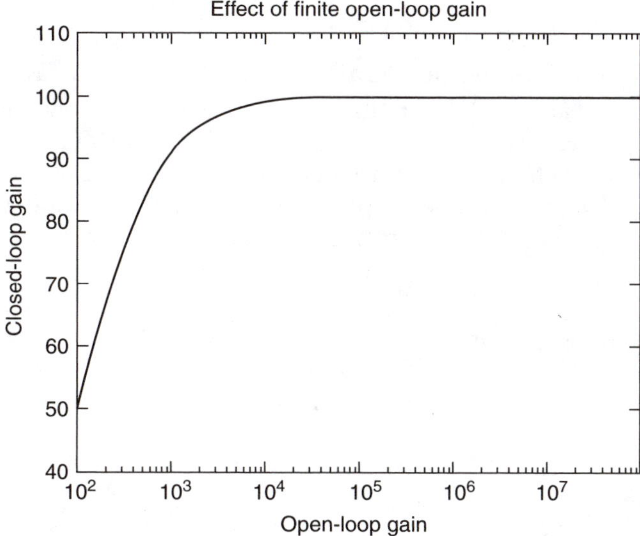

FIGURE 11.16
Closed-loop gain vs. open-loop gain.

For the voltage follower with gain shown in Figure 11.12, it can be shown that the closed-loop gain of the amplifier with finite open-loop gain is

$$\frac{V_O}{V_{IN}} = -\frac{(1+R_2/R_1)}{1+(1+R_2/R_1)/A} \tag{11.37}$$

11.5 Frequency Response of Op Amps

The simplified block diagram of the internal structure of the operational amplifier is shown in Figure 11.17.

Each of the individual sections of the operational amplifier contains a lowpass RC section, with its corner (pole) frequency. Thus, an op amp will have an open-loop gain with frequency that can be expressed as

$$A(s) = \frac{A_O}{(1+s/w_1)(1+s/w_2)(1+s/w_3)} \tag{11.38}$$

where $w_1 < w_2 < w_3$ and A_O is gain at dc.

For most operational amplifiers, w_1 is very small (approx. 20π radians/sec) and w_2 might be in the range of 2 to 6 megaradians/sec.

Example 11.4 Open-Loop Gain Characteristics of an Op Amp

The constituent parts of an operational amplifier have the following internal characteristics: the pole of the difference amplifier is at 200 Hz and the gain is –500. The pole of the voltage amplifier and level shifter is 400 kHz and has a gain of 360. The pole of the output stage is 800 kHz and the gain is 0.92. Sketch the magnitude response of the operational amplifier open-loop gain.

Solution

The lowpass filter response can be expressed as

$$\frac{V_O}{V_{IN}}(jw) = -\frac{C_{rstage}}{1+jf/f_p} \tag{11.39}$$

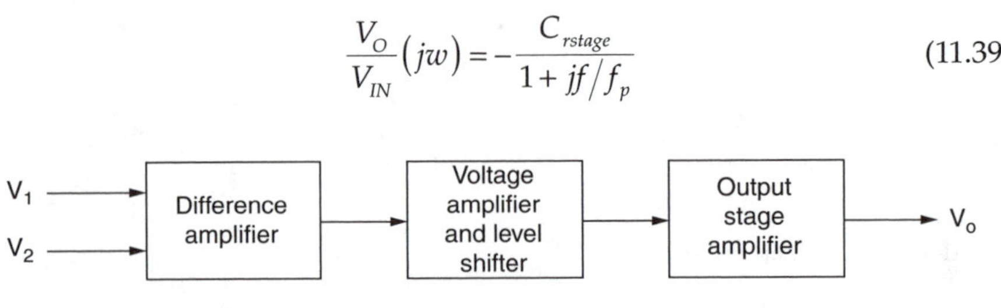

FIGURE 11.17
Internal structure of an operational amplifier.

or

$$\frac{V_O}{V_{IN}}(s) = \frac{C_{rstage}}{1 + s/w_p} \tag{11.40}$$

The transfer function of the amplifier is given as

$$A(s) = \frac{-500}{(1 + s/400\pi)} \frac{360}{(1 + s/8\pi 10^5)} \frac{0.92}{(1 + s/1.6\pi 10^6)} \tag{11.41}$$

The above expression simplifies to

$$A(s) = \frac{2.62 \times 10^{21}}{(s + 400\pi)(s + 8\pi 10^5)(s + 1.6\pi 10^6)} \tag{11.42}$$

MATLAB script

```
% Open Loop Gain of an op amp
% poles are
p1 = 400*pi; p2 = 8e5*pi; p3 = 1.6e6*pi;
p = [p1 p2 p3];
% zeros
z = [0];
const = 2.62e21;
% convert to poles and zeros  and
% find the frequency response
a3 = 1;
a2 = p1 + p2 + p3;
a1 = p1*p2 + p1*p3 + p2*p3;
a0 = p1*p2*p3;
den = [a3 a2 a1 a0];
num = [const];
w = logspace(1,8);
h = freqs(num,den,w);
f = w/(2*pi);
g_db = 20*log10(abs(h));

% plot the magnitude response
semilogx(f,g_db)
title('Magnitude Response')
xlabel('Frequency, Hz')
ylabel('Gain, dB')
```

The frequency response of the operational amplifier is shown in Figure 11.18.

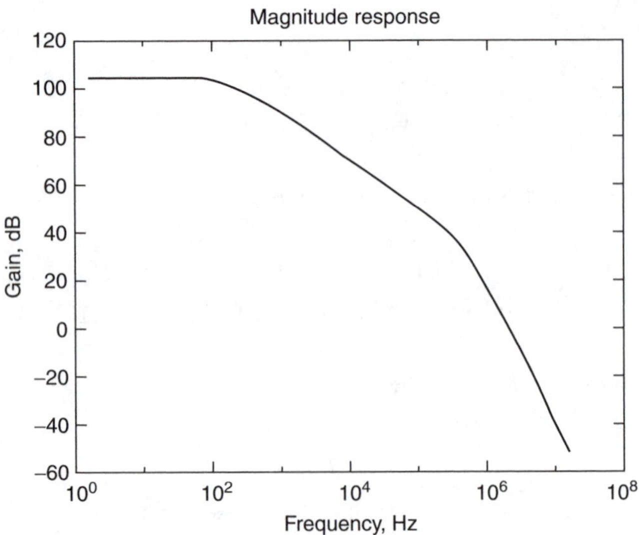

FIGURE 11.18
Open-loop gain characteristics of an op amp.

For an internally compensated op amp, there is a capacitor included on the integrated circuit (IC) chip. This causes the op amp to have a single pole lowpass response. The process of making one pole dominant in the open-loop gain characteristics is called frequency compensation, and it is done to ensure the stability of the op amp. For an internally compensated op amp, the open-loop gain $A(s)$ can be written as

$$A(s) = \frac{A_O}{(1 + s/w_b)} \tag{11.43}$$

where A_O is dc open-loop gain and w_b is break frequency.

For the 741 op amp, $A_O = 10^5$ and $w_b = 20\pi$ rad/sec. At physical frequencies $s = jw$, Equation (11.43) becomes

$$A(jw) = \frac{A_O}{(1 + jw/w_b)} \tag{11.44}$$

For frequencies $w > w_b$, Equation (11.44) can be approximated by

$$A(jw) = \frac{A_O w_b}{jw} \tag{11.45}$$

The unity gain bandwidth, w_t (the frequency at which the gain goes to unity), is given as

$$w_t = A_O w_b \tag{11.46}$$

For the inverting amplifier shown in Figure 11.5, if we substitute Equation (11.43) into Equation (11.35), we get a closed-loop gain

$$\frac{V_O}{V_{IN}}(s) = -\frac{R_2/R_1}{1 + (1 + R_2/R_1)/A_O + \dfrac{s}{w_t/(1 + R_2/R_1)}} \tag{11.47}$$

In the case of the noninverting amplifier shown in Figure 11.12, if we substitute Equation (11.43) into Equation (11.37), we get the closed-loop gain expression

$$\frac{V_O}{V_{IN}}(s) = \frac{1 + R_2/R_1}{1 + (1 + R_2/R_1)/A_o + \dfrac{s}{w_t/(1 + R_2/R_1)}} \tag{11.48}$$

From Equation (11.47) and Equation (11.48), it can be seen that the break frequency for the inverting and noninverting amplifiers is given by the expression

$$w_{3dB} = \frac{w_t}{1 + R_2/R_1} \tag{11.49}$$

The following example illustrates the effect of the ratio

$$\frac{R_2}{R_1}$$

on the frequency response of an op amp circuit.

Example 11.5 Effect of Closed-Loop Gain on the Frequency Response of an Op Amp

An op amp has an open-loop dc gain of 10^7, the unity gain bandwidth of 10^8 Hz. For an op amp connected in an inverting configuration (Figure 11.5), plot the magnitude response of the closed-loop gain if

$$\frac{R_2}{R_1} = 100, 600, 1100$$

Solution

Equation (11.47) can be written as

$$\frac{V_O}{V_{IN}}(s) = \frac{\dfrac{w_t R_2}{R_1\left(1 + {R_2}/{R_1}\right)}}{s + \dfrac{w_t}{A_O} + \dfrac{w_t}{\left(1 + {R_2}/{R_1}\right)}} \tag{11.50}$$

MATLAB script

```
% Inverter closed-loop gain versus frequency
w = logspace(-2,10);     f = w/(2*pi);
r12 = [100 600 1100];
a =[];     b = [];    num = [];    den = [];   h = [];
for i = 1:3
 a(i) = 2*pi*1.0e8*r12(i)/(1+r12(i));
 b(i) = 2*pi*1.0e8*((1/(1+r12(i))) + 1.0e-7);
 num = [a(i)];
 den = [1 b(i)];
 h(i,:) = freqs(num,den,w);
end
semilogx(f,abs(h(1,:)),'b',f,abs(h(2,:)),'b',f,
abs(h(3,:)),'b')
title('Op Amp Frequency Characteristics')
xlabel('Frequency, Hz')
ylabel('Gain')
axis([1.0e-2,1.0e10,0,1200])
text(1.5e-2, 150, 'Resistance ratio of 100')
text(1.5e-2, 650, 'Resistance ratio of 600')
text(1.50e-2, 1050, 'Resistance ratio of 1100')
```

Figure 11.19 shows the plots obtained from the MATLAB program.

11.6 Slew Rate and Full-Power Bandwidth

Slew rate (*SR*) is a measure of the maximum possible rate of change of the output voltage of an op amp. Mathematically, it is defined as

$$SR = \left.\frac{dV_O}{dt}\right|_{\max} \tag{11.51}$$

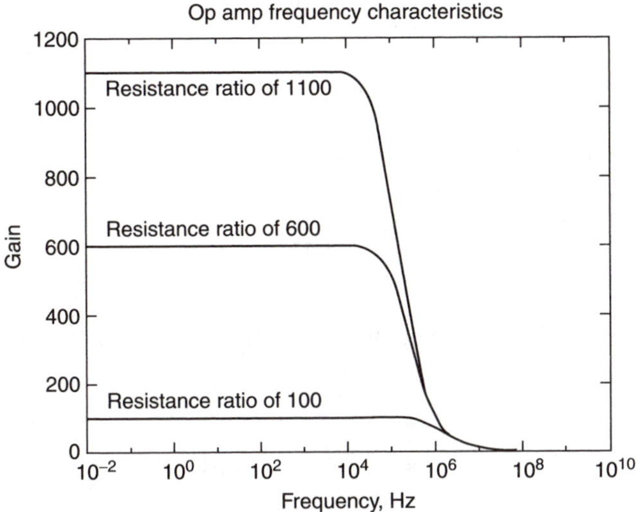

FIGURE 11.19
Frequency response of an op amp inverter with different closed-loop gains.

The slew rate is often specified on the op amp data sheets in V/μsec. Poor op amps might have slew rates of around 1 V/μsec. Good op amps with slew rates up to 1000 V/μs are available but are relatively expensive.

Slew rate is important when an output signal must follow a large input signal that is rapidly changing. If the slew rate is lower than the rate of change of the input signal, then the output voltage will be distorted. The output voltage will become triangular and attenuated. However, if the slew rate is higher than the rate of change of the input signal, no distortion occurs, and the input and output of the op amp circuit will have similar wave shapes.

As mentioned in the Section 11.5, a frequency-compensated op amp has an internal capacitance that is used to produce a dominant pole. In addition, the op amp has a limited output current capability, due to the saturation of the input stage. If we designate I_{max} as the maximum possible current that is available to charge the internal capacitance of an op amp, the charge on the frequency-compensation capacitor is

$$CdV = Idt$$

Thus, the highest possible rate of change of the output voltage is

$$SR = \left. \frac{dV_O}{dt} \right|_{max} = \frac{I_{max}}{C} \tag{11.52}$$

For a sinusoidal input signal given by

$$v_i(t) = V_m \sin wt \tag{11.53}$$

the rate of change of the input signal is

$$\frac{dv_i(t)}{dt} = wV_m \cos wt \tag{11.54}$$

Assuming that the input signal is applied to a unity gain follower, then the output rate of change

$$\frac{dV_O}{dt} = \frac{dv_i(t)}{dt} = wV_m \cos wt \tag{11.55}$$

The maximum value of the rate of change of the output voltage occurs when $\cos(wt) = 1$, i.e., $wt = 0, 2\pi, 4\pi, \ldots$, the slew rate

$$SR = \left.\frac{dV_O}{dt}\right|_{\max} = wV_m \tag{11.56}$$

Equation (11.56) can be used to define full-power bandwidth. The latter is the frequency at which a sinusoidal rated output signal begins to show distortion due to slew rate limiting. Thus

$$w_m V_{o,rated} = SR \tag{11.57}$$

Thus

$$f_m = \frac{SR}{2\pi, V_{o,rated}} \tag{11.58}$$

The full-power bandwidth can be traded for output rated voltage, thus, if the output rated voltage is reduced, the full-power bandwidth increases. The following example illustrates the relationship between the rated output voltage and the full-power bandwidth.

Example 11.6 Output Voltage vs. Full-Power Bandwidth

The LM 741 op amp has a slew rate of 0.5 V/μsec. Plot the full-power bandwidth vs. the rated output voltage if the latter varies from ±1 to ±10 V.

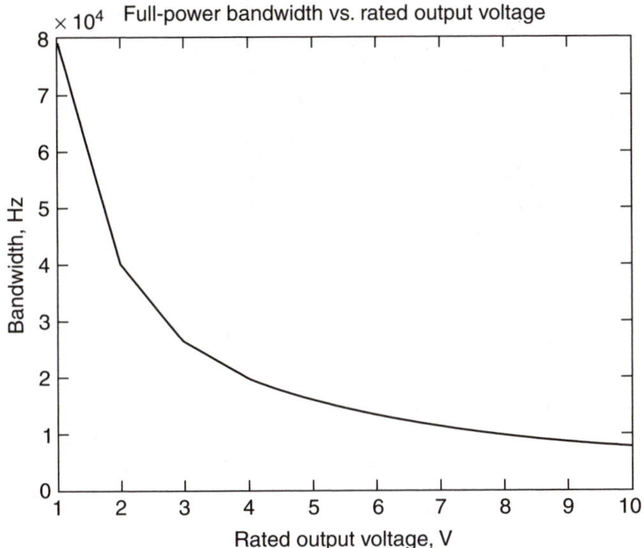

FIGURE 11.20
Rated output voltage vs. full-power bandwidth.

Solution

```
% Slew rate and full-power bandwidth
sr = 0.5e6;
v0 = 1.0:10;
fm = sr./(2*pi*v0);
plot(v0, fm)
title('Full-power Bandwidth vs. Rated Output Voltage')
xlabel('Rated Output Voltage, V')
ylabel('Bandwidth, Hz')
```

Figure 11.20 shows the plot for Example 11.6.

11.7 Common-Mode Rejection

For practical op amps, when two inputs are tied together and a signal applied to the two inputs, the output will be nonzero. This is illustrated in Figure 11.21a, where the common-mode gain, A_{cm}, is defined as

$$A_{cm} = \frac{v_o}{v_{i,cm}} \tag{11.59}$$

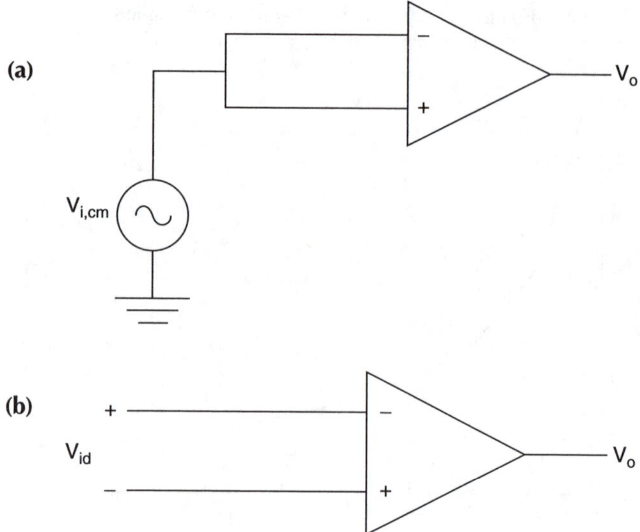

FIGURE 11.21
Circuits showing the definitions of (a) common-mode gain and (b) differential-mode gain.

The differential-mode gain, A_d, is defined as

$$A_d = \frac{v_o}{v_{id}} \tag{11.60}$$

For an op amp with arbitrary input voltages, V_1 and V_2 (see Figure 11.21b), the differential input signal, v_{id}, is

$$v_{id} = V_2 - V_1 \tag{11.61}$$

and the common mode input voltage is the average of the two input signals,

$$V_{i,cm} = \frac{V_2 + V_1}{2} \tag{11.62}$$

The output of the op amp can be expressed as

$$V_O = A_d v_{id} + A_{cm} v_{i,cm} \tag{11.63}$$

The common-mode rejection ratio (CMRR) is defined as

$$\text{CMRR} = \left| \frac{A_d}{A_{cm}} \right| \tag{11.64}$$

The CMRR represents the op amp's ability to reject signals that are common to the two inputs of an op amp. Typical values of CMRR range from 80 to 120 dB. CMRR decreases as frequency increases.

For an inverting amplifier as shown in Figure 11.5, because the noninverting input is grounded, the inverting input will also be approximately 0 V due to the virtual short circuit that exists in the amplifier. Thus, the common-mode input voltage is approximately zero and Equation (11.63) becomes

$$V_O \cong A_d V_{id} \tag{11.65}$$

The finite CMRR does not affect the operation of the inverting amplifier.

A method normally used to take into account the effect of finite CMRR in calculating the closed-loop gain is as follows: The contribution of the output voltage due to the common-mode input is A_{cm} $V_{i,cm}$. This output voltage contribution can be obtained if a differential input signal, V_{error}, is applied to the input of an op amp with zero common-mode gain.

Thus

$$V_{error} A_d = A_{cm} V_{i,cm} \tag{11.66}$$

$$V_{error} = \frac{A_{cm} V_{i,cm}}{A_d} = \frac{V_{i,cm}}{\text{CMRR}} \tag{11.67}$$

Figure 11.22 shows how to use the above technique to analyze a noninverting amplifier with a finite CMRR.

From Figure 11.22b, the output voltage is given as

$$V_O = V_i\left(1 + R_2/R_1\right) + \frac{V_i}{\text{CMRR}}\left(1 + R_2/R_1\right) \tag{11.68}$$

The following example illustrates the effect of a finite CMRR on the closed-loop gain of a noninverting amplifier.

Example 11.7 Effect of CMRR on the Closed-Loop Gain

For the amplifier shown in Figure 11.22, if R_2 = 50 kΩ and R_1 = 1 kΩ, plot the closed-loop gain vs. CMRR for the following values of the latter: 10^4, 10^5, 10^6, 10^7, 10^8, and 10^9.

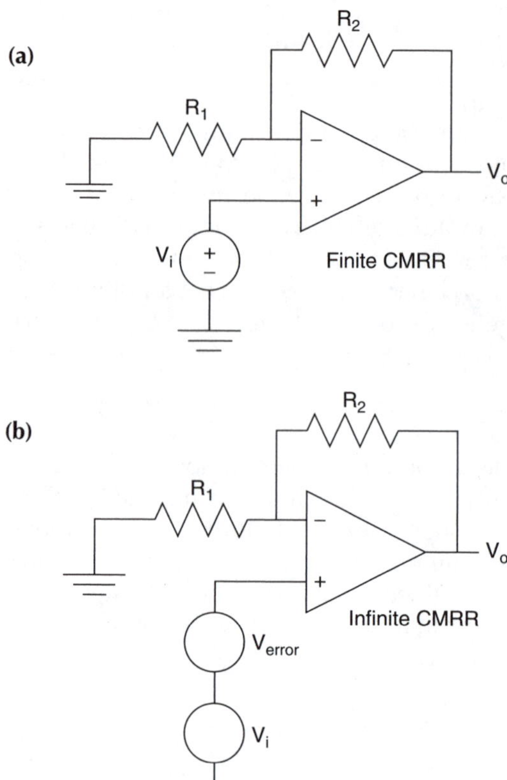

FIGURE 11.22
Noninverting amplifier: (a) finite CMRR; (b) infinite CMRR.

Solution

MATLAB script

```
% Non-inverting amplifier with finite CMRR
r2 = 50e3; r1 = 1.0e3; rr = r2/r1;
cmrr = logspace(4,9,6);  gain = (1+rr)*(1+1./cmrr);
semilogx(cmrr, gain ,cmrr, gain, 'bo')
xlabel('Common-mode Rejection Ratio')
ylabel('Closed Loop Gain')
title('Gain versus CMRR')
axis([1.0e3,1.0e10,50.998, 51.008])
```

Figure 11.23 shows the effect of CMRR on the closed loop of a noninverting amplifier.

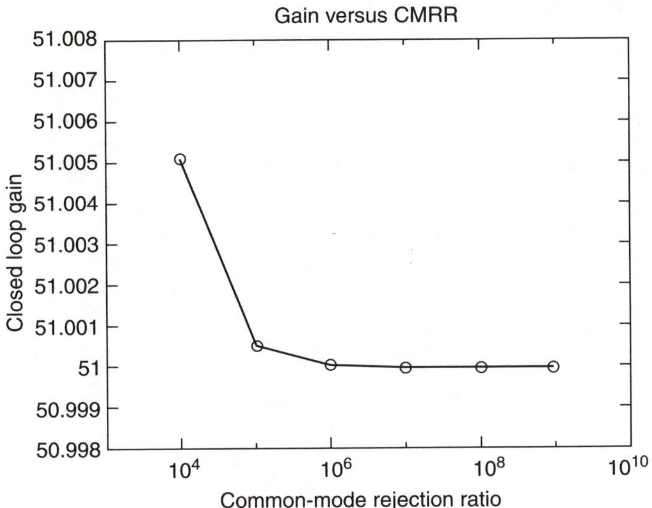

FIGURE 11.23
Effect of finite CMRR on the gain of a noninverting amplifier.

Bibliography

1. Belanger, P.R., Adler, E.L., and Rumin, N.C., *Introduction to Circuits with Electronics: An Integrated Approach*, Holt, Rinehart and Winston, New York, 1985.
2. Ferris, C.D., *Elements of Electronic Design*, West Publishing Co., St. Paul, MN, 1995.
3. Howe, R.T. and Sodini, C.G., *Microelectronics, An Integrated Approach*, Prentice Hall, Upper Saddle River, NJ, 1997.
4. Irvine, R.G., *Operational Amplifiers — Characteristics and Applications*, Prentice Hall, Upper Saddle River, NJ, 1981.
5. Rashid, M.H., *Microelectronic Circuits, Analysis and Design*, PWS Publishing Company, Boston, 1999.
6. Sedra, A.S. and Smith, K.C., *Microelectronic Circuits*, 4th ed., Oxford University Press, New York, 1998.
7. Wait, J.V., Huelsman, L.P., and Korn, G.A., *Introduction to Operational Amplifiers — Theory and Applications*, 2nd ed., McGraw-Hill, New York, 1992.
8. Warner, R.M., Jr. and Grung, B.L., *Semiconductor Device Electronics*, Holt, Rinehart and Winston, New York, 1991.

Problems

Problem 11.1

For the circuit shown in Figure P11.1:

(a) Derive the transfer function

$$\frac{V_O}{V_{IN}}(s)$$

(b) If $R_1 = 1\ \text{k}\Omega$, obtain the magnitude response.

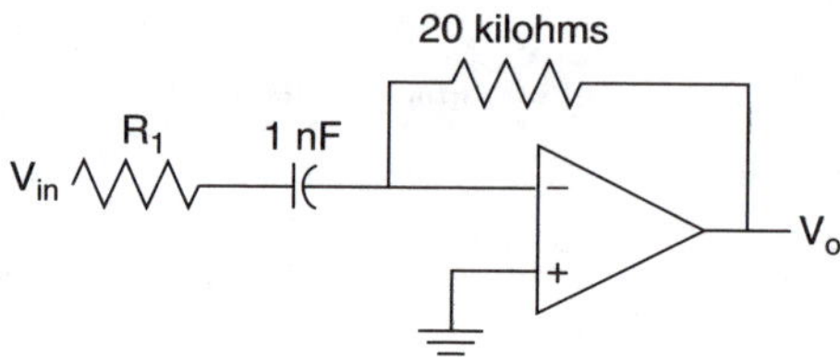

FIGURE P11.1
An op amp filter.

Problem 11.2

For Figure 11.4, $Z_1 = 2$ K, and Z_1 consists of a series combination of a resistor of 10 K and a capacitor of 5 nF. Plot the magnitude response assuming the op amp is ideal.

Problem 11.3

Find the poles and zeros of the circuit shown in Figure P11.3. Use MATLAB to plot the magnitude response. The resistance values are in kilohms.

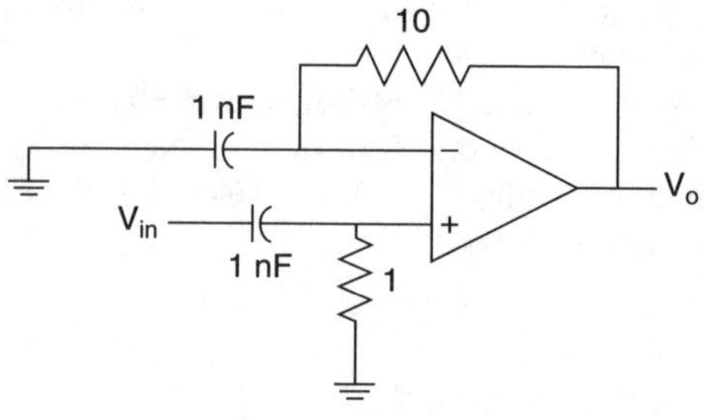

FIGURE P11.3
An op amp circuit.

Problem 11.4

For the inverting amplifier, shown in Figure 11.5, plot the 3-dB frequency vs. resistance ratio

$$\frac{R_2}{R_1}$$

for the following values of the resistance ratio: 10, 100, 1000, 10,000, and 100,000. Assume that $A_O = 10^6$ and $f_t = 10^7$ Hz.

Problem 11.5

For the inverting amplifier, shown in Figure 11.5, plot the closed-loop gain vs. resistance ratio

$$\frac{R_2}{R_1}$$

for the following open-loop gains: $A_O = 10^3$, 10^5, and 10^7. Assume a unity gain bandwidth of $f_t = 10^7$ Hz and that the resistance ratio,

$$\frac{R_2}{R_1}$$

has the following values: 10, 100, and 1000.

Problem 11.6

For the amplifier shown in Figure 11.12, if the open-loop gain is 106, R_2 = 24K, and R_1 = 1K, plot the frequency response for a unity gain bandwidth of 10^6, 10^7, and 10^8 Hz.

Problem 11.7

For Figure 11.12, if the open-loop gain is finite, (a) show that the closed-loop gain is given by the expression shown in Equation (11.37). (b) If R_2 = 100 K and R_1 = 0.5 K, plot the percentage error in the magnitude of the closed-loop gain for open-loop gains of 10^2, 10^4, 10^6, and 10^8.

Problem 11.8

An uncompensated op amp has the following characteristics: the pole of the difference amplifier is at 300 Hz and the gain is –450. The pole of the voltage amplifier and level shifter is 700 kHz and has a gain of 150. The pole of the output stage is 600 kHz and the gain is 0.88. Sketch the magnitude response of the operational amplifier open-loop gain.

Problem 11.9

An op amp with a slew rate of 1 V/μsec is connected in the unity gain follower configuration. A square wave of zero dc voltage and a peak voltage of 1 V and a frequency of 100 kHz is connected to the input of the unity gain follower. Write a MATLAB program to plot the output voltage of the amplifier.

Problem 11.10

For the noninverting amplifier, if $R_{icm} = 400$ MΩ, $R_{id} = 50$ MΩ, $R_1 = 2$ kΩ, and $R_2 = 30$ kΩ, plot the input resistance vs. the dc open-loop gain A_O. Assume the following values of the open-loop gain: 10^3, 10^5, 10^7, and 10^9.

Problem 11.11

For the amplifier shown in Figure 11.22, R_2 and R_1 were chosen such that

$$\left(1+\frac{R_2}{R_1}\right)$$

takes on the values of 20, 30, and 40. For each of the values of

$$\left(1+\frac{R_2}{R_1}\right)$$

plot the closed-loop gain vs. CMRR for the following values of the CMRR: 10^5, 10^6, 10^7, 10^8, 10^9, and 10^{10}.

12

Transistor Circuits

In this chapter, MATLAB will be used to solve problems involving the metal-oxide semiconductor field effect and bipolar junction transistors. The general topics to be discussed in this chapter are the dc model of BJT and MOSFET, biasing of discrete and integrated circuits, and frequency response of amplifiers.

12.1 Bipolar Junction Transistors

A bipolar junction transistor (BJT) consists of two pn junctions connected back to back. The operation of the BJT depends on the flow of both majority and minority carriers. There are two types of BJT: npn and pnp transistors. The electronic symbols of the two types of transistors are shown in Figure 12.1.

The dc behavior of the BJT can be described by the Ebers-Moll Model. The equations for the model are

$$I_F = I_{ES}\left[\exp\left(\frac{V_{BE}}{V_T}\right) - 1\right] \tag{12.1}$$

$$I_R = I_{CS}\left[\exp\left(\frac{V_{BC}}{V_T}\right) - 1\right] \tag{12.2}$$

and

$$I_C = \alpha_F I_F - I_R \tag{12.3}$$

$$I_E = -I_F + \alpha_R I_R \tag{12.4}$$

and

$$I_B = (1 - \alpha_F) I_F + (1 - \alpha_R) I_R \tag{12.5}$$

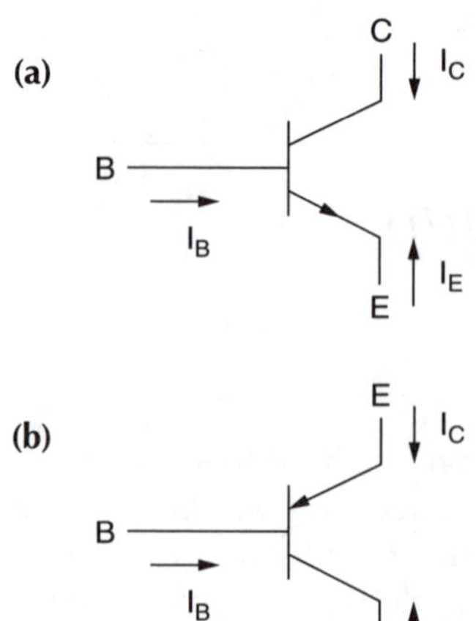

FIGURE 12.1
(a) NPN transistor; (b) PNP transistor.

where
- I_{ES} and I_{CS} are the base-emitter and base-collector saturation currents, respectively,
- α_R is the large signal reverse current gain of a common-base configuration, and
- α_F is the large signal forward current gain of the common-base configuration,

and

$$V_T = \frac{kT}{q} \tag{12.6}$$

where
- k is the Boltzmann's constant ($k = 1.381 \times 10^{-23}$ V.C/K)
- T is the absolute temperature in Kelvin
- q is the charge of an electron ($q = 1.602 \times 10^{-19}$ C)

The forward and reverse current gains are related by the expression

$$\alpha_R I_{CS} = \alpha_F I_{ES} = I_S \tag{12.7}$$

where I_S is the BJT transport saturation current.

The parameters α_R and α_F are influenced by impurity concentrations and junction depths. The saturation current, I_S, can be expressed as

$$I_S = J_S A \tag{12.8}$$

where A is the area of the emitter and J_S is the transport saturation current density, and it can be further expressed as

$$J_S = \frac{qD_n n_i^2}{Q_B} \tag{12.9}$$

where

D_n is the average effective electron diffusion constant

n_i is the intrinsic carrier concentration in silicon (n_i= 1.45 × 10^{10} atoms/cm^3 at 300 K)

Q_B is the number of doping atoms in the base per unit area

The dc equivalent circuit of the BJT is based upon the Ebers-Moll model. The model is shown in Figure 12.2. The current sources $\alpha_R I_R$ indicate the interaction between the base-emitter and base-collector junctions due to the narrow base region.

In the case of a pnp transistor, the directions of the diodes in Figure 12.2 are reversed. In addition, the voltage polarities of Equation (12.1) and Equation (12.2) are reversed. The resulting Ebers-Moll equations for pnp transistors are

$$I_E = I_{ES}\left[\exp\left(\frac{V_{EB}}{V_T}\right) - 1\right] - \alpha_R I_{CS}\left[\exp\left(\frac{V_{CB}}{V_T}\right) - 1\right] \tag{12.10}$$

$$I_C = -\alpha_F I_{ES}\left[\exp\left(\frac{V_{EB}}{V_T}\right) - 1\right] + I_{CS}\left[\exp\left(\frac{V_{CB}}{V_T}\right) - 1\right] \tag{12.11}$$

The voltages at the base-emitter and base-collector junctions will define the regions of operation. The four regions of operation are forward-active, reverse-active, saturation, and cut-off. Figure 12.3 shows the regions of operation based on the polarities of the base-emitter and base-collector junctions.

Forward-Active Region

The forward-active region corresponds to forward biasing the emitter-base junction and reverse biasing the base-collector junction. It is the normal operational region of transistors employed for amplifications. If $V_{BE} > 0.5$ V and $V_{BC} < 0.3$ V, then Equation (12.1) to Equation (12.4) and Equation (12.6) can be rewritten as

$$I_C = I_S \exp\left(\frac{V_{BE}}{V_T}\right) \tag{12.12}$$

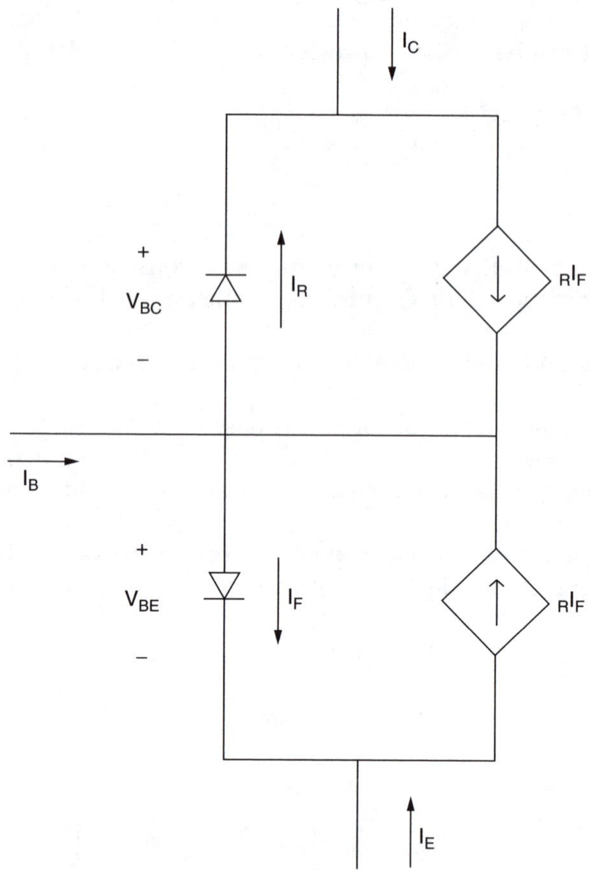

FIGURE 12.2
Ebers-Moll static model for an NPN transistor (injection version).

$$I_E = -\frac{I_S}{\alpha_F}\exp\left(\frac{V_{BE}}{V_T}\right) \tag{12.13}$$

From Figure 12.1,

$$I_B = -(I_C + I_E) \tag{12.14}$$

Substituting Equation (12.12) and Equation (12.13) into Equation (12.14), we have

$$I_B = I_S\frac{(1-\alpha_F)}{\alpha_F}\exp\left(\frac{V_{BE}}{V_T}\right) \tag{12.15}$$

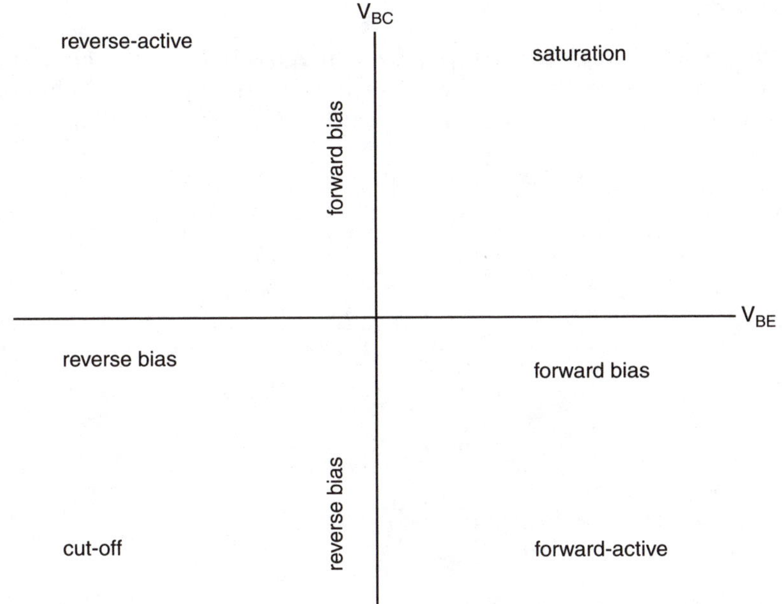

FIGURE 12.3
Regions of operation for a BJT as defined by the bias of V_{BE} and V_{BC}.

$$I_B = \frac{I_S}{\beta_F} \exp\left(\frac{V_{BE}}{V_T}\right) \tag{12.16}$$

where β_F is the large signal forward current gain of common-emitter configuration:

$$\beta_F = \frac{\alpha_F}{1-\alpha_F} \tag{12.17}$$

From Equation (12.12) and Equation (12.16), we have

$$I_C = \beta_F I_B \tag{12.18}$$

We can also define β_R, the large signal reverse current gain of the common-emitter configuration, as

$$\beta_R = \frac{\alpha_R}{1-\alpha_R} \tag{12.19}$$

Reverse-Active Region

The reverse-active region corresponds to reverse biasing the emitter-base junction and forward biasing the base-collector junction. The Ebers-Moll model in the reverse-active region ($V_{BC} > 0.5$ V and $V_{BE} < 0.3$ V) simplifies to

$$I_E = I_S\left[\frac{V_{BC}}{V_T}\right] \tag{12.20}$$

$$I_B = \frac{I_S}{\beta_R}\exp\left[\frac{V_{BC}}{V_T}\right] \tag{12.21}$$

Thus,

$$I_E = \beta_R I_B \tag{12.22}$$

The reverse-active region is seldom used.

Saturation and Cut-Off Regions

The saturation region corresponds to forward biasing both base-emitter and base-collector junctions. A switching transistor will be in the saturation region when the device is in the conducting or "ON" state.

The cut-off region corresponds to reverse biasing the base-emitter and base-collector junctions. The collector and base currents are very small compared to those that flow when transistors are in the active-forward and saturation regions. In most applications, it is adequate to assume that $I_C = I_B = I_E = 0$ when a BJT is in the cut-off region. A switching transistor will be in the cut-off region when the device is not conducting or is in the "OFF" state.

Example 12.1 Input Characteristics of a BJT

Assume that a BJT has an emitter area of 5.0 mil^2, $\beta_F = 120$, $\beta_R = 0.3$, transport current density $J_S = 2 \times 10^{-10}$ $\mu A/\text{mil}^2$, and $T = 300$ K. Plot I_E vs. V_{BE} for $V_{BC} = -1$ V. Assume $0 < V_{BE} < 0.7$ V.

Solution

From Equation (12.1), Equation(12.2), and Equation (12.4), we can write the following MATLAB program.

MATLAB script

```
%Input characteristics of a BJT
k=1.381e-23; temp=300; q=1.602e-19;
cur_den=2e-10; area=5.0; beta_f=120; beta_r=0.3;
vt=k*temp/q; is=cur_den*area;
```

```
alpha_f=beta_f/(1+beta_f);
alpha_r = beta_r/(1+beta_r);
ies=is/alpha_f;
vbe=0.3:0.01:0.65;
ics=is/alpha_r;
m=length(vbe);
for i = 1:m
ifr(i) = ies*exp((vbe(i)/vt)-1);
ir1(i) = ics*exp((-1.0/vt)-1);
ie1(i) = abs(-ifr(i) + alpha_r*ir1(i));
end
plot(vbe,ie1)
title('Input characteristics')
xlabel('Base-emitter voltage, V')
ylabel('Emitter current, A')
```

Figure 12.4 shows the input characteristics.

Experimental studies indicate that the collector current of the BJT in the forward-active region increases linearly with the voltage between the collector-emitter V_{CE}. Equation 12.12 can be modified as

$$I_C \cong I_S \exp\left[\frac{V_{BE}}{V_T}\right]\left(1+\frac{V_{CE}}{V_{AF}}\right) \tag{12.23}$$

where V_{AF} is a constant dependent on the fabrication process.

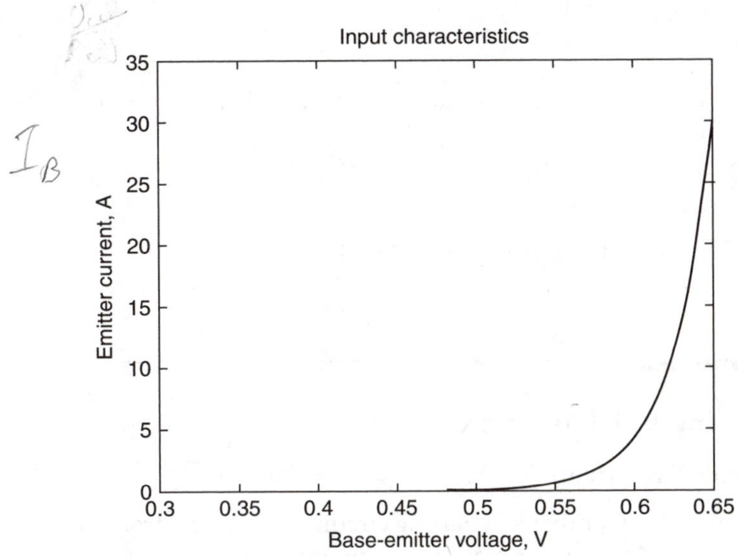

FIGURE 12.4
Input characteristics of a bipolar junction transistor.

Example 12.2 Output Characteristics of an npn Transistor

For an npn transistor with an emitter area of 5.5 mil², $\alpha_F = 0.98$, $\alpha_R = 0.35$, $V_{AF} = 250$ V, and transport current density is 2.0×10^{-9} μA/mil². Use MATLAB to plot the output characteristic for $V_{BE} = 0.65$ V. Neglect the effect of V_{AF} on the output current I_C. Assume a temperature of 300 K.

Solution

MATLAB script

```
%output characteristic of an npn transistor
k=1.381e-23; temp=300; q=1.602e-19;
cur_den=2.0e-15; area=5.5; alpha_f=0.98;
alpha_r=0.35; vt=k*temp/q; is=cur_den*area;
ies=is/alpha_f; ics=is/alpha_r;
vbe= [0.65];
vce=[0 0.07 0.1 0.2 0.3 0.4 0.5 0.6 0.7 1 2 4 6];
n=length(vbe);
m=length(vce);
for i=1:n
  for j=1:m
    ifr(i,j)= ies*exp((vbe(i)/vt) - 1);
    vbc(j) = vbe(i) - vce(j);
    ir(i,j) = ics*exp((vbc(j)/vt) - 1);
    ic(i,j) = alpha_f*ifr(i,j) - ir(i,j);
  end
end
ic1 = ic(1,:);
plot(vce, ic1,'b')
title('Output Characteristic')
xlabel('Collector-emitter Voltage, V')
ylabel('Collector Current, A')
text(3,3.1e-4, 'Vbe = 0.65 V')
axis([0,6,0,4e-4])
```

Figure 12.5 shows the output characteristic.

12.2 Biasing BJT Discrete Circuits

12.2.1 Self-Bias Circuit

One of the most frequently used biasing circuits for discrete transistor circuits is the self-bias of the emitter-bias circuit shown in Figure 12.6.

The emitter resistance, R_E, provides stabilization of the bias point. If V_{BB} and R_B are the Thevenin equivalent parameters for the base bias circuit, then

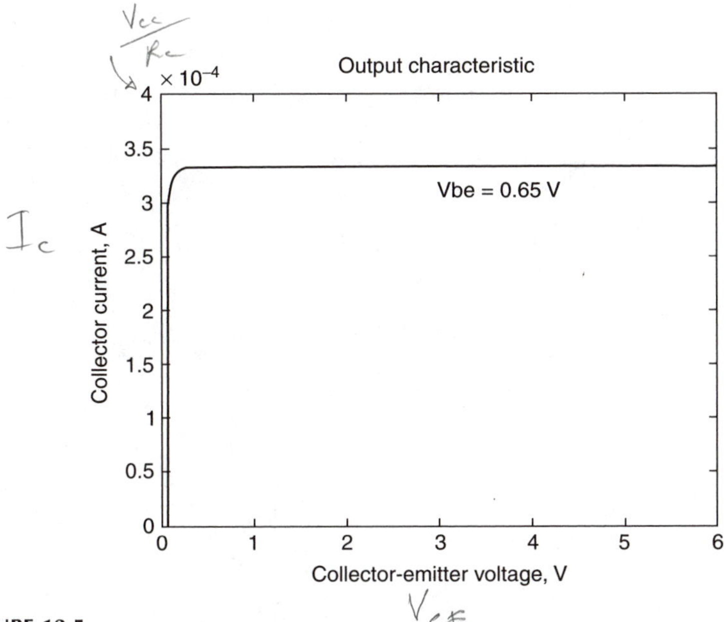

FIGURE 12.5
Output characteristic on an npn transistor.

$$V_{BB} = \frac{V_{CC} R_{B2}}{R_{B1} + R_{B2}} \tag{12.24}$$

$$R_B = R_{B1} \| R_{B2} \tag{12.25}$$

Using Kirchhoff's voltage law (KVL) for the base circuit, we have

$$V_{BB} = I_B R_B + V_{BE} + I_E R_E \tag{12.26}$$

Using Equation (12.18) and Figure 12.6b, we have

$$I_E = I_B + I_C = I_B + \beta_F I_B = (\beta_F + 1) I_B \tag{12.27}$$

Substituting Equation (12.18) and Equation (12.27) into Equation (12.26), we have

$$I_B = \frac{V_{BB} - V_{BE}}{R_B + (\beta_F + 1) R_E} \tag{12.28}$$

or

$$I_C = \frac{V_{BB} - V_{BE}}{\frac{R_B}{\beta_F} + \frac{(\beta_F + 1)}{\beta_F} R_E} \tag{12.29}$$

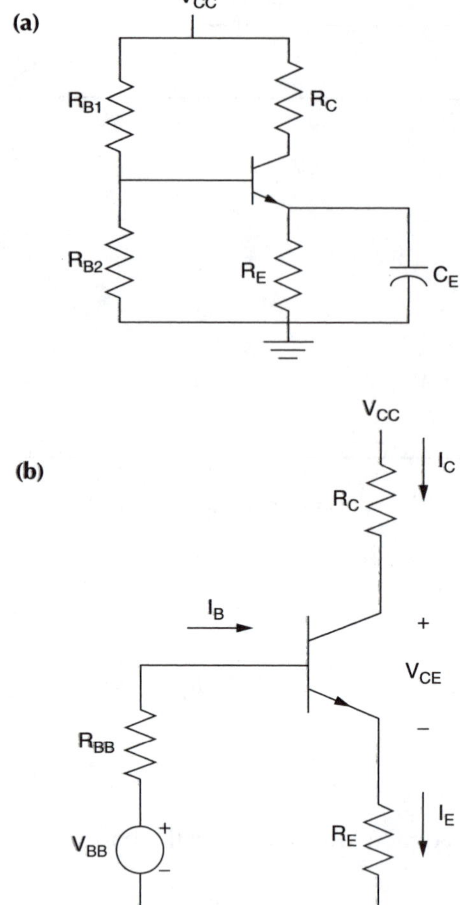

FIGURE 12.6
(a) Self-bias circuit, (b) DC equivalent circuit of (a).

Applying KVL at the output loop of Figure 12.6b gives

$$V_{CE} = V_{CC} - I_C R_C - I_E R_E \tag{12.30}$$

$$= V_{CC} - I_C \left(R_C + \frac{R_E}{\alpha_F} \right) \tag{12.31}$$

12.2.2 Bias Stability

Equation (12.30) gives the parameters that influence the bias current I_C. The voltage V_{BB} depends on the supply voltage V_{CC}. In some cases, V_{CC} would

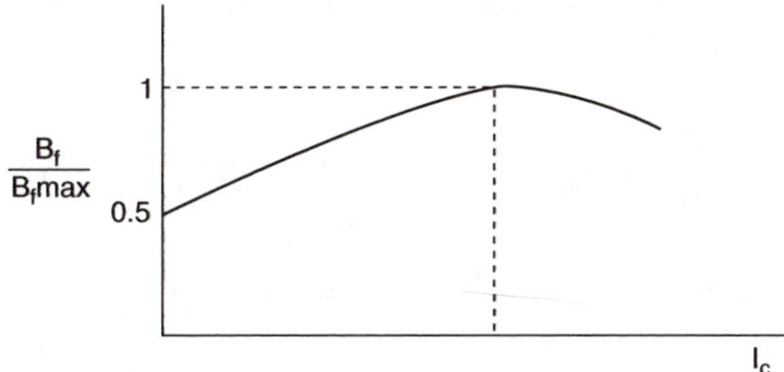

FIGURE 12.7
Normalized plot of β_F as a function of collector current.

vary with I_C, but by using a stabilized voltage supply we can ignore the changes in V_{CC}, and hence V_{BB}. The changes in the resistances R_{BB} and R_E are negligible. There is a variation of β_F with respect to changes in I_C. A typical plot of β_F vs. I_C is shown in Figure 12.7.

Temperature changes cause two transistor parameters to change. These are (1) base-emitter voltage (V_{BE}) and (2) collector leakage current between the base and collector (I_{CBO}). The variation on V_{BE} with temperature is similar to the changes of the pn junction diode voltage with temperature. For silicon transistors, the voltage V_{BE} varies almost linearly with temperature as

$$\Delta V_{BE} \cong -2(T_2 - T_1)\ \text{mV} \tag{12.32}$$

where T_1 and T_2 are in degrees Celsius.

The collector-to-base leakage current, I_{CBO}, approximately doubles every 10 degree temperature rise. As discussed in Section 9.1, if I_{CBO1} is the reverse leakage current at room temperature (25 °C), then

$$I_{CBO2} = 2^{((T_2-25)/10)} I_{CBO1}$$

and

$$\Delta I_{CBO} = I_{CBO2} - I_{CBO1} = I$$

$$= I_{CBO}\left[2^{((T_2-25)/10)} - 1\right] \tag{12.33}$$

Since the variations in I_{CBO} and V_{BE} are temperature dependent, but changes in V_{CC} and β_F are due to factors other than temperature, the information about the changes in V_{CC} and β_F must be specified.

From the above discussion, the collector current is a function of four variables: V_{BE}, I_{CBO}, β_F, V_{CC}. The change in collector current can be obtained using partial derivatives. For small parameter changes, a change in collector current is given as

$$\Delta I_C = \frac{\partial I_C}{\partial V_{BE}} \Delta V_{BE} + \frac{\partial I_C}{\partial V_{CBO}} \Delta I_{CBO} + \frac{\partial I_C}{\partial \beta_F} \Delta \beta_F + \frac{\partial I_C}{\partial V_{CC}} \Delta V_{CC} \tag{12.34}$$

The stability factors can be defined for the four variables as

$$S_\beta = \frac{\partial I_C}{\partial \beta_F} \cong \frac{\Delta I_C}{\Delta \beta_F}$$

$$S_v = \frac{\partial I_C}{\partial V_{BE}} \cong \frac{\Delta I_C}{\Delta V_{BE}}$$

$$S_I = \frac{\partial I_C}{\partial I_{CBO}} \cong \frac{\Delta I_C}{\Delta I_{CBO}}$$

and

$$S_{VCC} = \frac{\partial I_C}{\partial V_{CC}} \cong \frac{\Delta I_C}{\Delta V_{CC}} \tag{12.35}$$

Using the stability factors, Equation (12.34) becomes

$$\Delta I_C = S_V \Delta V_{BE} + S_\beta \Delta \beta_F + S_I \Delta I_{CBO} + S_{VCC} \Delta V_{CC} \tag{12.36}$$

From Equation (12.30),

$$S_V = \frac{dI_C}{dV_{BE}} = -\frac{1}{R_B/\beta_F + R_E\left(\beta_F + 1/\beta_F\right)} \tag{12.37}$$

From Equation (12.31),

$$I_C = \frac{V_{CC} - V_{CE}}{R_C + R_E/\alpha_F} \tag{12.38}$$

Thus, the stability factor S_{VCC} is given as

$$S_{VCC} = \frac{dI_C}{dV_{CC}} = \frac{1}{R_C + R_E/\alpha_F} \tag{12.39}$$

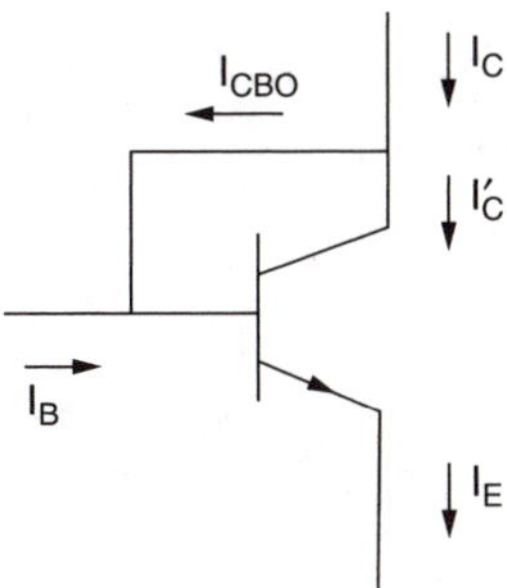

FIGURE 12.8
Current in transistor including I_{CBO}.

To obtain the stability factor S_I, an expression for I_C involving I_{CBO} needs to be derived. The derivation is assisted by referring to Figure 12.8.

The current

$$I_C = I'_C + I_{CBO} \tag{12.40}$$

and

$$I'_C = \beta_F(I_B + I_{CBO}) \tag{12.41}$$

From Equation (12.40) and Equation (12.41), we have

$$I_C = \beta_F I_B + (\beta_F + 1)I_{CBO} \tag{12.42}$$

Assuming that $\beta_F + 1 \cong \beta_F$, then

$$I_C = \beta_F I_B + \beta_F I_{CBO} \tag{12.43}$$

so

$$I_B = \frac{I_C}{\beta_F} - I_{CBO} \tag{12.44}$$

The loop equation of the base-emitter circuit of Figure 12.6(b) gives

$$V_{BB} - V_{BE} = I_B R_{BB} + R_E(I_B + I_C)$$

$$= I_B(R_{BB} + R_E) + R_E I_C \tag{12.45}$$

Assuming that $\beta_F + 1 \cong \beta_F$ and substituting Equation (12.44) into Equation (12.45), we get

$$V_{BB} - V_{BE} = (R_{BB} + R_E)\left(\frac{I_C}{\beta_F} - I_{CBO}\right) + I_C R_E \tag{12.46}$$

Solving for I_C, we have

$$I_C = \frac{V_{BB} - V_{BE} + (R_{BB} + R_E)I_{CBO}}{(R_{BB} + R_E)/\beta_F + R_E} \tag{12.47}$$

Taking the partial derivative,

$$S_I = \frac{\partial I_C}{\partial I_{CBO}} = \frac{R_{BB} + R_E}{(R_{BB} + R_E)/\beta_F + R_E} \tag{12.48}$$

The stability factor involving β_F and S_β can also be found by taking the partial derivative of Equation (12.47). Thus,

$$S_\beta = \frac{\partial I_C}{\partial \beta} = \frac{(R_B + R_E)[V_{BB} - V_{BE} + (R_B + R_E)I_{CBO}]}{(R_B + R_E + \beta R_E)^2} \tag{12.49}$$

The following example shows the use of MATLAB for finding the changes in the quiescent point of a transistor due to variations in temperature, base-to-emitter voltage, and common emitter current gain.

Example 12.3 Self-Bias Circuit — Stability Factors and Collector Current as a Function of Temperature

The self-bias circuit of Figure 12.6 has the following element values: R_{B1} = 50 K, R_{B2} = 10 K, R_E = 1.2 K, R_C = 6.8 K, β_F varies from 150 to 200, and V_{CC} is 10 ± 0.05 V. I_{CBO} is 1 μA at 25°C. Calculate the collector current at 25°C, and plot the change in collector current for temperatures between 25 and 100°C. Assume V_{BE} and β_F at 25°C are 0.7 V and 150, respectively.

Solution

Equation (12.25), Equation (12.26), and Equation (12.30) can be used to calculate the collector current. At each temperature, the stability factors are calculated using Equation (12.37), Equation (12.39), Equation (12.48), and Equation (12.49). The changes in V_{BE} and I_{CBO} with temperature are obtained using Equation (12.32) and Equation (12.33), respectively. The change in I_C for each temperature is calculated using Equation (12.36).

MATLAB script

```
% Bias stability
%
rb1=50e3; rb2=10e3; re=1.2e3; rc=6.8e3;
vcc=10; vbe=0.7; icbo25=1e-6; beta=(150+200)/2;
vbb=vcc*rb2/(rb1+rb2);
rb=rb1*rb2/(rb1+rb2);
ic=beta*(vbb-vbe)/(rb+(beta+1)*re);

%stability factors are calculated
svbe=-beta/(rb+(beta+1)*re);
alpha=beta/(beta+1);
svcc=1/(rc + (re/alpha));
svicbo=(rb+re)/(re+(rb+re)/alpha);
sbeta=((rb+re)*(vbb-vbe+icbo25*(rb+re))/
(rb+re+beta*re)^2);
% Calculate changes in Ic for various temperatures

t=25:1:100;
len_t = length(t);
dbeta = 50; dvcc=0.1;
for i=1:len_t
       dvbe(i)= -2e-3*(t(i)-25);
       dicbo(i)=icbo25*(2^((t(i)-25)/10)-1);
       dic(i)=svbe*dvbe(i)+svcc*dvcc...
      +svicbo+dicbo(i)+sbeta*dbeta;
end
plot(t,dicbo)
title('Change in Collector Current vs. Temperature')
xlabel('Temperature, degree C')
ylabel('Change in Collector Current, A')
```

Figure 12.9 shows I_C vs. temperature.

12.3 Integrated Circuit Biasing

Biasing schemes for discrete electronic circuits are not suitable for integrated circuits (IC) because of the large number of resistors and the large coupling and bypass capacitors required for biasing discrete electronic circuits. It is uneconomical to fabricate IC resistors since they take a disproportionately large area on an IC chip. In addition, it is almost impossible to fabricate IC inductors. Biasing of ICs is performed mostly using transistors that are connected to create constant current sources. Examples of integrated circuit biasing schemes are discussed in this section.

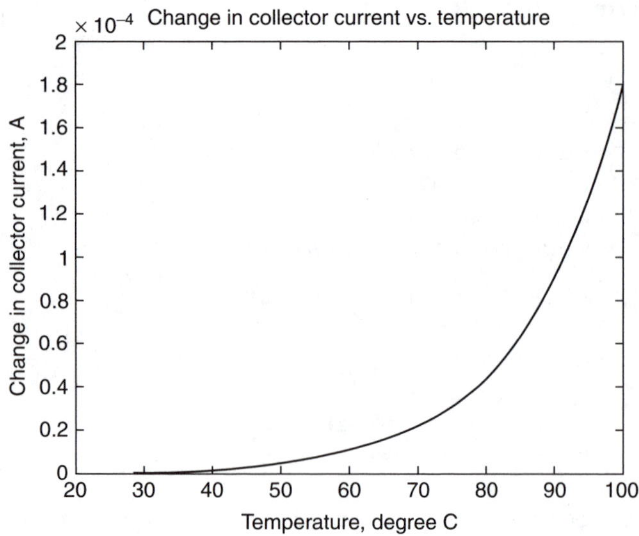

FIGURE 12.9
I_C vs. temperature.

12.3.1 Simple Current Mirror

A simple current mirror is shown in Figure 12.10. The current mirror consists of two matched transistors Q_1 and Q_2 with their bases and emitters connected. The transistor Q_1 is connected as a diode by shorting the base to its collector.

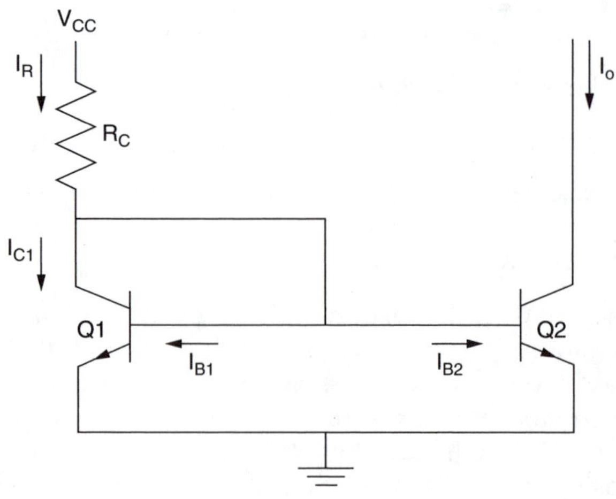

FIGURE 12.10
Simple current mirror.

From Figure 12.10, we observe that

$$I_R = \frac{V_{CC} - V_{BE}}{R_C} \tag{12.50}$$

Using Kirchhoff's current law (KCL), we get

$$I_R = I_{C1} + I_{B1} + I_{B2}$$

$$= I_{E1} + I_{B2} \tag{12.51}$$

But

$$I_{B2} = \frac{I_{E2}}{\beta + 1}$$

Assuming matched transistors

$$I_{B1} \cong I_{B2}$$

$$I_{E1} \cong I_{E2} \tag{12.52}$$

From Equation (12.51) and Equation (12.52), we get

$$I_R = I_{E1} + \frac{I_{E2}}{\beta + 1} \cong I_{E2}\left[1 + \frac{1}{\beta + 1}\right] = \left[\frac{\beta + 2}{\beta + 1}\right] I_{E2} \tag{12.53}$$

and

$$I_O = I_{C2} = \beta I_{B2} = \frac{\beta I_{E2}}{\beta + 1}$$

Therefore,

$$I_O = \left[\frac{\beta}{\beta + 1}\right]\left[\frac{\beta + 1}{\beta + 2}\right] I_R = \frac{\beta}{\beta + 2} I_R \tag{12.54}$$

$$I_O \cong I_R \text{ if } \beta >> 1 \tag{12.55}$$

Equation (15.55) is true provided Q_2 is in the active mode. In the latter mode of transistor operation, the device Q_2 behaves as a current source. For Q_2 to be in the active mode, the following relation should be satisfied:

$$V_{CE2} > V_{CEsat}$$

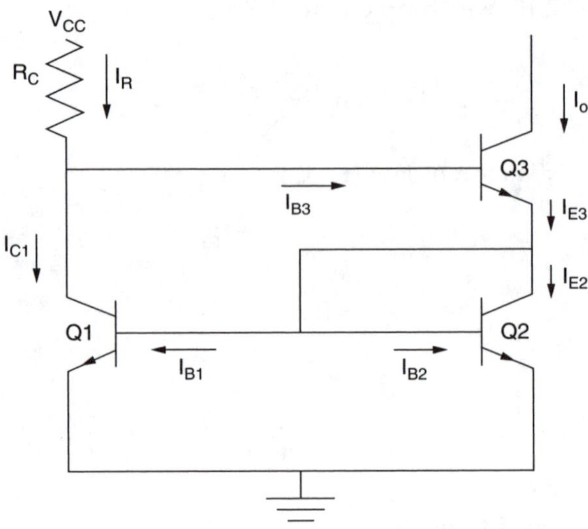

FIGURE 12.11
Wilson current source.

12.3.2 Wilson Current Source

The Wilson current source, shown in Figure 12.11, achieves high output resistance and an output current that is less dependent on transistor β_F. To obtain an expression for the output current, we assume that all three transistors are identical. Thus

$$I_{C1} = I_{C2}$$

$$V_{BE1} = V_{BE2}$$

$$\beta_{F1} = \beta_{F2} = \beta_{F3} = \beta_F \tag{12.56}$$

Using KCL at the collector of transistor Q_3, we get

$$I_{C1} = I_R - I_{B3} = I_R - \frac{I_O}{\beta_F}$$

therefore,

$$I_O = \beta_F (I_R - I_{C1}) \tag{12.57}$$

Using KCL at the emitter of Q_3, we obtain

$$\begin{aligned} I_{E3} &= I_{C2} + I_{B1} + I_{B2} = I_{C1} + 2I_{B1} \\ &= I_{C1}\left(1 + \frac{2}{\beta_F}\right) \end{aligned} \tag{12.58}$$

But

$$I_O = \alpha_F I_{E3} = \frac{\beta_F}{\beta_F + 1} I_{E3} \tag{12.59}$$

Substituting Equation (12.58) into Equation (12.59), we have

$$I_O = \left(\frac{\beta_F}{\beta_F + 1}\right)\left(1 + \frac{2}{\beta_F}\right) I_{C1} \tag{12.60}$$

Simplifying Equation (12.60), we get

$$I_{C1} = \left(\frac{\beta_F + 1}{\beta_F + 2}\right) I_O \tag{12.61}$$

Combining Equation (12.57) and Equation (12.61), we obtain

$$I_O = \beta_F\left[I_R - \left(\frac{\beta_F + 1}{\beta_F + 2}\right) I_O\right] \tag{12.62}$$

Simplifying Equation (12.62), we get

$$\begin{aligned} I_O &= \left(\frac{\beta_F^2 + 2\beta_F}{\beta_F^2 + 2\beta_F + 2}\right) I_R \\ &= \left(1 - \frac{2}{\beta_F^2 + 2\beta_F + 2}\right) I_R \end{aligned} \tag{12.63}$$

For reasonable values of β_F,

$$\left(\frac{2}{\beta_F^2 + 2\beta_F + 2}\right) << 1$$

and Equation (12.63) becomes

$$I_O \cong I_R$$

Thus, β has little effect on the output current, and

$$I_R = \frac{V_{CC} - V_{BE3} - V_{BE1}}{R_C} \tag{12.64}$$

Example 12.4 Comparison of Simple Current Mirror and Wilson Current Source

For Figure 12.10 and Figure 12.11, what are the percentage differences between the reference and output currents for the β_F from 40 to 200. Assume that for both figures, $V_{CC} = 10$ V, $R_C = 50$ kΩ, and $V_{BE} = 0.7$ V.

Solution

We use Equation (12.50) to calculate I_R and Equation (12.53) to find I_0 of the simple current mirror. Similarly, we use Equation (12.64) to find I_R and Equation (12.63) to calculate I_0 of the Wilson current source.

MATLAB script

```
% Integrated circuit Biasing
vcc=10; rc=50e3; vbe=0.7;
beta =40:5:200; ir1=(vcc-vbe)/rc;
ir2=(vcc-2*vbe)/rc; m=length(beta);
for i=1:m
   io1(i) = beta(i)*ir1/(beta(i) + 2);
   pd1(i)=abs((io1(i)-ir1)*100/ir1);
   io2(i)=(beta(i)^2+2*beta(i))/
   (beta(i)^2+2*beta(i)+2);
   pd2(i)=abs((io2(i)*ir2-ir2)*100/ir2);
end
subplot(211), plot(beta,pd1)
xlabel('Transistor Beta')
ylabel('Percentage Error')
text(90,4.0,'Error for Simple Current Mirror')
subplot(212),plot(beta,pd2)
xlabel('Transistor Beta')
ylabel('Percentage Error')
text(90, 0.15, 'Error for Wilson Current Source')
```

Figure 12.12 shows the percentage errors obtained for the simple current mirror and Wilson current source.

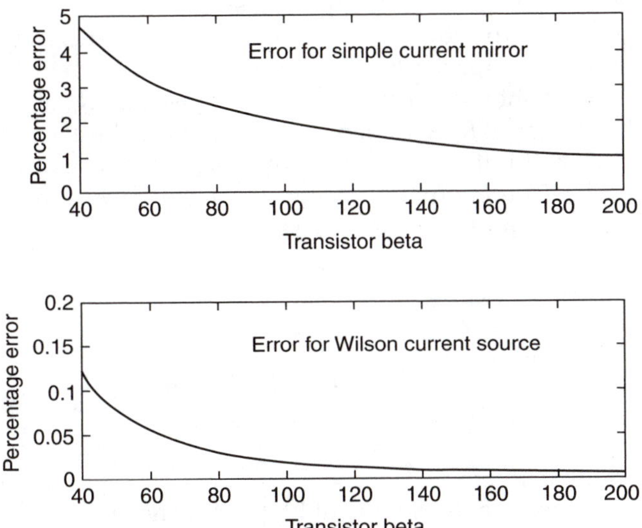

FIGURE 12.12
Percentage error between reference and output currents for simple current mirror and Wilson current source.

12.4 Frequency Response of Common-Emitter Amplifier

The common-emitter amplifier, shown in Figure 12.13, is capable of generating a relatively high current and voltage gains. The input resistance is medium and is essentially independent of the load resistance R_L. The output resistance is relatively high and is essentially independent of the source resistance.

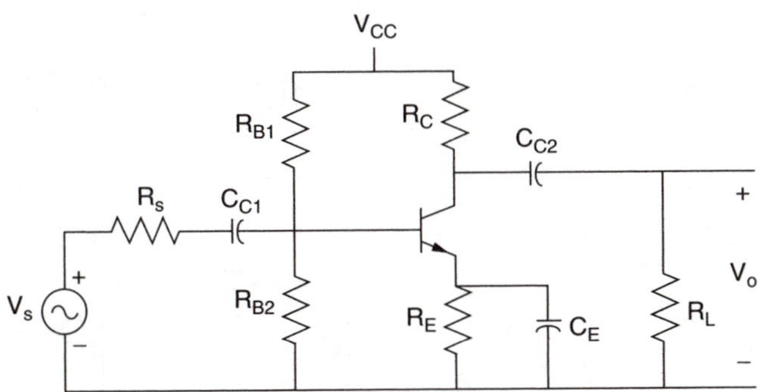

FIGURE 12.13
Common-emitter amplifier.

The coupling capacitor, C_{C1}, couples the source voltage v_S to the biasing network. Coupling capacitor C_{C2} connects the collector resistance R_C to the load R_L. The bypass capacitance C_E is used to increase the midband gain, since it effectively short circuits the emitter resistance R_E at midband frequencies. The resistance R_E is needed for bias stability. The external capacitors C_{C1}, C_{C2}, and C_E will influence the low-frequency response of the common-emitter amplifier. The internal capacitances of the transistor will influence the high frequency cut-off. The overall gain of the common-emitter amplifier can be written as

$$A(s) = \frac{A_m s^2 (s + w_z)}{(s + w_{L1})(s + w_{L2})(s + w_{L3})(1 + s/w_H)} \tag{12.65}$$

where

A_M is the midband gain.
w_H is the frequency of the dominant high-frequency pole.
w_{L1}, w_{L2}, and w_{L3} are low frequency poles introduced by the coupling and bypass capacitors.
w_Z is the zero introduced by the bypass capacitor.

The midband gain is obtained by short circuiting all the external capacitors and open circuiting the internal capacitors. Figure 12.14 shows the equivalent for calculating the midband gain.

From Figure 12.14, the midband gain, A_m, is

$$A_m = \frac{V_O}{V_S} = -\beta\left[r_{CE} \| R_C \| R_L\right]\left[\frac{R_B}{R_B + r_\pi}\right]\left[\frac{1}{R_S + \left[R_B \| r_\pi\right]}\right] \tag{12.66}$$

It can be shown that the low-frequency poles, w_{L1}, w_{L2}, and w_{L3}, can be obtained by the following equations:

$$\tau_1 = \frac{1}{w_{L1}} = C_{C1} R_{IN} \tag{12.67}$$

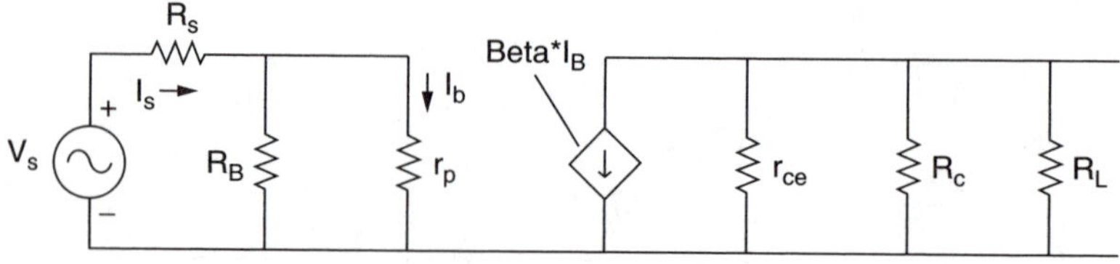

FIGURE 12.14
Equivalent circuit for calculating midband gain.

where

$$R_{IN} = R_S + \left[R_B \| r_\pi\right] \tag{12.68}$$

$$\tau_2 = \frac{1}{w_{L2}} = C_{C2}\left[R_L + \left(R_C \| r_{ce}\right)\right] \tag{12.69}$$

and

$$\tau_3 = \frac{1}{w_{L3}} = C_E R_E^{'} \tag{12.70}$$

where

$$R_E^{'} = R_E \left\| \left[\frac{r_\pi}{\beta_F + 1} + \left(\frac{R_B \| R_S}{\beta_F + 1}\right)\right]\right. \tag{12.71}$$

and the zero

$$w_Z = \frac{1}{R_E C_E} \tag{12.72}$$

Normally, $w_Z < w_{L3}$ and the low-frequency cut-off w_L is larger than the largest pole frequency. The low-frequency cut-off can be approximated as

$$w_L \cong \sqrt{\left(w_{L1}\right)^2 + \left(w_{L2}\right)^2 + \left(w_{L3}\right)^2} \tag{12.73}$$

The high-frequency equivalent circuit of the common-emitter amplifier is shown in Figure 12.15.

In Figure 12.15, C_μ is the collector-base capacitance, C_π is the emitter to base capacitance, and r_X is the resistance of silicon material of the base region between the base terminal B and an internal or intrinsic base terminal B′.

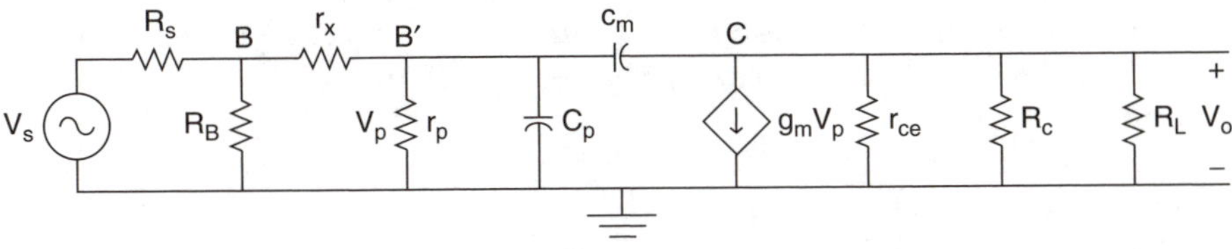

FIGURE 12.15
Equivalent circuit of CE amplifier at high frequencies.

Using the Miller theorem, it can be shown that the 3-dB frequency at high frequencies is approximately given as

$$w_H^{-1} = \left(r_\pi \middle\| \left[r_x + \left(R_B \| R_S\right)\right]\right) C_T \tag{12.74}$$

where

$$C_T = C_\pi + C_\mu\left[1 + g_m\left(R_L \| R_C\right)\right] \tag{12.75}$$

and

$$g_m = \frac{I_C}{V_T} \tag{12.76}$$

In the following example, MATLAB is used to obtain the frequency response of a common-emitter amplifier.

Example 12.5 Frequency Response of a Common-Emitter Amplifier

For the CE amplifier shown in Figure 12.13, $\beta = 150$, $R_L = 2\ \text{k}\Omega$, $R_C = 4\ \text{k}\Omega$, $C_\pi = 100$ pF, $C_\mu = 5$ pF, $V_{CC} = 10$ V, $r_{ce} = r_o = 60\ \text{k}\Omega$, $R_E = 1.5\ \text{k}\Omega$, $C_{C1} = 2\ \mu\text{F}$, $C_{C2} = 4\ \mu\text{F}$, $C_E = 150\ \mu\text{F}$, $R_{B1} = 60\ \text{k}\Omega$, $R_{B2} = 40\ \text{k}\Omega$, $R_S = 100\ \Omega$, and $r_x = 10\ \Omega$.
Use MATLAB to plot the magnitude response of the amplifier.

Solution

Equation (12.67), Equation (12.69), Equation (12.70), and Equation (12.74) are used to calculate the poles of Equation (12.65). The zero of the overall amplifier gain is calculated using Equation (12.66). The MATLAB program is as follows:

MATLAB script

```
%Frequency response of CE Amplifier
rc=4e3; rb1=60e3; rb2=40e3; rs=100; rce=60e3;
re=1.5e3; rl=2e3; beta=150; vcc=10; vt=26e-3; vbe =0.7;
cc1=2e-6; cc2=4e-6; ce=150e-6;, rx=10; cpi=100e-12;
cmu=5e-12;
% Ic is calculated
rb = (rb1 * rb2)/(rb1 + rb2);
vbb = vcc * rb2/(rb1 + rb2);
icq = beta * (vbb - vbe)/(rb + (beta + 1)*re);
% Calculation of low frequency poles
```

```
% using equations (12.67), (12.69) and (12.70)
rpi=beta * vt/icq;
rb_rpi=rpi * rb/(rpi + rb);
rin=rs + rb_rpi;
wl1=1/(rin * cc1);
rc_rce=rc * rce/(rc + rce);
wl2=1/(cc2 * (rl + rc_rce));
rb_rs=rb * rs/(rb + rs);
rx1=(rpi + rb_rs)/(beta + 1);
re_prime=re * rx1/(re + rx1);
wl3=1/(re_prime * ce);
% Calculate the low frequency zero using equation
(12.72)
wz = 1/(re*ce);
% Calculate the high frequency pole using equation
(12.74)
gm = icq/vt;
rbrs_prx = (rb * rs/(rb + rs)) + rx;
rt = (rpi * rbrs_prx)/(rpi + rbrs_prx);
rl_rc = rl * rc/(rl + rc);
ct = cpi + cmu * (1 + gm * rl_rc);
wh = 1/(ct * rt);
% Midband gain is calculated
rcercrl = rce * rl_rc/(rce + rl_rc);
am = -beta * rcercrl * (rb/(rb + rpi)) * (1/(rin));
% Frequency response calculation using equation (12.65)
a4 = 1; a3 = wl1 + wl2 + wl3 + wh;
a2 = wl1*wl2 + wl1*wl3 + wl2*wl3 + wl1*wh + wl2*wh +
wl3*wh;
a1 = wl1*wl2*wl3 +wl1*wl2*wh + wl1*wl3*wh + wl2*wl3*wh;
a0 = wl1*wl2*wl3*wh;
den=[a4 a3 a2 a1 a0];
b3 = am*wh;
b2 = b3*wz; b1 =0; b0 = 0;
num = [b3 b2 b1 b0];
w = logspace(1,10);
h = freqs(num,den,w);
mag = 20*log10(abs(h));
f = w/(2*pi);
% Plot the frequency response
semilogx(f,mag,'b')
title('Magnitude Response')
xlabel('Frequency, Hz')
ylabel('Gain, dB')
axis([1, 1.0e10, 0, 45])
```

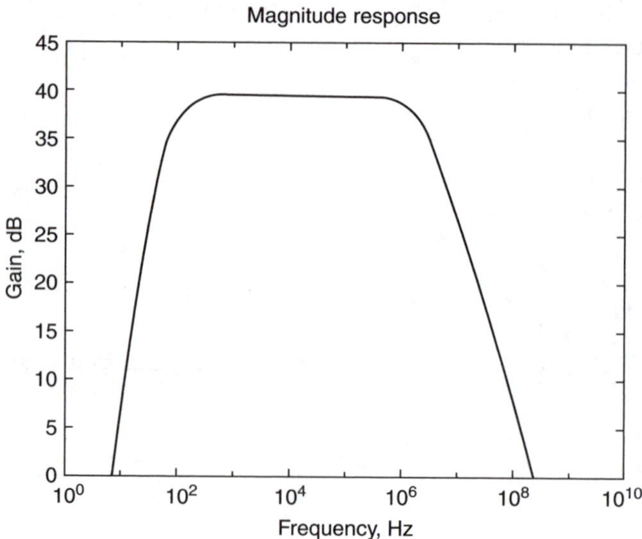

FIGURE 12.16
Frequency response of a CE amplifier.

The frequency response is shown in Figure 12.16.

12.5 MOSFET Characteristics

A metal-oxide-semiconductor field effect transistor (MOSFET) is a four-terminal device. The terminals of the device are the gate, source, drain, and substrate. There are two types of MOSFETs: the enhancement type and the depletion type. In the enhancement type MOSFET, the channel between the source and drain has to be induced by applying a voltage on the gate. In the depletion type MOSFET, the structure of the device is such that a channel exists between the source and drain. Because of the oxide insulation between the gate and the channel, MOSFETs have high input resistance. The electronic symbol of a MOSFET is shown in Figure 12.19.

Mosfets can be operated in three modes: cut-off, triode, and saturation regions. Because the enhancement mode MOSFET is widely used, the presentation in this section will be done using an enhancement-type MOSFET. In the latter device, the channel between the drain and source has to be induced by applying a voltage between the gate and source. The voltage needed to create the channel is called the threshold voltage, V_T. For an n-channel enhancement-type mosfet, V_T is positive and for a p-channel device it is negative.

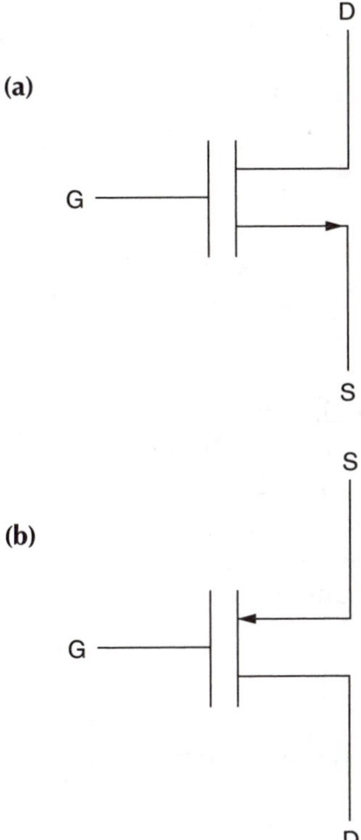

FIGURE 12.17
Circuit symbol of (a) N-channel and (b) P-channel MOSFETs.

Cut-Off Region

For an n-channel MOSFET, if the gate-source voltage V_{GS} satisfies the condition

$$V_{GS} < V_T \tag{12.77}$$

then the device is cut off. This implies that the drain current is zero for all values of the drain-to-source voltage.

Triode Region

When $V_{GS} > V_T$ and V_{DS} is small, the MOSFET will be in the triode region. In the latter region, the device behaves as a nonlinear voltage-controlled resistance. The I-V characteristics are given by

$$I_D = k_n\left[2(V_{GS} - V_T)V_{DS} - V_{DS}{}^2\right] \tag{12.78}$$

provided

$$V_{DS} \leq V_{GS} - V_T \tag{12.79}$$

where

$$k_n = \frac{\mu_n \varepsilon \varepsilon_{ox}}{2t_{ox}} \frac{W}{L} = \frac{\mu_n C_{ox}}{2} \left(\frac{W}{L} \right) \tag{12.80}$$

and

μ_n is surface mobility of electrons.
ε is permittivity of free space (8.85E–14 F/cm).
ε_{ox} is the dielectric constant of SiO_2.
t_{ox} is the thickness of the oxide.
L is the length of the channel.
W is the width of the channel.

Saturation Region

MOSFETs can operate in the saturation region. A MOSFET will be in saturation provided

$$V_{DS} \geq V_{GS} - V_T \tag{12.81}$$

and I-V characteristics are given as

$$I_D = k_n (V_{GS} - V_T)^2 \tag{12.82}$$

The dividing locus between the triode and saturation regions is obtained by substituting

$$V_{DS} = V_{GS} - V_T \tag{12.83}$$

into either Equation (12.78) or Equation (12.82), so we get

$$I_D = k_n V_{DS}^2 \tag{12.84}$$

In the following example, I-V characteristics and the locus that separates triode and saturation regions are obtained using MATLAB.

Example 12.6 I-V Characteristics of an n-Channel MOSFET

For an n-channel enhancement-type MOSFET with k_n = 1 mA/V^2 and V = 1.5 V, use MATLAB to sketch the I-V characteristics for V_{GS} = 4, 6, and 8 V and for V_{DS} between 0 and 12 V.

Solution

MATLAB script

```
% I-V characteristics of mosfet
%
kn=1e-3; vt=1.5;
vds=0:0.5:12;
vgs=4:2:8;
m=length(vds);
n=length(vgs);

for i=1:n
   for j=1:m
      if vgs(i) < vt
      cur(i,j)=0;
      elseif vds(j) >= (vgs(i) - vt)
      cur(i,j)=kn * (vgs(i) - vt)^2;
      elseif vds(j) < (vgs(i) - vt)
      cur(i,j)= kn*(2*(vgs(i)-vt)*vds(j) - vds(j)^2);
      end
        end
end
plot(vds,cur(1,:),'b',vds,cur(2,:),'b',vds,cur(3,:),'b')
xlabel('Vds, V')
ylabel('Drain Current,A')
title('I-V Characteristics of a MOSFET')
text(6, 0.009, 'Vgs = 4 V')
text(6, 0.023, 'Vgs = 6 V')
text(6, 0.045, 'Vgs = 8 V')
```

Figure 12.18 shows the I-V characteristics.

12.6 Biasing of MOSFET Circuits

A popular circuit for biasing discrete MOSFET amplifiers is shown in Figure 12.19. The resistances R_{G1} and R_{G2} will define the gate voltage. The resistance R_S improves operating point stability.

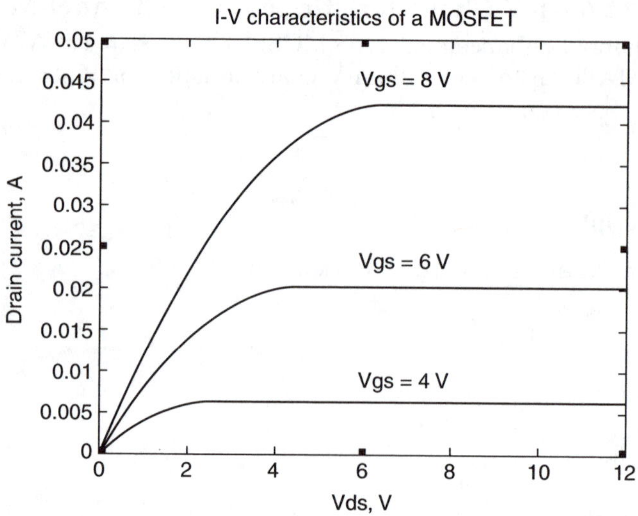

FIGURE 12.18
I-V characteristics of N-channel enhancement-type MOSFET.

Because of the insulated gate, the current that passes through the gate of the MOSFET is negligible. The gate voltage is given as

$$V_G = \frac{R_{G1}}{R_{G1} + R_{G2}} V_{DD} \tag{12.85}$$

The gate-source voltage V_{GS} is

$$V_{GS} = V_G - I_S R_S \tag{12.86}$$

For conduction of the MOSFET, the gate-source voltage V_{GS} should be greater than the threshold voltage of the MOSFET, V_T. Since $I_D = I_S$, Equation (12.86) becomes

$$V_{GS} = V_G - I_D R_S \tag{12.87}$$

The drain-source voltage is obtained by using KVL for the drain-source circuit

$$\begin{aligned} V_{DS} &= V_{DD} - I_D R_D - I_S R_S \\ &= V_{DD} - I_D (R_D + R_S) \end{aligned} \tag{12.88}$$

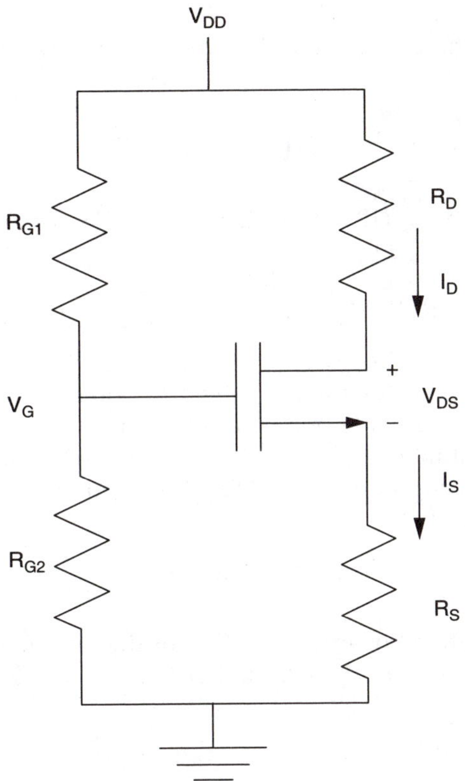

FIGURE 12.19
Simple biasing circuit for enhancement-type NMOS (n-channel MOSFET).

For proper operation of the bias circuit,

$$V_{GS} > V_T \tag{12.89}$$

When Equation (12.89) is satisfied, the MOSFET can either operate in the triode or saturation region. To obtain the drain current, it is initially assumed that the device is in saturation, and Equation (12.82) is used to calculate I_D. Equation (12.81) is then used to confirm the assumed region of operation. If Equation (12.82) is not satisfied, then Equation (12.78) is used to calculate I_D. The method is illustrated by the following example.

Example 12.7 Operating Point Calculation of an n-Channel MOSFET Biasing Circuit

For Figure 12.19, $V_T = 2$ V, $k_n = 0.5$ mA/V^2, $V_{DD} = 9$V, $R_{G1} = R_{G2} = 10$ MΩ, and $R_S = R_D = 10$ kΩ. Find I_D and V_{DS}.

Solution

Substituting Equation (12.86) into Equation (12.82), we have

$$I = k_n\left(V_g - I_D R_D - V_T\right)^2 \tag{12.90}$$

Simplifying Equation (12.90), we have

$$0 = k_n R_D^2 I_D^2 - k_n\left[1 + 2(V_g - V_T)R_D\right]I_D + k_n\left(V_g - V_T\right)^2 \tag{12.91}$$

The above quadratic equation is solved to obtain I_D. Two solutions of I_D are obtained. However, only one is sensible and possible. The possible one is the one that will make $V_{GS} > V_T$. With the possible value of I_D obtained, V_{DS} is calculated using Equation (12.88). It is then verified whether

$$V_{DS} > V_{GS} - V_T$$

The above condition ensures saturation of the device. If the device is not in saturation, then substituting Equation (12.86) into Equation (12.78), we obtain

$$I_D = k_n\left[2(V_g - I_D R_D - V_T)(V_{DD} - (R_D + R_S)I_D) - \left(V_{DD} - (R_D + R_S)I_D\right)^2\right] \tag{12.92}$$

Simplifying Equation (12.92), we obtain the quadratic equation

$$\begin{aligned} 0 = {} & I_D^2\left[(R_S + R_D)^2 + 2R_D(R_D + R_S)\right] \\ & + I_D\left[2V_{DD}(R_D + R_S) - 2V_{DD}R_D - 2(V_g - V_T)(R_D + R_S) - \frac{1}{k_n}\right] \\ & + 2(V_g - V_T)V_{DD} - V_{DD}^2 \end{aligned} \tag{12.93}$$

Two roots are obtained by solving Equation (12.93). The sensible and possible root is the one that will make

$$V_{GS} > V_T$$

The MATLAB program for finding I_D is shown below.

MATLAB script

```
%
% Analysis of MOSFET bias circuit
%
vt=2; kn=0.5e-3; vdd=9;
rg1=10e6; rg2=10e6; rs=10e3; rd=10e3;
vg=vdd * rg2/(rg1 + rg2);
% Id is calculated assuming device is in saturation
a1 = kn*(rd^2);
a2 = -(1 + 2*(vg - vt)*rd)*kn;
a3 = kn * (vg - vt)^2;
p1 = [a1, a2, a3];
irt = roots(p1);
% check for the sensible value of the drain current
vgs = vg - rs * irt(1);
 if vgs > vt
   id = irt(1);
   else
   id = irt(2);
% check for sensible value of the drain current
   vgs = vg - rs*irt(1);
   if vgs > vt
     id = irt(1);
   else
     id = irt(2);
   end
vds=vdd - (rs + rd)*id;
end
% print out results
fprintf('Drain current is%7.3e Amperes\n',id)
fprintf('Drain-source voltage is%7.3e Volts\n', vds)
```

The results are

```
Drain current is 2.484e-004 Amperes
Drain-source voltage is 4.032e+000 Volts
```

The circuit shown in Figure 12.20 is a MOSFET transistor with the drain connected to the gate. The circuit is normally referred to as a diode-connected enhancement transistor.

From Equation (12.88), the MOSFET is in saturation provided

$$V_{DS} > V_{GS} - V_T$$

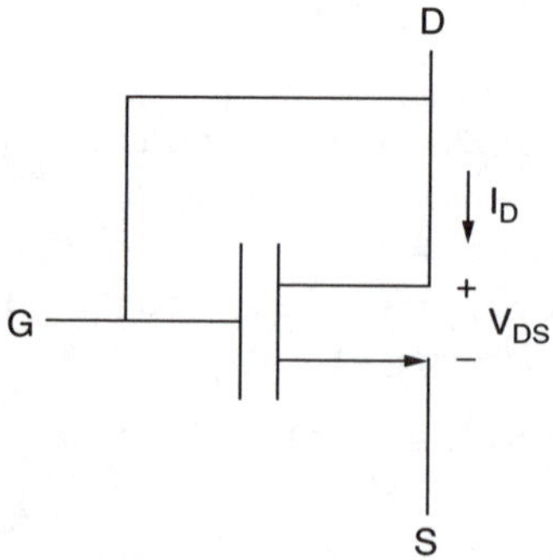

FIGURE 12.20
Diode-connected enhancement type MOSFET.

i.e.,

$$V_{DS} - V_{GS} > -V_T \text{ or } V_{DS} + V_{SG} > -V_T$$

or

$$V_{DG} > -V_T \tag{12.94}$$

Since $V_{DG} = 0$ and V_T is positive for n-channel MOSFET, the device is in saturation and

$$i_D = k_n (V_{GS} - V_T)^2 \tag{12.95}$$

But if $V_{GS} = V_{DS}$, Equation (12.95) becomes

$$i_D = k_n (V_{DS} - V_T)^2$$

The diode-connected enhancement MOSFET can also be used to generate dc currents for NMOS and complementary MOSFET (CMOS) analog integrated circuits. A circuit for generating dc currents that are constant multiples of a reference current is shown in Figure 12.21. It is a MOSFET version of the current mirror circuits discussed in Section 12.3.

Assuming the threshold voltages of the transistors of Figure 12.21 are the same, then since transistor T1 is in saturation,

$$I_{REF} = k_1 (V_{GS1} - V_T)^2 \tag{12.96}$$

Since transistors T1 and T2 are connected in parallel, we get

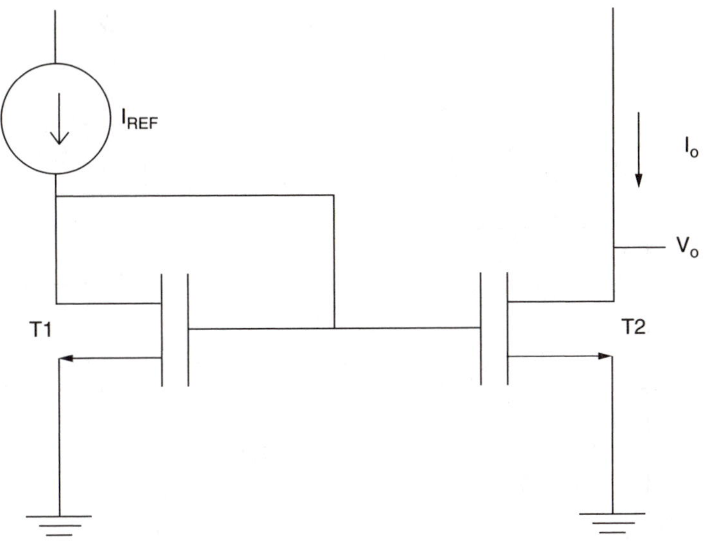

FIGURE 12.21
Basic MOSFET current mirror.

$$V_{GS1} = V_{GS2} = V_{GS} \tag{12.97}$$

and

$$I_0 = k_2 (V_{GS2} - V_T)^2$$

$$I_0 = k_2 (V_{GS} - V_T)^2 \tag{12.98}$$

Combining Equation (12.96) and Equation (12.98), the current

$$I_0 = I_{REF} \left(\frac{k_2}{k_1} \right) \tag{12.99}$$

and using Equation (12.74), Equation (12.99) becomes

$$I_0 = I_{REF} \left(\frac{(W/L)_2}{(W/L)_1} \right) \tag{12.100}$$

Thus, I_0 will be a multiple of I_{REF}, and the scaling constant is determined by the device geometry. In practice, because of the finite output resistance of transistor T2, I_0 will be a function of the output voltage v_0.

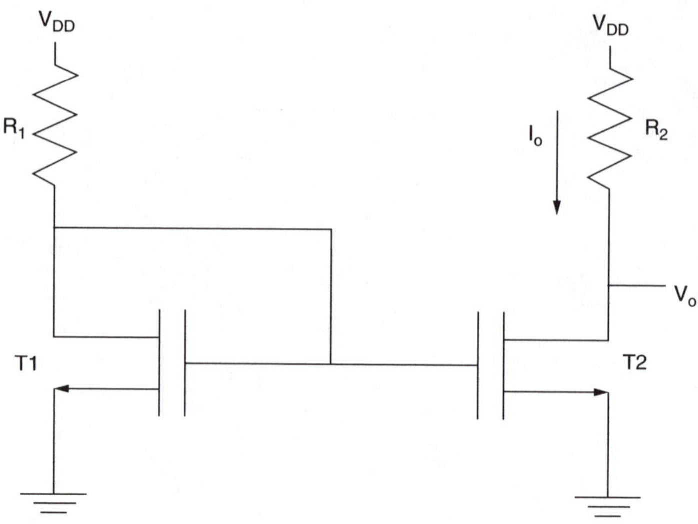

FIGURE 12.22
Circuit diagram for Example 12.8.

Example 12.8 Voltage and Current Calculations for a MOSFET Current Mirror

For the circuit shown in Figure 12.22, $R_1 = 1.5\ \text{M}\Omega$, $L_1 = L_2 = 6\ \mu\text{m}$, $W_1 = 12\ \mu\text{m}$, $W_2 = 18\ \mu\text{m}$, $V_T = 2.0$ V, and $V_{DD} = 5$ V. Find the output current I_{D1}, V_{GS1}, I_0 and R_2. Assume that $V_0 = 2.5$ V and $\mu C_{OX} = 30\ \mu\text{A/V}^2$. Neglect channel length modulation.

Solution

Since T1 is in saturation,

$$I_{D1} = k_{n1}\left(V_{GS} - V_T\right)^2 = k_{n1}(V_{DS} - V_T)^2 \tag{12.101}$$

$$V_{DS} = V_{DD} - I_{D1}R_1 \tag{12.102}$$

Substituting Equation (12.100) into Equation (12.99), we get

$$I_{D1} = k_{n1}\left(V_{DD} - V_T - R_1 I_{D1}\right)^2$$

$$\frac{I_{D1}}{k_{n1}} = \left(V_{DD} - V_T\right)^2 - 2(V_{DD} - V_T)R_1 I_{D1} + R_1^2 I_{D1}^2$$

$$0 = R_1^2 I_{D1}^2 - \left(2(V_{DD} - V_T)R_1 + \frac{1}{k_{n1}}\right)I_{D1} + \left(V_{DD} - V_T\right)^2 \tag{12.103}$$

The above quadratic equation will have two solutions, but only one solution of I_{D1} will be valid. The valid solution will result in $V_{GS} > V_T$.

Using Equation (12.100), we obtain

$$I_0 = I_{D1}\left(\frac{\left(W/L\right)_2}{\left(W/L\right)_1}\right) \tag{12.104}$$

and

$$R = \frac{5 - V_0}{I_0} \tag{12.105}$$

The MATLAB program is as follows:

MATLAB script

```
% Current mirror
%
ucox = 30e-6; l1 = 6e-6; l2 = 6e-6;
w1 = 12e-6; w2=18e-6;
r1=1.5e6; vt=2.0; vdd=5; vout=2.5;
% roots of quadratic equation(12.103) is obtained
kn = ucox * w1/(2 * l1);
a1 = r1^2;
a2 = -2*(vdd - vt)*r1 - (1/kn);
a3 = (vdd - vt)^2;
p = [a1,a2,a3];
i = roots(p);
% check for realistic value of drain current
vgs=vdd - r1*i(1);
if vgs > vt
   id1 = i(1);
  else
   id1 = i(2);
end
% output current is calculated from equation(12.100)
% r2 is obtained using equation (12.105)
iout = id1*w2*l1/(w1 * l2);
r2=(vdd - vout)/iout;
% print results
fprintf('Gate-source Voltage of T1 is%8.3e
Volts\n',vgs)
fprintf('Drain Current of T1 is%8.3e Ampers\n', id1)
fprintf('Drain Current Io is%8.3e Ampers\n', iout)
```

```
fprintf('Resistance R2 is%8.3e Ohms\n', r2)
```

The results are

```
Gate-source Voltage of T1 is 1.730e+000 Volts
Drain Current of T1 is 1.835e-006 Ampers
Drain Current Io is 2.753e-006 Ampers
Resistance R2 is 9.082e+005 Ohms
```

12.7 Frequency Response of Common-Source Amplifier

The common-source amplifier has characteristics similar to those of the common-emitter amplifier discussed in Section 12.4. However, the common-source amplifier has higher input resistance than that of the common-emitter amplifier. The circuit for the common-source amplifier is shown in Figure 12.23.

The external capacitors C_{C1}, C_{C2}, and C_S, will influence the low-frequency response. The internal capacitances of the field-effect transistor (FET) will affect the high-frequency response of the amplifier. The overall gain of the common-source amplifier can be written in a form similar to Equation (12.65).

The midband gain, A_m, is obtained from the midband equivalent circuit of the common-source amplifier. This is shown in Figure 12.24. The equivalent circuit is obtained by short-circuiting all the external capacitors and open-circuiting all the internal capacitances of the FET.

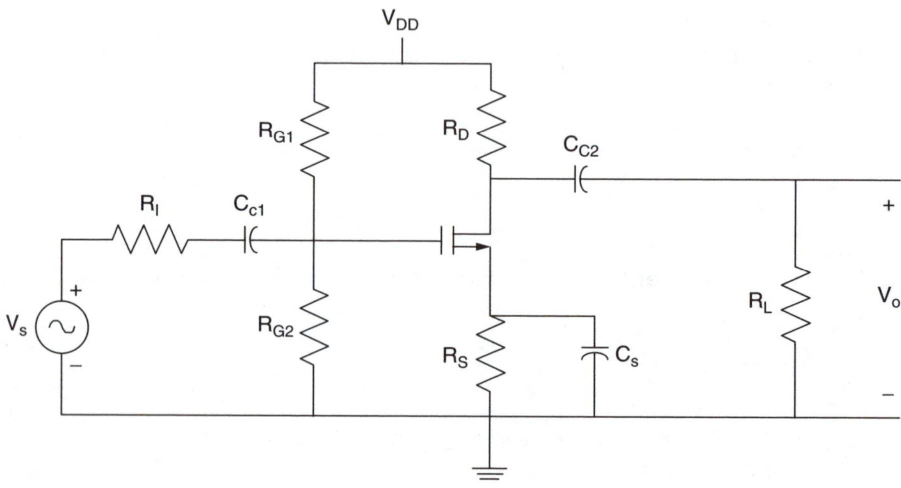

FIGURE 12.23
Common-source amplifier.

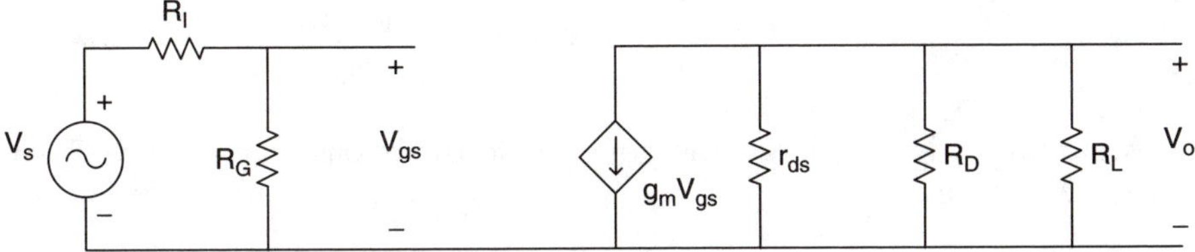

FIGURE 12.24
Midband equivalent circuit of common-source amplifier.

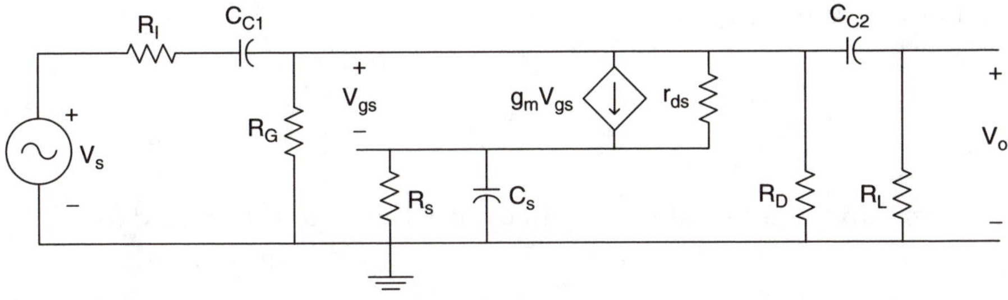

FIGURE 12.25
Equivalent circuit for obtaining the poles at low frequencies of common-source amplifier.

Using voltage division,

$$v_{gs} = \frac{R_G}{R_I + R_G} v_S \tag{12.106}$$

From Ohm's law,

$$v_0 = -g_m v_{gs} \left(r_{ds} \| R_D \| R_L \right) \tag{12.107}$$

Substituting Equation (12.106) into Equation (12.107), we obtain the midband gain as

$$A_m = \frac{v_0}{v_s} = -g_m \left(\frac{R_G}{R_G + R_I} \right) \left(r_{ds} \| R_D \| R_L \right) \tag{12.108}$$

At low frequencies, the small signal equivalent circuit of the common-source amplifier is shown in Figure 12.25.

It can be shown that the low-frequency poles due to C_{C1} and C_{C2} can be written as

$$\tau_1 = \frac{1}{w_{L1}} \cong C_{C1}(R_g + R_I) \tag{12.109}$$

$$\tau_2 = \frac{1}{w_{L2}} \cong C_{C2}\left(R_L + R_D \| r_{ds}\right) \tag{12.110}$$

Assuming r_d is very large, the pole due to the bypass capacitance C_S can be shown to be

$$\tau_3 = \frac{1}{w_{L3}} \cong C_S\left(\frac{R_S}{1 + g_m R_S}\right) \tag{12.111}$$

and the zero of C_S is

$$w_Z = \frac{1}{R_S C_S} \tag{12.112}$$

The 3-dB frequency at the low frequency can be approximated as

$$w_L \cong \sqrt{(w_{L1})^2 + (w_{L2})^2 + (w_{L3})^2} \tag{12.113}$$

For a single-stage common-source amplifier, the source bypass capacitor is usually the determining factor in establishing the low 3-dB frequency.

The high-frequency equivalent circuit of a common-source amplifier is shown in Figure 12.26. In the figure, the internal capacitances of the FET, C_{gs}, C_{gd}, and C_{ds}, are shown. The external capacitors of the common-source amplifier are short-circuited at high frequencies.

Using the Miller theorem, Figure 12.26 can be simplified. This is shown in Figure 12.27.

The voltage gain at high frequencies is

$$A_V = \frac{v_0}{v_s} \cong -\left(\frac{R_G}{R_G + R_I}\right)\left(\frac{g_m R_L'}{(1 + s(R_G \| R_I)C_1)(1 + sR_L'C_2)}\right) \tag{12.114}$$

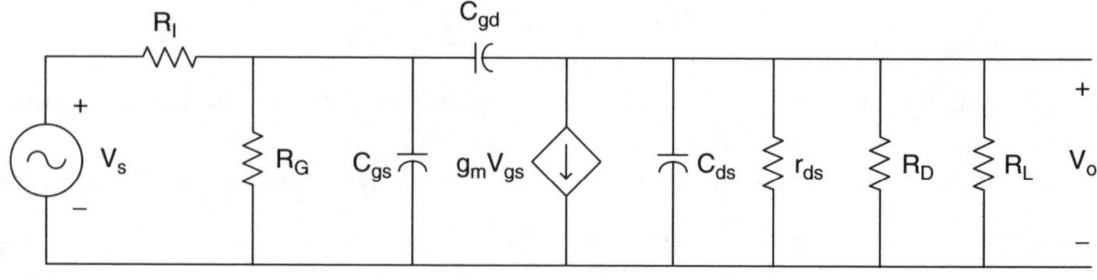

FIGURE 12.26
High-frequency equivalent circuit of common-source amplifier.

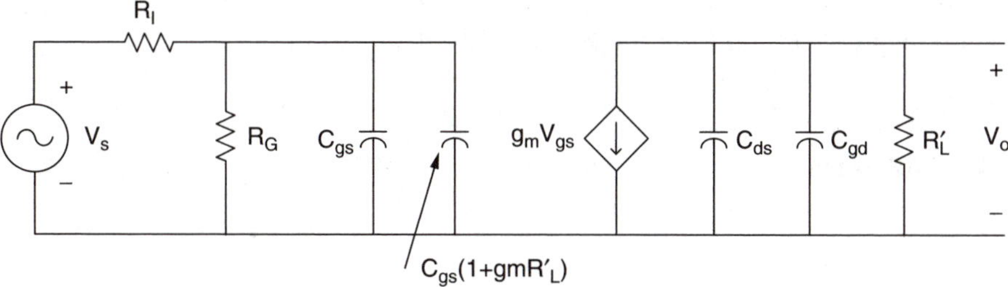

FIGURE 12.27
Simplified high-frequency equivalent circuit for common-source amplifier.

where

$$C_1 = C_{gs} + C_{gd}\left(1 + g_m R_L'\right) \tag{12.115}$$

and

$$C_2 = C_{ds} + C_{gd} \tag{12.116}$$

The high-frequency poles are

$$w_{H1} = \frac{1}{C_1\left(R_G \| R_I\right)} \tag{12.117}$$

$$w_{H2} = \frac{1}{C_2\left(R_L \| R_D \| r_{ds}\right)} \tag{12.118}$$

The approximate high-frequency cut-off is

$$w_H = \frac{1}{\sqrt{\left(\frac{1}{w_{H1}}\right)^2 + \left(\frac{1}{w_{H2}}\right)^2}} \tag{12.119}$$

In the following example, MATLAB is used to obtain the midband gain, cut-off frequencies, and bandwidth of a common-source amplifier.

Example 12.9 Common-Source Amplifier Gain, Cut-Off Frequencies, and Bandwidth

For the common-source amplifier, shown in Figure 12.23, $C_{C1} = C_{C2} = 1$ μF and $C_S = 50$ μF. The FET parameters are $C_{gd} = C_{ds} = 1$ pF, $C_{gs} = 10$ pF, $g_m = 10$ mA/V, and $r_{ds} = 50$ kΩ. $R_D = 8$ kΩ, $R_L = 10$ kΩ, $R_S = 2$ kΩ, $R_I = 50$ Ω, $R_{G1} = 5$ MΩ, and $R_{G2} = 5$ MΩ.

Determine (a) the midband gain, (b) the low-frequency cut-off, (c) the high-frequency cut-off, and (d) the bandwidth of the amplifier.

Solution

MATLAB script

```
%
% common-source amplifier
%
rg1=5e6; rg2=5e6; rd=8e3; rl=10e3;
ri=50; rs=2e3; rds=50e3;
cc1=1e-6; cc2=1e-6; cs=50e-6;
gm=10e-3; cgs=10e-12; cgd=1e-12; cds=1e-12;

% Calculate midband gain using equation (12.108)
a = (1/rds) + (1/rd) + (1/rl);
rlprime = 1/a;
rg = rg1*rg2/(rg1 + rg2);
gain_mb = -gm*rg*rlprime/(ri + rg);

% Calculate Low cut-off frequency using equation (12.113)
t1 = cc1*(rg + ri);
wl1 = 1/t1;
rd_rds = (rd*rds)/(rd + rds);
t2 = cc2 * (rl + rd_rds);
wl2=1/t2;
t3=cs * rs/(1 + gm * rs);
wl3=1/t3;
wl=sqrt(wl1^2 + wl2^2 + wl3^2);

% Calculate high frequency cut-off using equations
(12.115 to 12.119)
c1=cgs + cgd * (1 + gm * rlprime);
c2=cds + cgd;
rg_ri=rg * ri/(rg + ri);
wh1=1/(rg_ri * c1);
wh2=1/(rlprime * c2);
int_term = sqrt((1/wh1)^2 + (1/wh2)^2);
wh = 1/int_term;
bw = wh-wl;

% Print results
fprintf('Midband Gain is%8.3f\n', gain_mb)
fprintf('Low frequency cut-off is%8.3e\n', wl)
fprintf('High frequency cut-off is%8.3e\n', wh)
fprintf('Bandwidth is%8.3e Hz\n', bw)
```

The results are

```
Midband Gain is -40.816
Low frequency cut-off is 2.182e+002
High frequency cut-off is 1.168e+008
Bandwidth is 1.168e+008 Hz
```

Bibliography

1. Belanger, P.R., Adler, E.L., and Rumin, N.C., *Introduction to Circuits with Electronics: An Integrated Approach*, Holt, Rinehart and Winston, New York, 1985.
2. Ferris, C.D., *Elements of Electronic Design*, West Publishing Co., St. Paul, MN, 1995.
3. Geiger, R.L., Allen, P.E., and Strader, N.R., *VLSI Design Techniques for Analog and Digital Circuits*, McGraw-Hill Publishing Co., New York, 1990.
4. Ghausi, M.S., *Electronic Devices and Circuits: Discrete and Integrated*, Holt, Rinehart and Winston, New York, 1985.
5. Howe, R.T. and Sodini, C.G., *Microelectronics, An Integrated Approach*, Prentice Hall, Upper Saddle River, NJ, 1997.
6. Rashid, M.H., *Microelectronic Circuits, Analysis and Design*, PWS Publishing Company, Boston, 1999.
7. Sedra, A.S. and Smith, K.C., *Microelectronic Circuits*, 4th ed., Oxford University Press, New York, 1998.
8. Warner, R.M., Jr. and Grung, B.L., *Semiconductor Device Electronics*, Holt, Rinehart and Winston, New York, 1991.
9. Wildlar, R.J., Design techniques for monolithic operational amplifiers, *IEEE Journal of Solid State Circuits*, SC-3, 341–348, 1969.

Problems

Problem 12.1

For the data provided in Example 12.2, use MATLAB to sketch the output characteristics for $V_{BE} = 0.3$, 0.5, and 0.7 V. Do not neglect the effect of V_{AF} on the collector current.

Problem 12.2

For the self-bias circuit, shown in Figure 12.6, the collector current involving I_{CBO} is given by Equation (12.47). Assuming that $R_{B1} = 75$ kΩ, $R_{B2} = 25$ kΩ, $R_E = 1$ kΩ, $R_C = 7.5$ kΩ, $\beta_F = 100$, and at 25°C, $V_{BE} = 0.6$ V and $I_{CBO} = 0.01$ μA, determine the collector currents for temperatures between 25 and 85°C. If R_E is changed to 3 kΩ, what will be the value of I_C at 25°C?

Problem 12.3

For Figure 12.13, if R_{B1} = 50 kΩ, R_{B2} = 40 kΩ, r_S = 50 Ω, r_X = 10 Ω, R_L = 5 kΩ, R_C = 5 kΩ, r_{ce} = 100 kΩ, $C_{C1} = C_{C2}$ = 2 μF, C_π = 50 pF, C_μ = 2 pF, β_F = 100, and V_{CC} = 10 V, explore the low-frequency response for the following values of R_E: 0.1 kΩ, 1 kΩ, and 5 kΩ. Calculate the high-frequency cut-off for R_E = 0.1 kΩ.

Problem 12.4

For the Widlar current source, shown in Figure P12.4, determine the output current if R_C = 40 KΩ, V_{CC} = 10 V, V_{BE1} = 0.7 V, β_F = 100 and R_E = 25 EΩ.

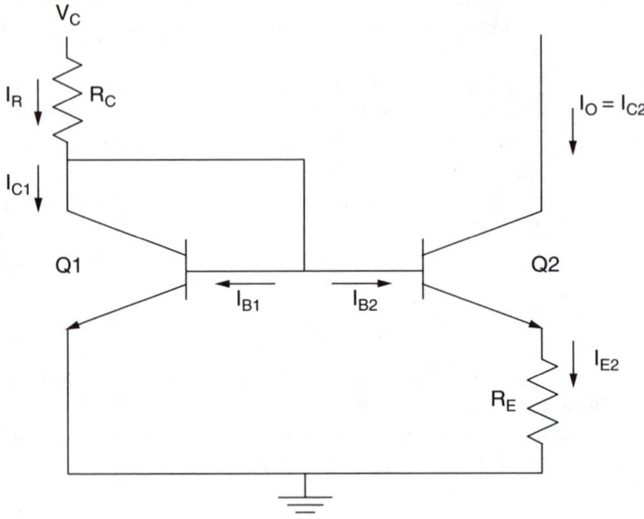

FIGURE P12.4
Widlar current source.

Problem 12.5

For an n-channel enhancement-type MOSFET with k_n = 2 mA/V² and V_T = 1 V, write a MATLAB program to plot the triode characteristics for V_{GS} = 2, 3, 4, and 5 V when V_{DS} < 1 V.

Problem 12.6

For Figure 12.19, V_T = 1.5 V, k_n = 0.5 mA/V², V_{DD} = 10 V, R_{G1} = 10 MΩ, R_{G2} = 12 MΩ, and R_D = 10 kΩ. Find I_D for the following values of R_S: 2, 4, 6, and 8 kΩ. Indicate the region of operation for each value of R_S.

Problem 12.7

For the common-source amplifier shown in Figure 12.23, R_D = 10 kΩ, R_L = 20 kΩ, R_I = 1.5 kΩ, R_S = 1000 Ω, R_{G1} = 10 MΩ, R_{G2} = 10 MΩ, $C_{C1} = C_{C2}$ = 2 μF, and C_S = 40 μF. The FET parameters are C_{gs} = 10 pF, $C_{gd} = C_{ds}$ = 1.5 pF, g_m = 5 mA/V, and r_{ds} = 100 kΩ. Use MATLAB to plot the frequency response of the amplifier.

13

Electronic Data Analysis

In this chapter, MATLAB functions needed for electronic data analysis are discussed. Several examples are worked out to illustrate the use of the functions.

13.1 Save, Load, and Textread Functions

In Chapters 1 and 3, some input/output commands of MATLAB were discussed. In this section, additional input/output commands, useful for data analysis, will be discussed. These are save, load, and textread functions.

13.1.1 Save and Load Functions

The **save** command saves data in MATLAB workspace to disk. The **save** command can store data either in a memory-efficient binary format, called a MAT-file, or an ASCII file. The general form of the **save** command is

save filename [List of variables] [options] (13.1)

where **save** (without filename, list of variables, and options) saves all the data in the current workspace to a file named **matlab.mat** in the current directory.

If a filename is included in the command line, the data will be saved in file "**filename.mat**".

If a list of variables is included, only those variables will be saved.

The options for the save command are shown in Table 13.1.

MAT-files are preferable for data that are generated and are going to be used by MATLAB. MAT-files are platform-independent. The files can be written and read by any computer that supports MATLAB. In addition, MAT-files preserve all the information about each variable in the workspace including its name, size, storage space in bytes, and class (structure array,

TABLE 13.1

Save Command Options

Option	Description
-mat	Save data in MAT-file format (default)
-ascii	Save data using 8-digit ASCII format
-ascii -double	Save data using 16-digit ASCII format
-ascii -double -tab	Save data using 16-digit ASCII format with Tabs
-append	Save data to an existing MAT-file

TABLE 13.2

Load Command Option

Option	Description
-mat	Load data from MAT-file (default in file extension is mat)
-ascii	Load data from space-separated file

double array, cell array, or character array). Furthermore, MAT-files have every variable stored in full precision.

The **ASCII files** are preferable if the data are to be exported or imported to programs other than MATLAB. It is recommended that if you save workspace content in ASCII format, *save only one variable at a time*. If more than one variable is saved, MATLAB will create ASCII data that might be difficult to interpret when loaded back into a MATLAB program.

The **load** command will load data from a MAT-file or ASCII file into the current workspace. The general format of the **load** command is

load filename [options] (13.2)

where **load** (by itself without filename and options) will load all the data in file matlab.mat into the current workspace and **load filename** will load data from the specified filename. The options for the load command are shown in Table 13.2.

It is strongly recommended that an ASCII data file that will be used with the MATLAB program should contain only numeric information and each row of the file should have the same number of data values. It is also recommended that an ASCII filename include the extension **.dat** to make it easier to distinguish m-files and MAT-files.

13.1.2 Textread Function

The **textread** command can be used to read ASCII files that are formatted into columns of data, where values in different columns might be of different types. The general form of the **textread** command is

[a, b, c,...] = textread(filename, format, n) (13.3)

where

filename is the name of file to open. The filename should be in quotes, i.e., 'filename'.

format is a string containing a description of the type of data in each column. The format descriptors are similar to those of fprintf. The format list should be in quotes. Supported functions include:

%d — Read a signed integer value.

%u — Read an integer value.

%f — Read a floating point value.

%s — Read a whitespace separated string.

%q — Read a (possibly double-quoted) string.

%c — Read characters (including white space; output is char array).

n is the number of lines to read. If n is missing, the command reads to the end of the file.

a, b, c are the output arguments. The number of output arguments must match the number of columns read.

The **textread** is much more than the **load** command. The **load** command assumes that all the data in the file being loaded are of a single type. The **load** command does not support different data types in different columns. In addition, the **load** command stores all the data in a single array. However, the **textread** command allows each column of data to go into a separate variable.

The following example illustrates the use of the **load** function.

Example 13.1 Voltage vs. Temperature of a Thermister

The data shown in Table 13.3 represent the corresponding temperature and voltage across a thermister. The data is stored in file ex13_1se.dat. Read data from the file and plot voltage as a function of temperature.

Solution

The **load** command will be used to read in the data on file.

MATLAB script

```
% data is stored in ex13_1se.dat
% read data using load command
%
load ex13_1se.dat -ascii;
temp = ex13_1se(:,1);
v1 = ex13_1se(:,2);
```

TABLE 13.3
Voltage vs. Temperature of a Thermister

Temperature, K	Voltage across Thermister, V
300	4.734
310	3.689
320	2.809
330	2.110
340	1.577
350	1.180
360	0.888
370	0.673
380	0.515
390	0.398
400	0.311

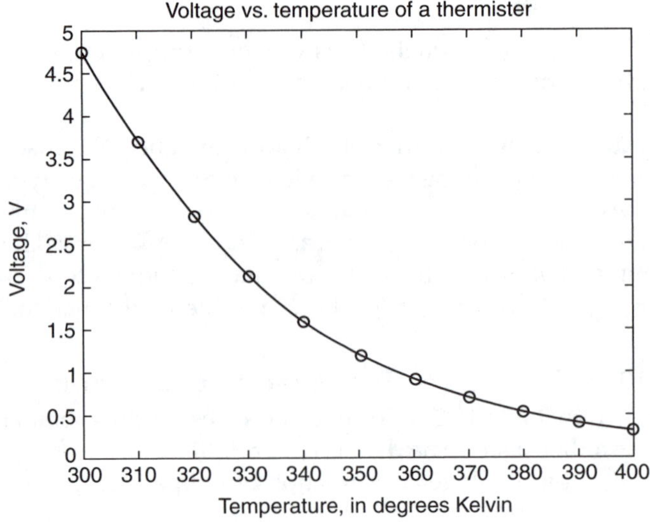

FIGURE 13.1
Voltage across a thermister as a function of temperature.

```
plot(temp, v1, 'ob', temp, v1, 'b');  % plot of voltage
vs. temperature
xlabel ('Temperature, in Degree Kelvin')
ylabel('Voltage, V')
title('Voltage vs. Temperature of a Thermister')
```

Figure 13.1 shows the voltage of the thermister as a function of temperature.

TABLE 13.4

Magnitude Characteristics of a Notch Filter

Frequency, Hz	Gain
2.51200E+06	2.23325E–02
2.51258E+06	1.94193E–02
2.51316E+06	1.65048E–02
2.51374E+06	1.35893E–02
2.51432E+06	1.06730E–02
2.51490E+06	7.75638E–03
2.51548E+06	4.83962E–03
2.51606E+06	1.92304E–03
2.51664E+06	9.93068E–04
2.51722E+06	3.90840E–03
2.51780E+06	6.82266E–03
2.51838E+06	9.73556E–03
2.51896E+06	1.26468E–02
2.52012E+06	1.84631E–02
2.52070E+06	2.13676E–02

Example 13.2 Notch Filter Characteristics

The data shown in Table 13.4 represent the frequency and voltage gain of a notch filter. The data are stored in file ex13_2se.dat. (a) Read data from the file and plot the magnitude characteristics of the filter. (b) Calculate the notch frequency.

Solution

MATLAB script

```
% data is stored in ex13_2se.dat
% read data using load command
%
load ex13_2se.dat -ascii
freq = ex13_2se(:,1);
gain = ex13_2se(:,2);
n = length(freq);
% Determination of center frequency
[vc, k] = min(gain);
fc = fre(k);
% Plot the frequency response
plot(freq, gain);% plot of  gain  vs. frequency
xlabel('Frequency, Hz')
ylabel('Gain')
title('Frequency Response of a Notch Filter')
fc
```

The center frequency of the notch filter is fc = 2516640 Hz. Figure 13.2 shows the magnitude characteristics of the notch filter.

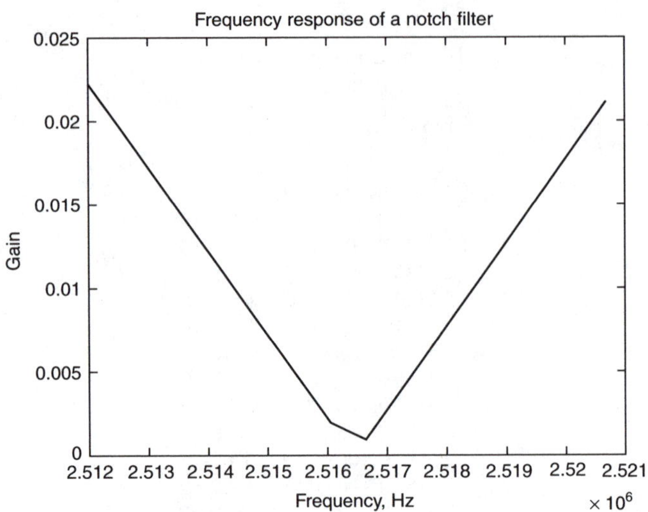

FIGURE 13.2
Magnitude response of a notch filter.

13.2 Statistical Analysis

In MATLAB, data analyses are performed on column-oriented matrices. Different variables are stored in the individual column cells, and each row represents a different observation of each variable. A set of data consisting of 10 samples in four variables would be stored in a matrix of size 10-by-4. Functions act on the elements in the column. Table 13.5 gives a brief description of various MATLAB functions for performing statistical analysis.

The following two examples illustrate the usage of some of the statistical analysis functions.

Example 13.3 Statistics of Resistors

Resistances of two bins containing 2-kΩ and 5-kΩ resistors, respectively, were measured using a multimeter. Ten resistors selected from the two bins have values shown in Table 13.5. For each resistor bin, determine the mean, median, and standard deviation.

Solution

The data shown in Table 13.5 are stored as a 2 × 10 matrix y.

MATLAB script

```
%  This program computes the mean, median, and standard
% deviation of resistors in bins
```

TABLE 13.5

Statistical Analysis Functions

Function	Description
corrcoef(x)	Determines correlation coefficients
hist(x)	Draws the histogram or the bar chart of x
max(x)	Obtains the largest value of x; if x is a matrix, max(x) returns a row vector containing the maximum elements of each column
[y, k] = max(x)	Obtains the maximum value of x and the corresponding locations (indices) of the first maximum value for each column of x
mean(x)	Determines the mean or the average value of the elements in the vector; if x is a matrix, mean(x) returns a row vector that contains the mean value of each column
median(x)	Finds the median value of the elements in the vector x; if x is a matrix, this function returns a row vector containing the median value of each column
min(x)	Finds the smallest value of x; if x is a matrix, min(x) returns a row vector containing the minimum values from each column
[y, k]=min(x)	Obtains the smallest value of x and the corresponding locations(indices) of the first minimum value from each column of x
sort(x)	Sorts the rows of a matrix a in ascending order
std(x)	Calculates and returns the standard deviation of x if it is a one-dimensional array; if x is a matrix, a row vector containing the standard deviation of each column is computed and returned

TABLE 13.6

Resistances in 2-kΩ and 5-kΩ Resistor Bins

Number	2-kΩ Resistor Bin	5-kΩ Resistor Bin
1	2050	5021
2	1992	5250
3	2021	4727
4	1980	5370
5	2070	4880
6	1940	5165
7	2005	4922
8	1998	5417
9	2021	4684
10	1987	5110

```
% the data is stored in matrix y
y = [2050  5021;
1992   5250;
2021  5250;
1980  5370;
2070  4880;
1940  5165;
2005  4922;
1998  5417;
2021  4684;
```

```
1987  5110];
%
% Calculate the mean
mean_r = mean(y);
% Calculate the median
median_r = median(y);
% Calculate the standard deviation
std_r = std(y);
% Print out the results
fprintf('Statistics of Resistor Bins\n\n')
fprintf('Mean of 2K, and 5K bins, respectively:%7.3e ,
%7.3e \n', mean_r)
fprintf('Median of 2K and 5K bins, respectively :%7.3e,
%7.3e\n',  median_r)
fprintf('Standard Deviation of 2K and 5K bins,
respectively:%7.8e,  %7.8e \n', std_r)
```

The results are

```
Statistics of Resistor Bins

Mean of 2K, and 5K bins, respectively:2.006e+003 ,
5.107e+003
Median of 2K and 5K bins, respectively :2.002e+003,
5.138e+003
Standard Deviation of 2K and 5K bins,
respectively:3.67187509e+001,  2.31329227e+002
```

Example 13.4 Correlation between Two Voltages of a Circuit

The voltages between two nodes of a circuit for different supply voltages are shown in Table 13.6. Plot V1 vs. V2. Determine the correlation coefficient between V1 and V2.

Solution

MATLAB script

```
% Solution to Example 13.4
% voltages V1and V2
v1 = [ 5.638  5.875  6.111  6.348  6.584 ...
   6.82  7.055  7.29  7.525  7.759 ...
   7.990  8.216  8.345];
v2 = [ 5.294  5.644  5.835  6.165  6.374 ...
   6.684  6.843  7.162  7.460  7.627  ...
   7.972  8.170  8.362];  %

% correlation coefficient between input and output
r = corrcoef(vs,vo);
```

TABLE 13.7

Voltages at Two Nodes

Voltage V1, V	Voltage V2, V
5.638	5.294
5.875	5.644
6.111	5.835
6.348	6.165
6.584	6.374
6.820	6.684
7.055	6.843
7.290	7.162
7.525	7.460
7.759	7.627
7.990	7.972
8.216	8.170
8.345	8.362

```
% coefficient
pfit = polyfit (v1, v2, 1);
% Linear equation is y = m*x + b
b = pfit(2)
m = pfit(1)
v2fit = m*v1 + b;
% Plot v versus ln(i) and best fit linear model
plot (v1, v2fit,'b', v1, v2, 'ob')
xlabel ('Voltage V1, V')
ylabel('Voltage, V2, V')
title('Correlation Between Voltages V1 and V2')
% print results
fprintf('Correlation Coefficient between input and
output voltages is %7.3f\n', r )
```

The plot of V1 and V2 is shown in Figure 13.3.

The MATLAB result is:

```
Correlation coefficient between input and output
voltages is 0.9992
```

13.3 Curve Fitting

The MATLAB **polyfit** function is used to compute the best fit of a set of data points to a polynomial with a specified degree. The polyfit function was discussed in Section 9.1. The general form of the function is

poly_xy = polyfit(x, y, n) (13.4)

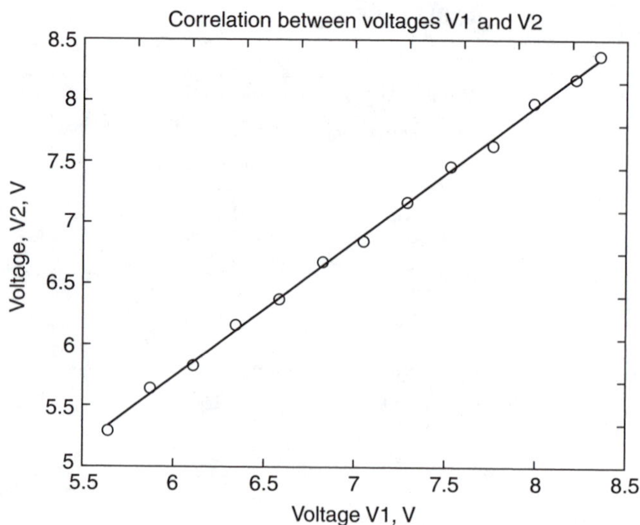

FIGURE 13.3
Correlation between V1 and V2.

where

- **x** and **y** are the data points.
- **n** is the nth degree polynomial that will fit the vectors x and y.
- **poly_xy** is a polynomial that fits the data in vector y to x in the least square sense. **poly_xy** returns (n + 1) coefficients in descending powers of x.

An application of the **polyfit** function is illustrated by the following example.

Example 13.5 Platinum Resistance Thermometer

In a platinum resistance thermometer with the American Alloy, the resistance, R, is related to the temperature, T, by the expression

$$R = a + bT + cT^2 \tag{13.5}$$

where R is the resistance in ohms and T is temperature in degrees Celsius. Using the data in Table 13.7, determine the coefficients a, b, and c.

MATLAB script

```
% Example 13.5
% Platinum Resistance thermometer
T = [ -100  -80  -60  -40 -20  0  ...
      20  40  60  80  100];
R = [ 160.3  168.34  176.32  184.26  192.16 ...
      200  207.79  215.54  223.24  230.89  238.5];
```

TABLE 13.8

Temperature vs. Resistance of a Thermometer

T, °C	R, Ω
–100	160.30
–80	168.34
–60	176.32
–40	184.26
–20	192.16
0	200
20	207.79
40	215.54
60	223.24
80	230.89
100	238.50

```
n = length(T);
%
% coefficient
pfit = polyfit (T, R, 2);
% Equation is of the form R = c*T*T + b*T + a
a = pfit(3)
b = pfit(2)
c = pfit(1)
%
for i = 1:n
   Rfit(i) = c*T(i)*T(i) + b*T(i) + a;
end
% Plot R versus T and best fit  model
plot (T, Rfit, 'b', T, R, 'ob')
xlabel ('Temperature, Centigrade')
ylabel('Resistance, Ohms')
title('Best Fit Model')
```

The best-fit model is plotted in Figure 13.4.
From MATLAB results, the constants a, b, and c are given as

```
a  =  199.9975
b  =  0.3910
c  =  -5.9819e-005
```

Example 13.6 Zener Voltage Regulator Circuit

The zener voltage regulator circuit has the corresponding input and output voltages shown in Table 13.8. (a) Determine the correlation coefficient between the two voltages. (b) What is the least mean square fit between the voltages (polynomial fit)?

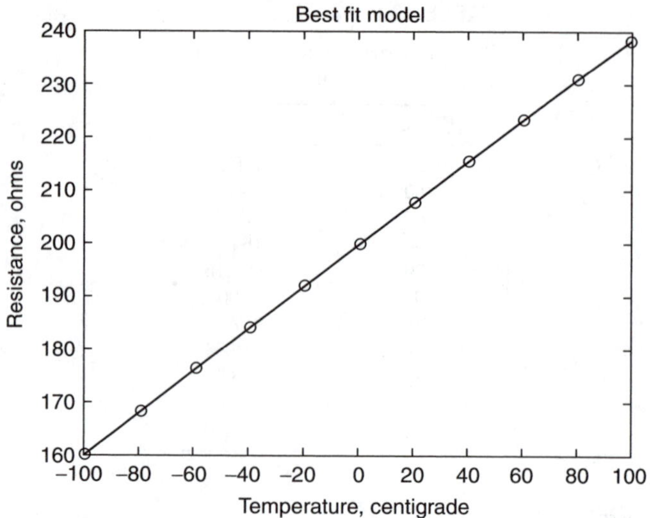

FIGURE 13.4
Resistance vs. temperature of a thermister.

TABLE 13.9
Input and Output Voltages of Zener Voltage Regulator

Input Voltage VS, V	Output Voltage VOUT, V
8.0	5.641
9.0	5.651
10.0	5.665
11.0	5.671
12.0	5.676
13.0	5.686
14.0	5.707
15.0	5.711
16.0	5.717
17.0	5.725
18.0	5.735

Solution

MATLAB script

```
% Solution to Example 13.6
% Input and output voltages
vs = [8  9  10  11  12  13  14  15  16  17  18];
vo = [5.641  5.651  5.665  5.671  5.676  5.686  ...
      5.707 5.711  5.717  5.725  5.735];
```

```
% correlation coefficient between input and output
r = corrcoef(vs,vo);

% Least square fit
pfit = polyfit (vs, vo, 1);
% Linear equation is y = m*x + b
b = pfit(2);
m = pfit(1);
vfit = m*vs + b;
% Plot vo versus vs and best fit linear model
plot (vs, vfit, 'b', vs, vo, 'ob')
xlabel ('Input Voltage, V')
ylabel('Output Voltage, V')
title('Best Fit Linear Model')
fprintf('Correlation Coefficent between input and
output voltages\n')
r
%
```

The MATLAB results are:

```
Correlation Coefficent between input and output
voltages

r =
     1.0000     0.9936
     0.9936     1.0000

The correlation coefficient is 0.9936
```

The plot is shown in Figure 13.5.

13.4 Other Functions for Data Analysis

Additional functions can be used for data analysis. Table 13.9 gives a brief description of various MATLAB functions for performing data analysis.

13.4.1 Integration Function (trapz)

The **quad** and **quad8** functions use an argument that is an analytic expression of the integrand. This facility allows the functions to reduce the integration subinterval automatically until a given precision is attained. However, if we require the integration of a function whose analytic expression is unknown, the MATLAB function **trapz** can be used to perform the numerical integration. The description of the function **trapz** follows.

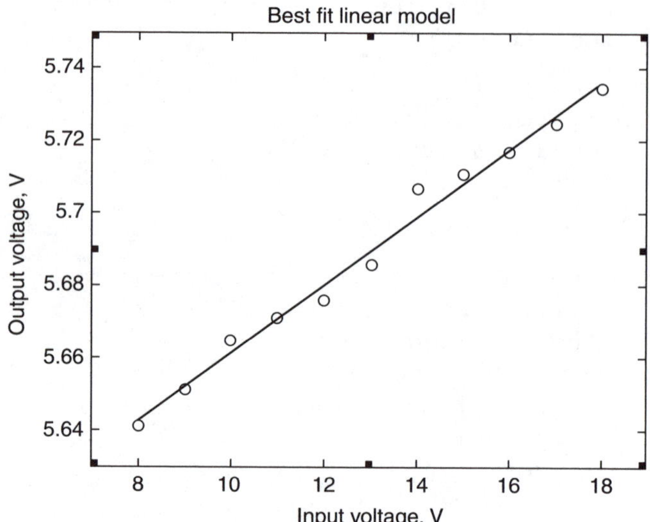

FIGURE 13.5
Best linear model of zener output voltage vs. input voltage.

The MATLAB function **trapz** is used to obtain the numerical integration of a function (with or without an analytic expression) by use of the trapezoidal rules. If a function f(x) has known values at x_1, x_2, ... x_n given as $f(x_1)$, $f(x_2)$, ... $f(x_n)$, respectively, the trapezoidal rule approximates the area under the function, i.e.,

$$A \cong (x_2 - x_1)\left[\frac{f(x_1) + f(x_2)}{2}\right] + (x_3 - x_2)\left[\frac{f(x_2) + (x_{3)}}{2}\right] + \ldots \tag{13.6}$$

For constant spacing where

$x_2 - x_1 = x_3 - x_2 = \ldots = h$, the above equation reduces to

$$A = h\left[\frac{1}{2}f(x_1) + f(x_2) + f(x_3) + \ldots f(x_{n-1}) + \frac{1}{2}f(x_n)\right] \tag{13.7}$$

The error in the trapezoidal method of numerical integration reduces as the spacing, h, decreases.

The general form of the **trapz** function is

S2 = trapz(x, y) (13.8)

where **trapz(x, y)** computes the integral of y with respect to x. X and y must be vectors of the same length.

TABLE 13.10

Data Analysis Functions

Function	Description
cov(x)	Obtains covariance matrix
cross(x, y)	Determines cross product of vectors x and y
cumprod(x)	Finds a vector of the same size as x containing the cumulative products of the values from x; if x is a matrix, cumprod(x) returns a matrix the same size as x containing cumulative products of values from the columns of x
cumsum(x)	Obtains a vector of the same size as x containing the cumulative sums of values from x; if x is a matrix, cumsum(x) returns a matrix the same size as x containing cumulative values from the columns of x
diff(x)	Computes the differences between elements of an array x; it approximates derivatives
dot(x, y)	Determines the dot product of vectors x and y
hist(x)	Draws the histogram or the bar chart of x
prod(x)	Calculates the product of the elements of x; if x is a matrix, prod(x) returns a row vector that contains the product of each column
rand(n)	Generates random numbers; if n = 1, a single random number is returned; if n > 1, an n-by-n matrix of random numbers is generated; rand(n) generates the random numbers uniformly distributed in the interval [0,1]
rand(m, n)	Generates an m-by-n matrix containing uniformly distributed random numbers between 0 and 1
rand('seed,' n)	Sets the seed number of the random number generator to n: if rand is called repeatedly with the same seed number, the sequence of random numbers becomes the same
rand('seed')	Returns the current value of the "seed" values of the random number generator
rand(m,n)	Generates an m-by-n matrix containing random numbers uniformly distributed between zero and one
randn(n)	Produces an n-by-n matrix containing normally distributed (Gaussian) random numbers with a mean of zero and variance of one
randn(m,n)	Produces an m-by-n matrix containing normally distributed (Gaussian) random numbers with a mean of zero and variance of one; to convert Gaussian random number r_n with mean value of zero and variance of one to a new Gaussian random number with mean of μ and standard deviation σ, we use the conversion formula: $$X = \sigma .\ r_n + \mu$$ Thus, random number data with 200 values, gaussian distributed with mean value of four and standard deviation of two can be generated with the equation $$\text{data\_g} = 2.\text{randn}(1,200) + 4$$
sum(x)	Calculates and returns the sum of the elements in x; if x is a matrix, this function calculates and returns a row vector that contains the sum of each column
trapz(x,y)	Trapezoidal integration of the function y = f(x); a detailed discussion of this function is presented in Section 13.4.2

Another form of the trapz function is

$$\textbf{S2 = trapz(Y)} \tag{13.9}$$

where **trapz(Y)** computes the trapezoidal integral of Y assuming unit spacing data points. If the spacing is different from one, assuming it is **h**, then trapz(Y) should be multiplied by **h** to obtain the numerical integration, i.e.,

$$S1 = (h)(S2) = (h).trapz(Y) \tag{13.10}$$

Example 13.7 Average Current Flowing through a Diode

The current flowing through a diode at specific instants of time is saved in file ex13_7se.dat. A sample of the data is shown in Table 13.10. (a) Plot the current with respect to time. (b) Find the average current passing through the diode.

Solution

MATLAB script

```
% Example 13.7
%
% The data is stored at ex13_7.dat Read the data using
load function
```

TABLE 13.11

Current Flowing through a Diode

Time, sec	Current, A
0.000E+00	–1.210E–11
1.000E–03	4.201E–09
2.000E–03	6.160E–09
3.000E–03	3.385E–02
4.000E–03	6.148E–02
5.000E–03	4.723E–02
6.000E–03	2.060E–03
7.000E–03	–4.583E–09
8.000E–03	–3.606E–09
9.000E–03	–2.839E–09
1.000E–02	–2.070E–09
1.100E–02	–1.268E–09
1.200E–02	–4.366E–10
1.300E–02	3.940E–10
1.400E–02	1.231E–09
1.500E–02	2.041E–09
1.600E–02	2.820E–09
1.700E–02	3.601E–09

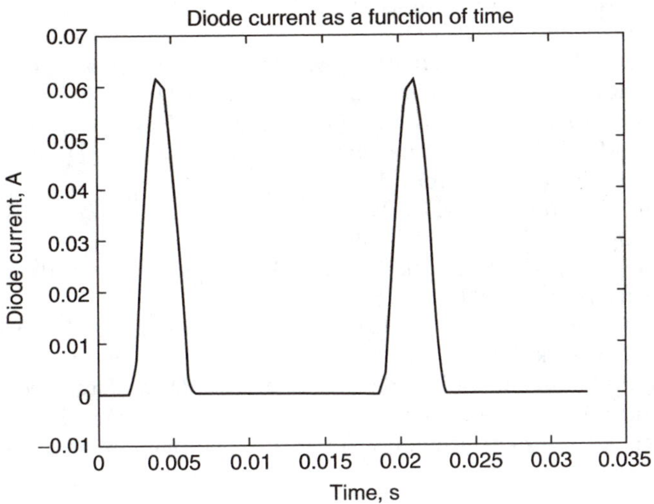

FIGURE 13.6
Diode current.

```
load 'ex13_7se.dat' -ascii; %
time = ex13_7se(:,1);
idiode = ex13_7se(:,2);
plot(time, idiode),   % plot of diode current
xlabel('Time, s')
ylabel('Diode Current, A')
title('Diode Current as a Function of Time')
i_ave = mean(idiode);  % average current
% Print out the results
fprintf('Average current is % 10.5e A\n',  i_ave)
```

The result from MATLAB is:

```
Average current is  8.78920e-003 A
```

The plot of the current vs. time is shown in Figure 13.6.

Bibliography

1. Attia, J.O., *PSPICE and MATLAB for Electronics: An Integrated Approach,* CRC Press, Boca Raton, FL, 2002
2. Belanger, P.R., Adler, E.L., and Rumin, N.C., *Introduction to Circuits with Electronics: An Integrated Approach,* Holt, Rinehart and Winston, New York, 1985.
3. Chapman, S.J., *MATLAB Programming for Engineers,* 2nd ed., Brook, Cole Thompson Learning, Pacific Grove, CA, 2002

4. Etter, D.M., *Engineering Problem Solving with MATLAB*, 2nd ed., Prentice Hall, Upper Saddle River, NJ, 1997.
5. Ferris, C.D., *Elements of Electronic Design*, West Publishing Co., St. Paul, MN, 1995.
6. Geiger, R.L., Allen, P.E., and Strader, N.R., *VLSI Design Techniques for Analog and Digital Circuits*, McGraw-Hill Publishing Co., New York, 1990.
7. Ghausi, M.S., *Electronic Devices and Circuits: Discrete and Integrated*, Holt, Rinehart and Winston, New York, 1985.
8. Howe, R.T. and Sodini, C.G., *Microelectronics, An Integrated Approach*, Prentice Hall, Upper Saddle River, NJ, 1997.
9. Rashid, M.H., *Microelectronic Circuits, Analysis and Design*, PWS Publishing Company, Boston, 1999.
10. Sedra, A.S. and Smith, K.C., *Microelectronic Circuits*, 4th ed., Oxford University Press, New York, 1998.
11. Warner, R.M., Jr. and Grung, B.L., *Semiconductor Device Electronics*, Holt, Rinehart and Winston, New York, 1991.
12. Wildlar, R.J., Design techniques for monolithic operational amplifiers, *IEEE Journal of Solid State Circuits*, SC-3, 341–348, 1969.

Problems

Problem 13.1

Table P13.1 shows the frequency response of a filter. Plot the magnitude and phase responses.

TABLE P13.1

Frequency Response of a Filter

Frequency, Hz	Magnitude, dB	Phase in Degrees
1.0 K	–14	107
1.9 K	–9.6	90
2.5 K	–5.9	72
4.0 K	–3.3	55
6.3 K	–1.6	39
10 K	–0.7	26
15.8 K	–0.3	17
25 K	–0.1	11
40 K	–0.05	7
63 K	–0.02	4
100 K	–0.008	3

Problem 13.2

The data in Table P13.2 show the frequency response of a twin-T notch filter. Determine the notch frequency and the bandwidth of the filter.

TABLE P13.2

Frequency Response of Twin-T Notch Filter

Frequency, Hz	Gain
1.000E+01	9.987E–01
5.012E+01	9.694E–01
1.000E+02	8.905E–01
5.012E+02	2.329E–01
1.000E+03	1.145E–01
5.012E+03	8.378E–01
1.000E+04	9.523E–01
5.012E+04	9.981E–01
1.000E+05	1.000E+00

Problem 13.3

The gains β of fifteen 2N3904 transistors, measured in the laboratory, are 210, 205, 225, 207, 215, 199, 230, 217, 228, 216, 229, 200, 219, 230, and 222.

(a) What is the mean value of β?

(b) What are the minimum and maximum values of β?

(c) What is the standard deviation of β?

Problem 13.4

Table P13.4 shows the corresponding source voltage and current flowing through a resistor of a circuit. Determine the correlation coefficient between the source voltage and the current.

TABLE P13.4

Source Voltage and Current Flowing through a Resistor

Voltage Source VS, V	Current IB, A
0.000	0.000E+00
2.000	3.81E–04
4.000	6.64E–04
6.000	1.12E–03
8.000	1.285E–03
10.000	1.66E–03

Problem 13.5

To the first order in $1/T$, the relationship between resistance R and temperature T of a thermister is given by

$$R(T) = R(T_0)\exp\left(\beta\left(\frac{1}{T} - \frac{1}{T_0}\right)\right)$$

where T is in Kelvin, T_0 is the reference temperature in Kelvin, $R(T_0)$ is the resistance at the reference temperature, and β is the temperature coefficient.

If $R(T_0)$ is 25,000 Ω at the reference temperature, use Table P13.5 to obtain the constants β and T_0.

TABLE P13.5

Resistance vs. Temperature of a Thermister

Temperature, K	Resistance, Ω
300	2.247E04
310	1.461E04
320	9.765E03
330	6.685E03
340	4.681E03
350	3.345E03
360	2.436E03
370	1.804E03
380	1.357E03
390	1.036E03
400	8.025E02

Problem 13.6

For a two-stage amplifier with feedback resistance, the output voltage of the amplifier was measured with respect to feedback resistance. Table P13.6 shows the output voltage as a function of feedback resistance. Draw the equation of best fit that models the output voltage with respect to the feedback resistance. What is the correlation coefficient between the output voltage and the feedback resistance?

TABLE P13.6

Output Voltage vs. Feedback Resistance

Feedback Resistance, Ohms	Output Voltage, V
10.0E03	3.000E–03
20.0E03	4.998E–03
30.0E03	6.996E–03
40.0E03	8.991E–03
50.0E03	1.098E–02
60.0E03	1.297E–02
70.0E03	1.496E–02
80.0E03	1.694E–02
90.0E03	1.892E–02
100.0E03	2.089E–02

Problem 13.7

The current flowing through a capacitor with respect to time is shown in Table P13.7. Find the charge stored in the capacitor.

TABLE P13.7

Current Flowing through a Capacitor

Time, S	**V(3), V**
0.000E+00	0.000E+00
4.000E–04	8.951E–01
8.000E–04	1.588E+00
1.200E–03	2.079E+00
1.600E–03	2.381E+00
2.000E–03	2.514E+00
2.400E–03	2.502E+00
2.800E–03	2.372E+00
3.200E–03	2.151E+00
3.600E–03	1.867E+00
4.000E–03	1.544E+00
4.400E–03	1.205E+00
4.800E–03	8.703E–01

Problem 13.8

Table P13.8 shows the characteristics of a zener diode. Plot the dynamic resistance as a function of reverse voltage of the zener diode.

TABLE P13.8

Zener Diode Characteristics

Reverse Voltages, V	**Reverse Current, A**
1	1.0E–13
3	1.2E–13
4	1.0E–12
5	1.0E–11
6	1.2E–9
7	1.0E–8
7.5	2.1E–8
7.7	15.0E–6
7.9	44.5E–6

Problem 13.9

Table P13.9 shows the drain current I_D vs. gate-source voltage V_{GS} of a MOSFET. Plot I_D vs. V_{GS} of the MOSFET. In addition, plot the transconductance as a function V_{GS}f.

TABLE P13.9

I_D vs. V_{GS} of a MOSFET

V_{GS}, V	I_D, A
0.0	5.010E–15
0.5	5.010E–15
1.0	5.010E–15
1.5	5.010E–15
2.0	5.010E–15
2.5	2.555E–08
3.0	2.183E–07
3.5	6.001E–07
4.0	1.171E–06
4.5	1.931E–06
5.0	2.879E–06

Index

C

D

E

F

G

H

I

N

O

P

Q

R

S

T

U

V

W

X

Y

Z